Silver Micro-Nanoparticles - Properties, Synthesis, Characterization, and Applications

Edited by Samir Kumar, Prabhat Kumar and Chandra Shakher Pathak

Published in London, United Kingdom

IntechOpen

Supporting open minds since 2005

Silver Micro-Nanoparticles - Properties, Synthesis, Characterization, and Applications
http://dx.doi.org/10.5772/intechopen.92480
Edited by Samir Kumar, Prabhat Kumar and Chandra Shakher Pathak

Contributors
Jaya T. T Varkey, Atamjit Singh, Komalpreet Kaur, Preet Mohinder Singh Bedi, Jaijeet Singh, Nitish Kumar, Manikandan Dhayalan, PriadharsinI Karikalan, Nalini Srinivasan, Su Mohammed Riyaz, William Da Silva, Leandro Rodrigues Rodrigues Oviedo, Pâmela Cristine Ladwig Muraro, Vinícius Rodrigues Oviedo, Daniel Moro Druzian, Olga V. Morozova, Dmitry V. Klinov, Srikanta Moharana, Ankita Subhrasmita Gadtya, Rozalin Nayak, Ram Naresh Mahaling, Guillermo Foladori, Noela Invernizzi, Hoang Nam Nguyen, Uday Dadwal, Rajendra Singh, Bipin Deochand Deochand Lade, Arti Sanjay Shanware, Ruchika M. Barapatre, Sunil T. Galatage, Aditya S. Hebalkar, Shradhey V. Dhobale, Omkar R. Mali, Pranav S. Kumbhar, Supriya V. Nikade, Suresh G. Killedar, Paulina De León Portilla, Enrique Sánchez Mora, Ana L. Lilia González, Ainil Hawa Jasni, Azirah Akbar Ali, Suresh Sagadevan, Zaharah Wahid, Lyubomir Lazov, Edmunds Teirumnieks, Risham Singh Ghalot, Kirandeep Kaur, Neena Bedi

First published in London, United Kingdom, 2021 by IntechOpen
IntechOpen is the global imprint of INTECHOPEN LIMITED, registered in England and Wales, registration number: 11086078, 5 Princes Gate Court, London, SW7 2QJ, United Kingdom
Printed in Croatia

British Library Cataloguing-in-Publication Data
A catalogue record for this book is available from the British Library

Additional hard and PDF copies can be obtained from orders@intechopen.com

Silver Micro-Nanoparticles - Properties, Synthesis, Characterization, and Applications
Edited by Samir Kumar, Prabhat Kumar and Chandra Shakher Pathak
p. cm.
Print ISBN 978-1-83968-659-7
Online ISBN 978-1-83968-660-3
eBook (PDF) ISBN 978-1-83968-661-0

For EU product safety concerns:
IN TECH d.o.o., Prolaz Marije Krucifikse Kozulić 3, 51000 Rijeka, Croatia, info@intechopen.com or visit our website at intechopen.com.

Meet the editors

Dr. Kumar is an experimental physicist with expertise in developing and studying sculptured thin films and interfaces. Dr. Kumar received his Ph.D. in Physics from the Indian Institute of Technology Delhi, India, in 2017. He has been working at Kyoto University since February 2018. His primary area of research is the synthesis of novel nanostructures by glancing angle deposition for plasmonics, surface-enhanced spectroscopy, photocatalysis, water repellent surfaces, and bio/chemical sensing applications.

Dr. Prabhat Kumar is a researcher in experimental material science with a specialization in transition metal oxides and dichalcogenides. He was awarded a Ph.D. in Physics from the Indian Institute of Technology Delhi, in 2019. He has an MTech in Nanotechnology and a BTech in Electronics and Communication Engineering. Dr. Kumar has experience in thin-film characterizations by X-ray photoelectron spectroscopy, electron microscopy, and thin films based terahertz modulators. He is currently working at the Institute of Physics of the Czech Academy of Sciences on atomic layer deposition technique for ultra-thin films of oxides and nitrides. He also has experience in energy storage and biomedical sensor applications of 3D graphene.

Dr. Chandra Shakher Pathak obtained his doctoral degree in Physics from the Indian Institute of Technology, Delhi, India. He obtained an MSc in Physics from Kumaun University Nainital, Uttarakhand, and an MTech (Gold medalist) in Advanced Materials Science and Technology from NIT Durgapur, India. Dr. Pathak's research work lies in the area of nanomaterials, thin-film growth, characterization, and electronic devices. He has research experience in nanomaterials, organic-inorganic heterojunctions, Schottky diodes, 2D materials, conducting atomic force microscopy, Kelvin probe force microscopy, and photovoltaics. Currently, he is working on the fabrication and stability of hybrid perovskite solar cells at Ben-Gurion National Solar Energy Center, Ben-Gurion University of the Negev, Israel. His bio-data has been published in Marquis Who's Who in the World 2014 (31st edition), Marquis Who's Who in the World 2015 (32nd Edition), and Who's Who in Science and Engineering 2016–2017 (12th Edition).

Contents

Preface XI

Section 1
Synthesis and Characterization of Silver Nanostructures 1

Chapter 1 3
Synthesis, Dielectric and Electrical Properties of Silver-Polymer Nanocomposites
by Srikanta Moharana, Ankita Subhrasmita Gadtya, Rozalin Nayak and Ram Naresh Mahaling

Chapter 2 25
Synthesis, Characterization of Dichlorofluorescein Silver Nanoparticles (DCF-SNPs) and Their Effect on Seed Germination of *Vigna radiata*
by Bipin Deochand Lade, Arti Sanjay Shanware and Ruchika M. Barapatre

Chapter 3 47
Multifunctional Silver Nanoparticles: Synthesis and Applications
by Nguyen Hoang Nam

Chapter 4 61
Silver Nanoparticles: Properties, Synthesis, Characterization, Applications and Future Trends
by Sunil T. Galatage, Aditya S. Hebalkar, Shradhey V. Dhobale, Omkar R. Mali, Pranav S. Kumbhar, Supriya V. Nikade and Suresh G. Killedar

Section 2
Biomedical Uses of Silver Nanostructures 81

Chapter 5 83
Biomedical Applications of Silver Nanoparticles
by Manikandan Dhayalan, Priadharsini Karikalan, Mohammed Riyaz Savaas Umar and Nalini Srinivasan

Chapter 6 95
Nanosilver in Biomedicine: Advantages and Restrictions
by Olga V. Morozova and Dmitry V. Klinov

Chapter 7 **107**
Silver Nanoparticles - Preparation Methods and Anti-Bacterial/Viral Remedy Impacts against COVID 19
by Lyubomir Lazov, Risham Singh Ghalot and Edmunds Teirumnieks

Chapter 8 **129**
Silver Nanoparticles Impregnated Wound Dressings: Recent Progress and Future Challenges
by Atamjit Singh, Kirandeep Kaur, Komalpreet Kaur, Jaijeet Singh, Nitish Kumar, Neena Bedi and Preet Mohinder Singh Bedi

Chapter 9 **147**
Silver Nanoparticles for Photocatalysis and Biomedical Applications
by William Leonardo da Silva, Daniel Moro Druzian, Leandro Rodrigues Oviedo, Pâmela Cristine Ladwig Muraro and Vinícius Rodrigues Oviedo

Section 3
Silver Nanostructures for SERS and Photoelectrochemical Applications **169**

Chapter 10 **171**
Theoretical and Experimental Study on the Functionalization Effect on the SERS Enhancement Factor of SiO_2-Ag Composite Films
by Paulina De León Portilla, Ana Lilia González Ronquillo and Enrique Sánchez Mora

Chapter 11 **191**
Silicon-Silver Dendritic Nanostructures Enabled Photoelectrochemical Solar Water Splitting for Energy Applications
by Uday Dadwal and Rajendra Singh

Chapter 12 **205**
Synthesis and Catalytic Activity Studies of Silver Nanoparticles Stabilized in Polymeric Hydro Gel
by Jaya T. Varkey

Chapter 13 **217**
Silver Nanoparticles in Various New Applications
by Ainil Hawa Jasni, Azirah Akbar Ali, Suresh Sagadevan and Zaharah Wahid

Section 4
Health Concerns **237**

Chapter 14 **239**
AgNano, the Construction of Occupational Health Standards: A Status Update
by Guillermo Foladori and Noela Invernizzi

Preface

Silver nanoparticles (AgNPs) have become one of the widely researched nanostructures in recent years because of their unique optical, physical, and chemical properties. Due to their structure, shape, and size, AgNPs can be modified to produce nanoparticles with exciting and unique properties. Materials based on AgNPs have a variety of interesting, challenging, and promising features suitable for various applications in biosensing, diagnostics, imaging, catalysis, and drug delivery. The size-dependent unique plasmonic characteristics of AgNPs, in particular, make them more attractive for biological and diagnostic applications.

This book provides a comprehensive overview of the significant applications and various methods for sustainable and green high-quality synthesis of AgNPs, and theoretical insight on the plasmonic properties of Ag composite films.

This book is divided into four sections. The first section explains the various methods for producing and characterizing AgNPs and their composites with dielectric and electrical properties. This section also covers a new class of multifunctional nanomaterials based on AgNP plasmonic activity and Fe_3O_4 superparamagnetic nanoparticles.

The second section delves into the biomedical applications of AgNPs in a detailed and systematic manner. This section discusses the photocatalytic activity of AgNPs, their potential use in the degradation of organic pollutants under ultraviolet and visible/solar radiation, as well as current strategies for developing AgNPs, functionalized wound dressings, and their potential outcomes, benefits, and future wound management challenges.

The third section discusses the surface-enhanced Raman spectroscopy (SERS) and photoelectrochemical (PEC) uses of AgNPs. It also discusses the production and catalytic characteristics of AgNPs stabilized in a hydrogel. This section gives an overview of the characteristics and conditions necessary to synthesize hydrogel-stabilized AgNPs. This section also looks into metal-assisted wet chemical etching of semiconductors (silicon) and semiconductor nanostructures-induced solar water splitting to obtain the best solar-to-hydrogen conversion efficiency.

The fourth and last section examines the different nanotechnology components and their relevance in today's environmental and human health fields.

The editors appreciate the authors' collaboration and contributions and those of the academic editorial board. We also thank IntechOpen, particularly Author Service

Manager Ms. Dolores Kuzelj and Commissioning Editor Ms. Anja Filipovic. We also want to express our gratitude to IntechOpen for allowing us to edit this fascinating book project.

Samir Kumar
Kyoto University,
Kyoto, Japan

Prabhat Kumar
Institute of Physics of the Czech Academy of Sciences,
Prague, Czech Republic

Chandra Shakher Pathak
Ben-Gurion University of the Negev,
Israel

Section 1

Synthesis and Characterization of Silver Nanostructures

Chapter 1

Synthesis, Dielectric and Electrical Properties of Silver-Polymer Nanocomposites

Srikanta Moharana, Ankita Subhrasmita Gadtya, Rozalin Nayak and Ram Naresh Mahaling

Abstract

Metallic nanoparticles and its composites have emerged as valuable asset in all phases of material science and engineering including electronic, optics and electromagnetic domains. Silver nanoparticles (Ag NPs) are one of the most vital and fascinating nanomaterials among several metallic nanoparticles due to its large surface ratio and outstanding properties with diverse field of potential applications. We demonstrated various synthesis techniques of nanocomposites, silver nanoparticles and composite based on these particles have shown great importance because of the remarkable properties (high electrical and thermal conductivity, good chemical stability and catalytic properties) of silver nanoparticles. This chapter provides various synthesis techniques for preparation of silver nanoparticles and their composites with dielectric and electrical properties in a lucid manner. The detail discussions of silver-polymer nanocomposites, emphasizing on each individual synthesis routes and properties have been carried out.

Keywords: Ag nanoparticles, composites, polymer manocomposites, properties

1. Introduction

In recent decade nanotechnology (nano signify very small that denotes to one billionth or 10^{-9} m in size) is a recognized as one of the most emerging fields of contemporary research deals with synthesis, manufacturing, strategy and tailoring of particle size approximately varying from 1 to 100 nm. The nanoparticles have unique magnetic, electronic and optical properties because of their high surface area to volume ratio and wide variety of applications including environmental health, optics, electronics; optoelectronics, catalysis and energy storage devices [1–6]. Nanoparticles possess small size; composition and shape have differences in their physical and chemical behaviors from their parent materials. Moreover, the smaller size of the nanomaterials also helps them to penetrate exact cellular locations and additional surface area facilitates increased absorption and targeted delivery of the substances [6–8]. A large number of nanomaterials have multitude of technological applicability in the field of engineering including electronics, biomedical, drug-gene delivery, environment, catalysis, light emitters, single electron transistors, non-linear optical or photo-electrochemical devices [9, 10]. The synthesis of nanomaterials by using chemical and physical methods is relatively expensive and potentially hazardous to the environment due to the effect of toxic chemicals and it is responsible for various biological risks [11]. Nanomaterials have rised as appropriate alternatives to overcome the shortcomings of micro-composites and monolithics, while challenges exist related to preparation for controlling elemental composition and stoichiometry in the nanocluster phase. Engineered nanomaterials produced with nanoscale dimension are commonly grouped into four kinds: carbon, metal oxides, dendrimers and composites [12]. However, composites are another powerful tool for the progress of specific materials according to our requirements. Nanocomposites are composites in which at least one of the phases exhibit dimensions in the nanometer range (1 nm = 10^{-9} m) and also shows high performance material with unusual performance and unique design possibilities [13]. The nanoparticles based composite by using various polymeric matrices and fillers such as metal powders (conductive filler) have made immense interest from academia and industry [14, 15]. These nanocomposites show various performances like high strength, stiffness at elevated temperature, corrosion resistance, low weight, low maintenance cost and attractive thermal coefficient [16]. Besides, composites generally divided into two stages such as continuous phase which is commonly uses as matrix and the other phases such as reinforcement which is embedded into the matrix. Moreover, the unique combination of matrix materials including polymer, carbon, metal, ceramics and different types of the reinforcements such as fibers, particles and layered materials have been widely utilized for the fabrication of composites [17]. It has been reported that, there is change in particle properties when size of the particle is less than a particular level called critical size [13]. In addition, as dimensions reach the nanometer level, interactions at phase interfaces become largely improved and this is important to enhance material properties. In view of this, the surface area-volume ratio of reinforced materials employed in the preparation of nanocomposites is the crucial role to the understanding of their structure–property co-relations. Furthermore, the discovery (1991) of carbon material especially carbon nanotube (CNTs) [18] and their subsequent use to fabricate composites with improved mechanical, thermal and electrical properties [13, 19] added a new and attractive dimension in the field of nanotechnology. Currently, nanocomposites offer new technology and business opportunities for all areas of industry making environmental friendly [20].

The synthesis and characterization of composites using organic and inorganic nanomaterials have been extensively investigated area of research with improved functional properties and wide range of potential applications such as coating, packaging materials, sensors, energy storage, etc. [21, 22]. The measurement of various properties of the prepared composite materials mostly depends on the characteristics of the original materials. There is a variety of properties of the matrix materials, for instance it combines the dispersed phase mutually. It protects the dispersed phase from chemical action and maintain in suitable position and its direction [23]. It is revealed that the properties of the polymer composites associated with chemistry extend of polymer chain and thermoset cure can change from the interface between reinforcement and matrix. The silver nanoparticle based polymer composites can be used as biocompatible materials with improved antimicrobial activity as well as good electrical conductivity and catalytic properties [24–26]. The polymer based nanocomposites using nanoparticles have been successfully synthesized by various researchers via free radical thermal polymerization [27] and photo polymerization [28]. There are several attempts to synthesize Ag based polymer composites: For example; Nikfarjam *et al.* [26] have fabricated Ag nanoparticle filled poly(methylmethacrylate) based composites by using in-situ photoiniferter mediated photo polymerization technique with good thermal stability and enhanced mechanical properties.

Urged by scientific interest and potential application for the various green synthetic techniques of silver nanoparticle related polymer based composites research has increased to a surprising scale, opening new challenges and opportunity for the further modulation of properties. This typical chapter is primarily divided into two major sections. The first section covers various synthetic routes of silver (Ag) nanoparticles. The second section covers the synthesis, dielectric and electrical properties of silver-polymer nanocomposites in the field of energy storage devices. We end this chapter with a few words on this new and exciting research area of the Ag nanoparticles followed by summary and perspective.

2. Synthesis of Ag nanoparticles

The new synthesis strategy for the material fabrications are of essential importance of nanostructure material in the field of nanotechnology. The applications of nanomaterials are possible only when nano-structured materials are made available with required size, morphology, crystal, chemical composition and their unique properties in the field of various technological applications [29–31]. Silver nanoparticles (Ag NPs) have outstanding microbial resistant ability. These nanoparticles have variety of applications in our daily life including clothes, household and personal care products and mostly owing to their antimicrobial properties. Moreover, the silver based nanostructured materials with specific physical, chemical and optical properties, especially altering sizes and shapes have been widely used in the field of electronic devices, paints, coatings, soaps, detergents, etc. [32]. In the above discussion, the following details of Ag based nanomaterials are significant to consider in their synthesis methods such as surface property, particle composition, size distribution, morphology and different types of reducing and capping agents used. Generally, the methods used for the preparation of metallic Ag Nps are classified into two categories, namely top-down or bottom-up approach [33]. The top down approach involves bulk materials and decrease them into nano-sized particles by using physical/chemical and mechanical processes [34]. The top-down approach is also used for the fabrication of many materials including semiconductor industry [35], in this approach metal oxide semiconductor field effect transistor (MOSFET) are imprinted onto a silica wafer by lithography based technique [36]. On the other hand, the preparation in bottom-up method requires single atoms and molecules into larger nanostructures to achieve nano-size particles [34]. Currently, the synthetic methods are divided into physical, chemical and biological green syntheses. In this respect, the physical and chemical synthesis method tend to more serious and hazardous as compared to the biological synthesis of Ag nanoparticles. This shows outstanding properties including high yield, solubility and stability [37]. The following segment discuss different synthesis methods in detail of Ag nanoparticle and their mechanisms, explaining how shape and size controlled Ag Nps can be achieved by proper selection of precursor chemicals, reducing and capping agent as well as concentration and molar ratio of chemicals.

2.1 Physical methods

The physical synthesis of Ag nanoparticles involves various processes such as evaporation-condensation process and the laser ablation technique [38, 39]. This technique is to synthesize large amount of Ag nanoparticles with high purity without use of chemicals that release toxic substances and expose human health and environment. However, there is a great challenge of agglomeration of nanoparticles because it is not used in the capping agents. Besides, both methods consume more power and require relatively longer duration of synthesis and difficult equipment, all of which increase their operating cost. The evaporation-condensation method commonly uses a gas phase technique that utilizes a tube furnace to synthesize nanospheres at atmospheric pressure. A variety of nanospheres using several materials including Au, Ag and PbS have been prepared by this method [40]. The centre of the tube furnace comprises a vessel carrying a base metal source which is evaporated into the carrier gas, permitting the final of nanoparticles. The shape, size and yield of the nanoparticles can be controlled by changing the plan of reaction facilities. Further, the synthesis of Ag nanoparticles by evaporation-condensation technique through tube furnace has various drawbacks. In this technique, the tube furnace occupies a huge space, consumes high energy elevating the surrounding temperature of the metal source and gives a longer period to maintain its thermal stability. To overcome these difficulties, Jung *et al.* [41] have synthesized Ag nanoparticles with higher concentration using ceramic heater and also utilizes in the efficiency.

Furthermore, another crucial approach for physical synthesis of Ag nanoparticle using laser ablation technique. The preparation of Ag nanoparticles through laser ablation of a bulk metal source placed in a liquid environment. Once irradiating with a pulse laser, the liquid environment may contain Ag nanoparticle of the base metal source, cleared from other compounds, ions or reducing agents [42]. However, there are various factors like laser power, duration of irradiation, type of base metal source and property of liquid media manipulate the features of the metal nanoparticles produced. Consequently, different chemical synthesis, the preparation of nanoparticles via laser ablation technique is pure and uncontaminated and in this method it uses mild surfactants in the solvent without using any other chemical reagents [43].

2.2 Chemical method

The synthesis of silver (Ag) nanoparticles by chemical method is the most commonly used technique and it is stable, colloidal dispersions in water or other appropriate organic solvents. The chemical process for synthesis of Ag nanoparticles in solution comprises the three major behaviors: (i) metal precursor, (ii) reducing agent and (iii) stabilizing or capping agent. The Ag nanoparticles are mainly chemically synthesized through Brust-Schiffrin synthesis (BSS) or the Turkevich method [43–46]. In order to achieve specific shape, size and various optical properties of the metal nanoparticles, it is essential to control the reducing agents and stabilizers are also taken into consideration. The uses of stabilizing agent during the preparation of metal nanoparticles are typically for avoiding aggregation [47]. So therefore, the following factors are needed to be considered for the safety and effectiveness of this technique, which includes the choice of appropriate solvent, use of environment friendly reducing agent and selection of non-toxic substances. There are various reducing agents used for the preparation of nanoparticles such as $NaBH_4$, N_2H_4, tri-sodium citrate (TSC), sodium citrate and N,N-dimethylformamide (DMF). Besides, in order to avoid aggregation between Ag nanoparticles, surfactant can be used such as sodium dodecyl sulphate (SDS), oleylamine and some polymeric materials like polyvinylpyrrolidone (PVP), polymethacrylic acid, polyethylene glycol (PEG) and polymethylmethacrylate have been reported to be the efficient protecting agents to stabilize as capping agents for stabilization [48–50]. This stabilization or capping agents can be accomplishing either through electrostatic or steric repulsion. For example, electrostatic stabilization is generally reached through anionic species like citrate, halide, carboxylates that adsorb or interact with Ag nanoparticles to impart a negative charge on the Ag nanoparticle surfaces. Thus, the surface charge of Ag nanoparticles can be controlled by coating with citrate ions to give a strong negative charge. Further, the polyethylene glycol (PEG) coated nanoparticles show good stability in highly concentrated salt solution, whereas lipoic acid coated particles with carboxylic group can also be used for bio conjugation. Meanwhile, the morphology of nanoparticles is strongly manipulated by the temperature variant were implemented during the preparation. The Ag nanoparticles exhibit much deviation in shape ranging from spherical to trigonal/hexagonal. It is also reported that ascorbic acid used as reducing agent for the formation of room temperature flower like silver nano architecture with average particle size is about 20 nm. Consequently, Ag nanoparticles have been synthesized by the polyol process with the support of supercritical carbon dioxide from nitrate salt of silver as the base material, polyvinylpyrrolidone (PVP) as the stabilizer for the silver clusters and ethylene glycol act as the reducing agent and solvent. However, polyvinylpyrrolidone (PVP) not only protected the nanosized silver particles from aggregation but also help nucleation phenomenon [51, 52].

2.3 Photochemical method

The Ag nanoparticle was successfully synthesized by using photo irradiation technique. In this technique, photo assisted synthesis of Ag used for the preparation of stable Ag nanoparticles by irradiation of a reaction mixture with a light source including laser or lamp in presence of photo reducing agents without introducing stabilizers or surfactants [53–55]. For instance, the laser irradiation of an aqueous solution of Ag salt and surfactant can made stable homogeneous dispersion of Ag nanoparticles with good distribution of shape and size of the particles. However, the syntheses of silver nanoparticles with narrow size distribution through ethylene glycol-water solvent system without use of a stabilizer. Further, it is also observed that Ag nanoparticles are synthesized by using per chlorate salt via pulse radiography technique. In this technique, there is a reduction of Ag^+ to Ag^0 was accomplished successfully using UV light as a substitute of chemical materials in a rubber matrix using photo reduction of film cast from natural rubber latex (NRL) comprising silver salt with an average size about 10 nm. It is also reported that other synthesis route such as microwave irradiation have also been utilized and this route is of much faster rate than that of the conventional heating via conduction and convection. Their size and preparation time of the nanoparticles is directly proportional to the irradiation power of the source of light [56]. In addition, photochemical processes also suggest a reasonable potential synthesis of controlled shape and size of Ag nanoparticles while multiple preparation steps might be required.

3. Synthesis of polymer-silver nanocomposites

3.1 In-situ polymerization

In-situ polymerization is a very efficient technique for the carbon based conductive fillers to be dispersed homogeneously in the polymer matrix, so it gives a strong interaction between the matrix and the filler particles. However, in-situ polymerization technique normally involves the addition of nanoparticle in a pristine monomer or a solution of monomer through polymerization in the presence of nanomaterials [57]. Many studies have been made to synthesize nanocomposites using in-situ polymerization techniques and also showed that the covalent linkages between the matrix and nanomaterials. Besides, the fabrication of silver nanoparticle is relatively simple, efficient process by using in-situ technique. It is a one step process for manufacture of nanoparticles which uses the corresponding precursors for synthesis and these nanoparticles can be directly grown using this technique. The most significant benefit of this technique is that it avoids particle agglomeration and maintains homogeneous distribution of the particles in the polymer matrix at the same time but, the main shortcoming of this technique is the slight possibility of left un-reacted educts in path of the reaction. These results may be control on the properties of the final product. For instance: Li *et al.* [58] have prepared polyvinylpyrrolidone (PVP) encapsulated stable Ag nanofluids with various concentrations of Ag via microwave synthesis technique. It is observed that the stable nanofluid comprising Ag nanoparticles of 30 to 60 nm and it is suitable for large scale production of nanofluids. Zhao *et al.* [59] have fabricated Ag deposited cellulose nanocomposites by using microwave assisted technique. It is low cost production and large scale production of prepared cellulose based nanocomposites. Singh *et al.* [60] have synthesized cellulose-Ag nanocomposites by using microwave assisted method and these nanocomposites have superb antimicrobial properties.

3.2 Ex-situ polymerization

Another important technique for preparation of nanoparticles based polymer composites by using ex-situ polymerization. This technique is more appropriate for large scale industrial applications. The main challenges of this method are preparation of nanoparticles with good homogeneity into the polymer matrix and have good thermal stability against aggregation [61]. In order to solve these difficulties sonication techniques were used to disperse the nanoparticles in the polymer matrix. For instance: Dhibar *et al.* [62] synthesized silver deposited polypyrrrole-graphene (Ag-PPy-Gr) nanocomposites used as an excellent candidate in the field of super capacitor applications [62]. In this work, it is observed that the graphene sheets are well uniformly coated in the polypyrrrole surface in the presence of silver nanoparticles. However, the specific capacitance of prepared Ag-PPy-Gr nanocomposites is 472 F/g at a current density 0.5 A/g. The development of these nanocomposites has enhanced electrochemical properties with the presence of both graphene and silver nanoparticles. In the ex-situ polymerization technique the uniform dispersion of nanoparticles in the polymer matrix becomes one important factor for preparation of transparent hybrid nanocomposites.

4. Different properties of polymer-silver nanocomposites

4.1 Dielectric properties

The dielectric properties of poly (vinylidene fluoride) (PVDF) based composites filled with nanosilver (nAg) deposited nickelate (Mg doped $La_{1.9}Sr_{0.1}NiO_4$) particles, which was reported by Thongbai *et al.* [63]. The composites (**Figure 1 (a,b)**) possessed high dielectric permittivity (ε') 62 and simultaneously achieved low dielectric loss ($\tan \delta$) of 0.027 at particular frequency region. Thus, these Ag-MLSNO/PVDF composites of greater dielectric properties with good mechanical flexibility and used in high technology future applications [63]. The nAg-MLNSO-PVDF composite were more thermally stable than that of the neat PVDF and it results these polymer composite systems has brilliant flexibility with lower concentration of fillers. These flexible dielectric polymer composites might be used in this modern period of electronic applications.

Zeng *et al.* [21] have reported silver nanoparticle-modified alumina microsphere hybrid composites for enhanced energy density and thermal conductivity. This

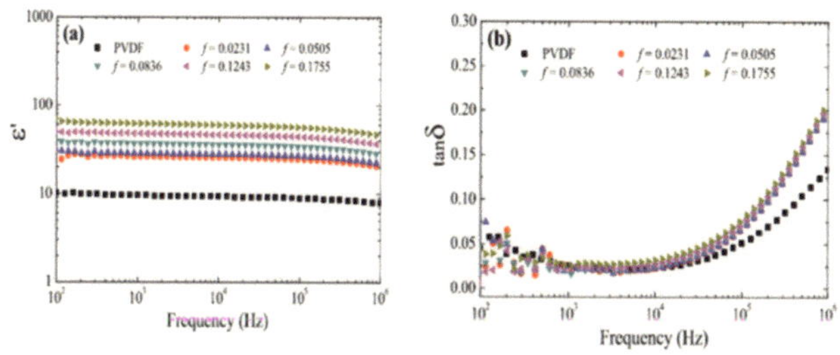

Figure 1.
Frequency dependence of (a) dielectric permittivity (ε') and (b) dielectric loss (tan δ) of nAg-MLSNO-PVDF composites. Reprinted with permission from ref. [63].

report tells us about the dielectric polymer composites which consisted of epoxy resins and silver nanoparticles decorated Al_2O_3 microsphere (Al_2O_3-AgNPs) with improved energy storage density and thermal conductivity. The dielectric loss of pure epoxy resins and Al_2O_3 -epoxy composites increases with increase in frequency. Moreover, the silver nanoparticle which is helpful to obtain Al_2O_3-AgNPs-epoxy resin composites with enhances the interfacial polarization and also leads to the high dielectric constant. For getting low dielectric loss and high breakdown strengths of the composite systems, the silver nanoparticles can be prevented from connecting into conductive pathways and is helpful for obtaining the polymer composites with high dielectric loss and high breakdown strengths [21]. It is observed that dielectric constant increases with Al_2O_3 AgNPs. With the increase in Al_2O_3 microsphere content, the motion of epoxy chain segments would be reserved and confined in the near vicinity, and the steric hindrance effect reduces the corresponding loss.

The enhancement of the thermal conductivity for Ag deposited alumina sphere through interfacial thermal resistance has been reported by Sun *et al.* [64]. In this work, the epoxy resin and commercial Al_2O_3 sphere was chosen as thermally conductive filler. The influence of silver deposition on the out-of-plane thermal conductivity of the composites was investigated. In this study, it is found that the composite with out-of-plane conductivity has increased with the increase in the Ag-deposited Al_2O_3 filler contents. The thermal conductivities of Al_2O_3-AgNPs-epoxy composites (**Figure 2(a)**) continuously improved [1.304 $Wm^{-1} K^{-1}$] until Ag NP content of 1.96 wt% and then decrease with increase in the filler content. This increase may be due to the bridge of Ag between the filler particles facilitate the heat flow through interfacial boundary, which in turn moderates the interfacial

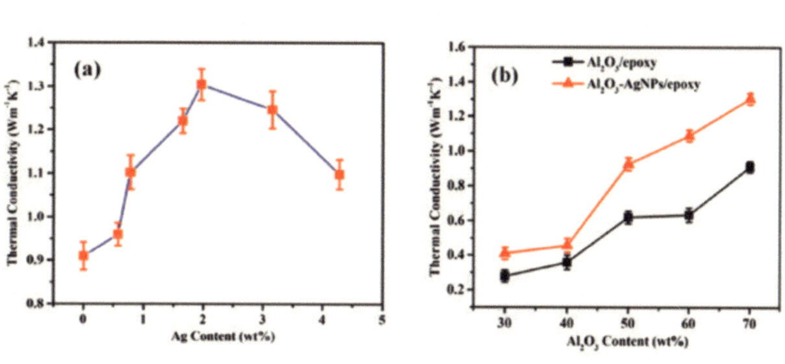

Figure 2.
Thermal conductivity of (a) Al_2O_3-Ag NPs-epoxy composites as a function of Ag contents, (b) variation of thermal conductivity of Al_2O_3-epoxy and Al2O3-AgNP-epoxy composites. Reprinted with permission from ref. [64].

thermal resistance and also help to improve thermal conductivity (**Figure 2(b)**) of the resultant composites [64, 65]. This method is helpful in the designing of thermally conductive polymer composites and can be used in the future generation electronic packaging applications.

Wu *et al.* [22] have reported about dielectric properties and thermal conductivity of the quantum-sized Ag decorated core or polydopamine (PDA)-epoxy based composites. The dopamine molecules are coated and attached to the surface of organic or inorganic matter which forms into supra-molecular polymer chain through the interaction of covalent bond and non-covalent bond [22]. However, the PDA is considerably improved the interfacial interaction between filler particles and polymer matrices, resulting in higher dielectric permittivity and breakdown strength [66, 67]. In this work they have found that, the composite has a low dielectric loss, high resistance and breakdown strength due to the Coulomb barrier formed by the Ag nanoparticles in the polymer. The novel core shell gray powder (AO*) sphere was prepared by the surface treatment of raw AO powder using dopamine. The polydopamine (PDA) shell helps to improve the interaction between AO fillers and epoxy matrix, which shows increase in dielectric permittivity and breakdown strength of the composites. Moreover, the strawberry like core shells AO* in Ag nanoparticles were prepared by hydrolysing silver nitrate and reducing the silver ion. The amalgamation of strawberry based core shell particles showed improved thermo mechanical, thermo stability and storage modulus. The AO* in Ag nanoparticles or epoxy composites indicates higher thermal conductivity values than that of the AO* or epoxy. It is accredited to the exceptional intrinsic thermal conductivity of Ag, mainly contributed from electronic transfer and phonon lattice vibration. Further, the enhanced dielectric, electromechanical and hydrophobic behaviors of core-shell Ag nano-wires (Ag NWs) in SiO_2 filler into polydimethylsiloxane (PDMS) based composites were prepared by sol gel technique and it is reported by Zhang *et al.* [68] and his co-workers. The polydimethylsiloxane (PDMS) is act as a polymer resin which is used in the field of insulators, coatings, medical devices, actuators and adhesives. Because it is due to the unique properties such as high electric insulation, good mechanical flexibility, mechanical stability under different ranges of temperature and high response speed [69].

Zang *et al.* [70] have reported that excellent energy storage density and efficiency in blend polymer poly (vinylidene fluoride) (PVDF) based composites by design of core-shell structured inorganic fibers and sandwich structured films. The dielectric material having great storage density is estimated to be obtained to minimize the size and weight of the dielectric capacitors, which is helpful in the pulsed power system and electric vehicles [70, 71]. Here they have taken into account the energy storage density and the energy storage efficiency for the polymer based composites using silver as nanofillers. These composites exhibited high polarization, low breakdown strength and high remnant polarization, which limits its energy storage density and efficiency [71, 72]. The silver (Ag) nanoparticle encapsulated $0.5Ba(Zr_{0.2}Ti_{0.8})O_3$-$0.5(Ba_{0.7}Ca_{0.3})TiO_3$ (BCZT)- PVDF composites with improved dielectric constant. Further, the Al_2O_3 fibers outer layer not only discharge the dielectric difference between the BCTZ and polymer matrix, which will weaken the distortion of electric field but also confines the mobility of electronics by using Ag particles and the interfacial charges. In this study both energy storage density and efficiency have been improved with the sandwich-structured PVDF films. Luo *et al.* [73] have been synthesized an Ag deposited $BaTiO_3$ hybrid and fabricated BT-Ag-PVDF composites. The effect of Ag content on the dielectric properties of the composites were analyzed based on the diffused electrical double layer theory. The experimental results revealed that, the dielectric constant of BT-Ag-PVDF composites significantly enhanced with the increase in the filler content about 613 (61 wt%) and reduced the dielectric loss of about 0.29 at 1 KHz. Furthermore, It was revealed that significantly enhancement of dielectric constant appeared when the inter particle distance decreased to a critical value of about 20 nm [73]. The facile method for the synthesis of Ag nanoparticles was successfully reinforced into the epoxy matrix *via* in-situ photochemical method. The Al-epoxy and Al-Ag-epoxy composites were prepared by mixing methods. The result showed that the dielectric constant of the Al-Ag-epoxy composite increases as compared to the Al-epoxy composites with simultaneously achieved negligible dielectric loss. Thus, these metal-polymer nanocomposites by in-situ photochemical method can be applied for the development of high-K polymer matrix to various fillers of host [74]. However, it is reported that the composites with high dielectric constant of core shell Ag@TiO_2 fillers successfully embedded into the polytetrafluoroethylene (PTFE) matrix, which provides the thickness of cell about 8–10 nm. It is observed that the composite of Ag@TiO_2 with rutile shells containing high dielectric constant than the anatase shells of Ag@TiO_2. The resultant dielectric constant maximum up to 200 at 100 Hz, which was 100 times higher than the pristine PTFE matrix with well suppressed dielectric loss of 0.005. Thus, the composites have potential applications in embedded devices [75].

Kuang *et al.* have been reported core shell Ag@C nanoparticles with diameter of 100–120 nm silver shell and 60–80 nm of carbon shell by hydrothermal method. The composite of Ag@C-PVDF composites was synthesized by solution casting method. The result of the composite illustrated that remarkably higher dielectric constant with low loss of 0.08 at 1000 Hz as compared to neat PVDF matrix due to the effect of the interfacial polarization. Furthermore, the permittivity and the tan δ of the composites increased with the increase of temperature [76]. The facile method for the synthesis of nanocomposites films of PVA reinforced into the silver nanoparticles and graphene oxide (GO) via solvent casting technique. It is observed that the composites have decrease in the real and imaginary part of impedance along with bulk resistivity on increasing GO content. Thus, on account of these above results of the GO-Ag-PVA nanocomposites can be employed as better promising candidate in the charge storage devices applications [77]. Xu et al. [78] have been demonstrated a new approach of ternary P(VDF-HFP) composites of Ag nanoparticles was decorated on TiO_2 nanowires with various aspects ratio *via* hydrothermal process. The result showed that the composites (**Figure 3(a,b)**) containing high aspect ratio of TiO_2 nanowires (i.e., pa-TiO_2) exhibits high dielectric constant (33.4) with simultaneously well suppressed dielectric loss of 0.023 at 100 Hz, which was 42.1% higher than that of Ag/pr-TiO_2/P(VDF-HFP) composites and also 227% greater than the pristine P (VDF-HFP) matrix. Hence, this method

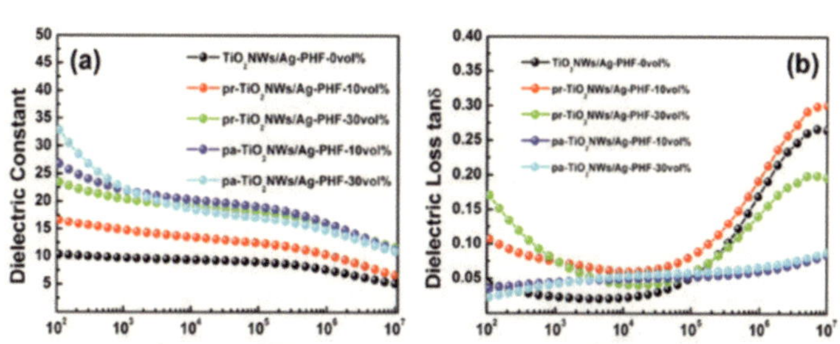

Figure 3.
Frequency dependence of (a) dielectric constant and (b) dielectric loss (tan δ) of Ag-TiO_2-P(VDF –HFP) ternary composites. Reprinted with permission from ref. [78].

could be provide a green route for the fabrication of high performance polymer composites for electric and electronic storage devices applications [78].

Phromviyo *et al.* [79] has been provided Ag@Ale–NPs composites were successfully fabricated on PVDF matrix via hydrothermal process using *Aloe vera* plant extracted solution as surface stabilizer and reducing agent. The composite of 0.18 volume fraction Ag@Ale-NP content exhibited high dielectric permittivity of 92.5 with low loss tangent of 0.049 at 1 KHz at room temperature. However, with the increase in filler content up to 0.22 the dielectric permittivity dramatically enhanced of ≈ 257.2 with very low loss of 0.26. Thus, the Ag@Ale-NPs-PVDF nanocomposites have greatly improved dielectric permittivity, due to the interfacial polarization effect and also the formation of micro-capacitor in the PVDF matrix [79]. Ghosh *et al.* have been prepared a CCTO ceramic fillers were prepared by sol–gel method and they successfully decorated by Ag nanoparticles embedded into the PVDF matrix to formed P/C-Ag: 30 composites via modified seeding method. These composites have high dielectric constant and negligible dielectric loss, which was 20% higher than that of the unmodified P/C: 30 composites. Thus, these types of work could be provided a better promising candidate of high performance polymer composites for energy storage applications [80].

The enhancement of both dielectric constant and breakdown strength of the modified Ag-OMMT nanoparticles incorporated in to the P(VDF-HFP) matrix to form P(VDF-HFP)-Ag-OMMT composites. However, the dielectric constant and breakdown strength of the P(VDF-HFP)-Ag-OMMT composites significantly increased with 4 vol% of Ag-OMMT contents and showed an energy density of 10.51 J cm^{-3} at 400 MV m^{-1}, which is ~2.25 times higher than that of pristine P(VDF-HFP) film. Thus, it is a simple and facile method to increase the energy density of polymer dielectric films by the addition of very low loading hybrid fillers [81]. Zou *et al.* [82] have been demonstrated silver (Ag) nanoparticles were dispersed homogeneously in P(VDF-TrFE) matrix to fabricate the composites by solvent casting method. The results showed that the enhancement of ferroelectric and dielectric properties of the composites with an increase in 38% of remnant polarization and 47% of dielectric constant than that of the pure P(VDF-TrFE) films [82]. Wang *et al.* have synthesized a PVDF based nanocomposites of Ag particles embedded in to the $BaTiO_3$ nanofillers via coaxial electro spinning technology followed by calcinations. The result indicated that, the composite of PVDF-Ag@BT exhibits high dielectric constant of 38.87 with well suppressed dielectric loss of 0.02 at 1 KHz. This remarkably enhancement of the dielectric constant may be due to the uniform distribution of the Ag nanoparticles on to the BT surface and the interfacial polarization effect. Furthermore, the Ag nanoparticles on the $BaTiO_3$ nanofibers exhibit the Coulomb blockade effect in the composites, which inhibits the electron migration and suppress the dielectric loss for better electronics applications [83].

The new strategy of core–shell graphene@polydopamine-Ag nanoplatelets and corresponding thermoplastic polyurethane (TPU)/Gns@PDA-Ag nanocomposites was reported by Zhu *et al.* [84]. This core shell nanocomposites were demonstrated by using dopamine oxidation polymerization and chemical reduction of silver nitride. The finding results indicates that, the dielectric constant for the 4 wt% of the TPU-Gns@PDA-Ag nanocomposites was 117.81, which was about 14 times greater than that of the pristine TPU matrix. Meanwhile, PDA shell not only promoted the homogeneous dispersion of Gns in the TPU matrix, but also enhanced the interface bonding between the nanoplatelets and TPU matrix. However, the Ag nanoparticles were uniformly embedded on the PDA shell, and the Coulomb blocking effect was successfully introduced into the composite. As a result, TPU-Gns@PDA-Ag nanocomposites exhibits low dielectric loss and high breakdown field strength compares with TPU-Gns nanocomposites. Thus, the obtained TPU-Gns@

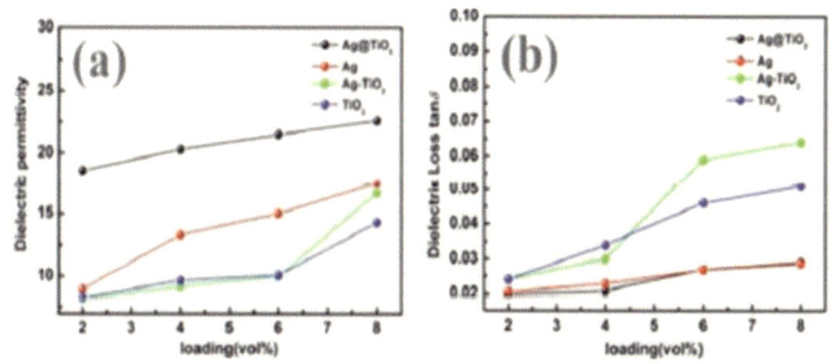

Figure 4.
Comparison of (a) dielectric permittivity and (b) dielectric loss of P(VDF-HFP)-Ag-TiO2 composites. Reprinted with permission from ref. [85].

PDA-Ag composite shows a better material for the potential applications in the field of energy storage [84]. Xiao *et al.* [85] have prepared the filler of Ag@TiO_2 core shell nanoparticles by vapor-thermal method and successfully embedded into the P(VDF-HFP) matrix to formed the P(VDF-HFP)-Ag@TiO_2 composites. The results revealed that the composite of P(VDF-HFP)-Ag@TiO_2 (**Figure 4(a,b)**) exhibits high dielectric constant (≈36) at 13 vol% of filler loading which was 4 times larger than that of the pristine P(VDF-HFP) matrix and well suppressed dielectric loss 0.03 at 1 KHz due to the effect of duplex polarization. Thus, this facile method of composites fabrication could be used as a smart functional material for embedded devices in the field of electronic industries [85].

Yang *et al.* reported the ternary composites system of polyimide (PI) embedded $CaCu_3Ti_4O_{12}$-Ag nanoparticles (CCTO@Ag) to fabricate PI-CCTO@Ag composites. The results showed that the Ag nanoparticle functionalized CCTO-PI composites having high dielectric constant value about 103 at 100 Hz at 3 vol% of filler content. This high dielectric constant based composite materials have well agreed with the percolation theory, space charge polarization and Maxwell-Wagner-Sillars effect. Furthermore, the dielectric loss value was reduced about 0.018 at 100 Hz due to blockage of charge transfer by insulating polyimide chains. Thus, the easy process ability, good flexibility and excellent dielectric properties of the composite have drawn potential applications in the field of charge-storage capacitors and embedded devices in electronic industry [86]. Chen *et al.* have been synthesized Ag-TiO_2-core shell nanoparticles (NPs) modified by two surfactants, i.e. octyl phosphonic acid (OPA) and pentafluorobenzyl phosphonic acid (PFBPA) reinforced polytetrafluroethylene (PTFE) based composites. Thus, the results showed that the nanocomposites by PFBPA exhibits higher permittivity than the OPA because of its uniform dispersion stability of inorganic fillers into the PTFE polymer matrix and results increase the interfacial polarization effect. Therefore, nanocomposites could be used as better promising materials for applications of embedded devices and high frequency fields in electronics [87].

Zhu *et al.* [88] have prepared a flexible polymer composite by using solution casting method with core-shell nanostructure (Ag cores coated by SiO_2 shell) and incorporating P(VDF-TrFE) matrix. The Ag@SiO_2 nanoparticles were prepared by a simple Stober method. The resultant composite showed increased in dielectric constant with lower dielectric loss along with improved breakdown field of 10 wt% and thickness of the SiO_2 shell about 10 and 20 nm. The enhancement of the dielectric performance suggested a promising way to fabricate composites for potential applications in electronic devices [66, 67, 88]. Fang *et al.* [89] have been reported BT-Ag hybrid particles with variation of Ag mass fractions of 0.31, 0.49, and 0.61 were synthesized by chemically depositing nano Ag on the surface of 100 nm BT particles. The results showed that the enhancement of the dielectric permittivity of the BT-Ag-PVDF (0.31, 0.49, 0.61 wt% of filler contents) composites exhibits 283, 350, and 783, respectively. On the other hand, the dielectric loss about <1 at 100 Hz, which is approximately three times higher than that of the untreated BT nanoparticles and it may be attributed to the low conductivity of the composites. Thus this method provided a way to fabricate BT-Ag hybrid particles into the polymer matrix for desired dielectric performance. Furthermore, the synthesized composites have attractive for potential applications in the high density packing technology and electronic systems [83, 84, 89]. Similarly, Wageh *et al.* [90] have successfully demonstrated a nano-silver decorated on the surface of the reduced graphene oxide (Ag-RGO) and incorporated into the polar poly (vinylidene fluoride) (PVDF) matrix. The results of these composites (**Figure 5(a,b)**) have lower electrical conductivity as well as increase in dielectric constant of 97 at 1 kHz of 1.5 vol% of Ag-RGO contents, while at the same time maintain the low loss. This enhancement in dielectric constant can be attributed to interfacial polarization or Maxwell-Wagner-Sillars (MWS) effect [79]. Thus, these results of the Ag–RGO–PVDF composite systems have drawn great promise for use as dielectric material for embedded capacitor applications [90].

4.2 Electrical properties

The simple method for the synthesis of electrically conductive composites comprising silver nanoparticles decorated on the carbon nanotubes (CNT) (Ag@CNT) incorporated into the epoxy resins by using wet chemistry reaction. The obtained result of the composites showed that higher electrical conductivity 0.10 wt% of Ag@CNTs, which was four times higher than the pure and functionalized CNTs. Thus, this enhancement of electrical conductivity of the Ag@CNTs-epoxy based composites was used as a potential application in the field of electronic package industries [91]. Sahu *et al.* [92] have been prepared polyaniline (PANI) coated functionalized multiwalled carbon nanotubes (f-CNTs) and silver nanoparticles (AgNPs) (PANI@f-CNTs and PANI@AgNPs hybrid nanofillers) based polymethylmethacrylate (PMMA) composites by in-situ polymerization technique. It is observed that the AC electrical conductivity (σ_{ac}), dielectric permittivity and dielectric loss (tan δ) of the resultant nanocomposites have significantly enhanced. Although the real impedance decreases by the incorporation of the hybrid nanofillers into the PMMA matrix, which concludes that the synthesized polymeric materials have most promising dielectric application [92]. Yusof *et al.* [93] have been synthesized a MWNT-Ag hybrids via reducing aqueous silver salt with the

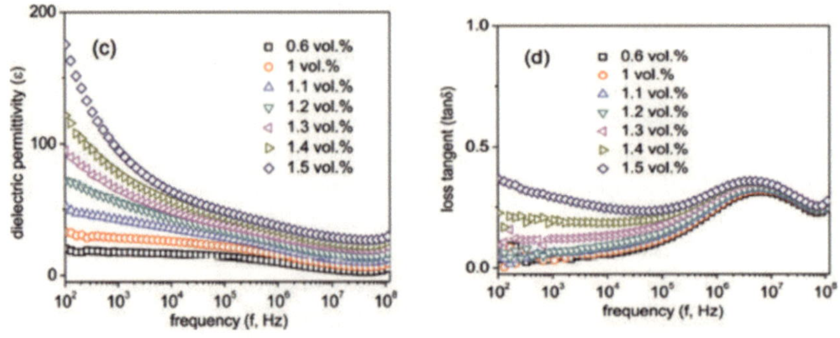

Figure 5.
Frequency dependence of (a) dielectric constant and (b) dielectric loss of Ag-RGO-PVDF composites with various percentage of filler contents. Reprinted with permission from ref. [90].

presence of SDS as stabilizing agent to enhance the deposition of the Ag NP on to the MWNT surface. The result obtained that MWNT-AgNP hybrids remarkably improved the structural, electrical and thermal properties, which could be a better material for the enhancement of electrical properties in the field-effect transistor (FET) applications at lower threshold voltage [93]. A facile and effective method of silver nanoparticles decorated on reduced graphene oxide (R-GO) was incorporated in to the PVDF matrix to formed Ag-RGO-PVDF composites via solution mixing method. The results indicated that the electrical conductivity of the composites increases at above the percolation threshold (0.17 vol %) due to the homogeneous dispersion of the Ag-RGO into the polymer matrix. Thus, this enhancement of electrical properties have drawn an attractive materials for antistatic, electrostatic dissipative and electromagnetic/radio frequency interference shielding applications. Furthermore, the resistivity of the composite system increased with increasing temperature, generating a pronounced positive temperature coefficient effect of resistivity [94]. Hussain *et al.* [95] have synthesized conducting PANI-Ag-MWCNTs composites comprising pure multi-walled carbon nanotubes (MWCNTs) and silver anchored with MWCNTs (Ag-MWCNTs) by using in situ polymerization approach. The electrical conductivity of the Ag-MWCNT (act as filler) is more strong interactions between PANI and Ag-MWCNTs and results improved electrical conductivity of the PANI. Thus, the PANI-Ag-MWCNTs composites were found to be the most stable composites [95]. Hsiao *et al.* [96] have prepared a silver nanoparticle deposited reduced graphene oxide-polyurethane (PU) composites (AgNP@RGO-PU) by electro spinning technique. The experimental result showed that the GO sheets can be attached to the PU surface through hydrogen bonding between GO and PU fibers. They also reported that the AgNPs deposited RGO-PU composites, which were modified by thiophenol through π-π interaction between the functional groups of AgNPs and RGO by dip-coating method. The 2 wt% of AgNP@RGO-PU composite (**Figure 6**) exhibited remarkably enhanced electrical properties (˂10 Ω/sq) [96]. Similarly, Nayak *et al.* [97] have reported expanded graphite (EG) decorated Ag nanoparticles using solid-state decomposition by a simple 'mix and heat' technique and prepared nanoparticles were deposited into the epoxy via stir casting method. It is revealed that the EG-Ag-epoxy composites with improved electrical conductivity. The EG-Ag composites exhibits better dispersion and lower porosity in the epoxy matrix which corresponding to high performance of thermal and electrical properties. Therefore, this method provided a green route to fabricate high

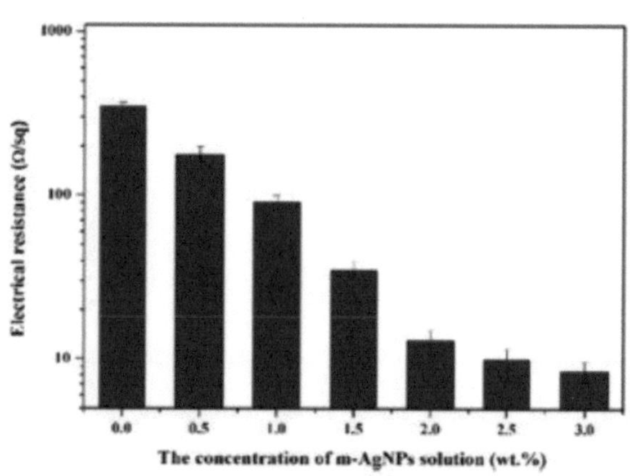

Figure 6.
Electrical resistance of AgNP@RGO-PU composites with various concentrations of filler content. Reprinted with permission from ref. [96].

efficiency and multifunctional epoxy based composite for advanced electronics and optoelectronics applications [97].

5. Conclusion

In summary, the syntheses of Ag nanoparticles have significant aspects of nanotechnology and it is used as nanofillers for fabricating nanocomposites. The Ag nanoparticle is a highly efficient, reliable high yielding and low cost technique. These nanoparticles have gained immense interest due to their unique physical and chemical properties as well as confirmed applicability in diverse fields such as electronics, catalysis, biotechnology and medicine. However, the shape and size distribution of silver nanoparticles can be controlled by adjusting reaction conditions such as reducing agent, stabilizing agent or using various synthesis techniques. The use of silver in the polymer based nanocomposites shows the enhancement in various properties such as dielectric and electrical in the field of energy storage devices. The Ag based nanocomposites have an enormous research interest in recent few times and potentially applicable in various fields especially in embedded high energy storage devices.

Acknowledgements

The authors gratefully acknowledge DST-SERB for financial support obtained through project grant of (CRG/2018/004101), New Delhi, Government of India.

Conflicts of interest

The authors declare no conflict of interest.

Author details

Srikanta Moharana[1]*, Ankita Subhrasmita Gadtya[1], Rozalin Nayak[2] and Ram Naresh Mahaling[2]*

1 School of Applied Sciences, Centurion University of Technology and Management, Odisha, India

2 Laboratory of Polymeric and Materials Chemistry, School of Chemistry, Sambalpur University, Burla, Odisha, India

*Address all correspondence to: srikantanit@gmail.com and rnmahaling@suniv.ac.in

References

[1] Narayanan KB, Sakthivel N. Biological synthesis of metal nanoparticles by microbes. Adv. Colloid Interface Sci. 2010; 156: 1-13.

[2] Kausar A, Anwar Z, Muhammad B. Recent Developments in Epoxy/Graphite, Epoxy/Graphene, and Epoxy/Graphene Nanoplatelet Composites: A Comparative Review. Polym. Plast Technol Eng. 2016; 55:1192-1210.

[3] Kausar A. Effect of Halloysite Nanoclay on Polymerization and Properties of Poly (3, 4-(2, 2-dimethylpropylenedioxy)-thiophene-co-aniline). Am. J. Polym. Sci. 2015; 5: 30-34.

[4] Shenashen MA, El-Safty SA, Elshehy EA. Synthesis, Morphological Control, and Properties of Silver Nanoparticles in Potential Applications. Part.Part. Syst. Charact. 2014; 31: 293-316.

[5] Khan, ZU, Kausar A, Ullah H. A Review on Composite Papers of Graphene Oxide, Carbon Nanotube, Polymer/GO, and Polymer/CNT: Processing Strategies, Properties, and Relevance. Polym. Plast Technol Eng. 2016; 55: 559-581.

[6] Iravani S, Korbekandi H, Mirmohammadi, SV, Zolfaghari, B. Synthesis of silver nanoparticles: chemical, physical and biological methods. Res Pharm Sci. 2014; 9: 385-406.

[7] Bilal M, Rasheed T, Iqbal HMN, Hu H, Zhang X. Silver nanoparticles: Biosynthesis and antimicrobial potentialities. Int. J. Pharmacol. 2017; 13: 832-845.

[8] Hamouda RA, Hussein MH, Abo-elmagd RA, Bawazir SS. Synthesis and biological characterization of silver nanoparticles derived from the cyanobacterium Oscillatoria limnetica. Scientific Reports, 2019; 9:13071.

[9] Khalil KA, Fouad H, Elsarnagawy T, Almajhdi FN. Preparation and characterization of electrospun PLGA/silver composite nanofibers for biomedical applications. Int. J. Electrochem. Sci. 2013; 8: 3483-3493.

[10] Kaviya SSJ, Viswanathan B. Green synthesis of silver nanoparticles using Polyalthia longifolia leaf extract along with D-sorbitol. J. Nanotech. 2011; 15:1-5.

[11] Shah M, Fawcett D, Sharma S, Tripathy SK, Eddy G, Poinern J. Green Synthesis of Metallic Nanoparticles via Biological Entities. Materials 2015; 8: 7278-7308.

[12] Awad MA, Hendi AA, Ortashi KMO, Alanazi AB, ALZahrani BA, Soliman DA. Greener Synthesis, Characterization, and Antimicrobiological Effects of *Helba* Silver Nanoparticle-PMMA Nanocomposite. Int. J. Polym. Sci., 2019, 1-7.

[13] Camargo PHC, Satyanarayan KG, Wypych F. Nanocomposites: Synthesis, Structure, Properties and New Application Opportunities. Mater. Res. 2009; 12: 1-39.

[14] Muzaleva PA, Kosobudskiia ID, Kulbatskiia DM, Ushakov NM. Polymer Composites Based on Polymethylmethacrylate with Silver Nanoparticles, Synthesis and Optical Properties. Inorg. Mater. Appl. Res. 2012; 3: 40-43.

[15] Yuan S, Shen F, Chua CK, Zhou K. Polymeric composites for powder-based additive manufacturing: Materials and applications. Prog. Polym. Sci. 2019; 91: 141-168.

[16] Azwa ZN, Yousif BF, Manalo AC, Karunasena W. A review on the degradability of polymeric composites based on natural fibres. Mater. Des. 2013; 47: 424-442.

[17] Yigezu BS, Jha PK, Mahapatra MM. The Key Attributes of Synthesizing Ceramic Particulate Reinforced Al-Based Matrix Composites through Stir Casting Process: A Review. Mater. Manuf. Process 2013; 28: 969-979.

[18] Iijima S. Helical microtubes of graphitic carbon. Nature 1991; 354:56-58.

[19] Weisenberger MC, Grulke EA, Jacques D, Ramtell T, Andrews R. Enhanced mechanical properties of polyacrylonitrile: multiwall carbon nanotube composite fibers. J. Nanosci. Nanotechnol., 2003; 3:535-539.

[20] Choa YH, Yang JK, Kim BH, Jeong YK, Lee JS, Nakayama T. Preparation and characterization of metal: ceramic nanoporous nanocomposite powders. J. Magn. Magn. Mater., 2003; 266:12-19.

[21] Ren L, Zeng X, Zhang X, Sun R, Tian X, Zeng Y, Xu JB, Wong CP, Silver nanoparticle-modified alumina microsphere hybrid composites for enhanced energy density and thermal conductivity. Compos Part A: Appl. Sci. Manuf. 2019; 119: 299-309.

[22] Wang Z,Yang M, Cheng Y, Liu J, Xiao B, Chen S, Huang J, Xie Q, Wu G, Wu H. Dielectric properties and thermal conductivity of epoxy composites using quantum-sized silver decorated core/shell structured alumina/polydopamine. Compos. Part A: Appl. Sci. Manuf. 2019; 118: 302-311.

[23] Ajayan PM, Schadler LS, Braun PV. Nanocomposite science and technology. Wiley; 2003.

[24] Grinou A, Bak H, Yun YS, Jin HJ. Polyaniline/Silver Nanoparticle-Doped Multiwalled Carbon Nanotube Composites. J. Disper. Sci. Technol. 2011; 33: 750-755.

[25] Afanasev DS, Yakovina OA, Kuznetsova AS, Listsyn AS. High activity CO oxidation of Ag nanoparticles supported on fumed silica. Catal. Commun. 2012; 22: 43-47.

[26] Makvandi P, Nikfarjam N, Sanjani NS, Qazvini NT. Effect Of Silver Nanoparticle On The Properties Of Poly(Methyl Methacrylate) Nanocomposite Network Made By In Situ Photoiniferter-Mediated Photopolymerization. Bull. Mater. Sci. 2015: 1625-1631.

[27] Boer B, Simon HK, Werts MPL, Vander Vegte EW, Hadziioannou G. Living Free radical Photopolymerization Initiated from Surface-Grafted Iniferter Monolayers. Macromolecules 2000; 33: 349-356.

[28] Balan L, Schneider R, Lougnot DJ. A new and convinent route to polyacrylate/silver nanocomposites by light-induced cross-linking polymerization. Prog. Org. Coat. 2008; 62: 351-357.

[29] Syafiuddin A, Salmiati, Salim MR, Kueh ABH, Hadibarata T, Nur H. A Review of silver nanoparticles: Research trends, global consumption, synthesis, properties, and future Challenges. J. Chin. Chem. Soc. 2017; 64:732-756.

[30] Desireddy A, Conn BE, Guo J, Yoon B, Barnett RN, Monahan BM, Kirschbaum K, Griffith WP, Whetten RL, Landman U, Bigioni TP. Ultrastable silver nanoparticles. Nature 2013; 501: 399-402.

[31] Atwater HA, Polman A. Plasmonics for improved photovoltaic devices. Nat. Mater. 2010; 9: 205-213.

[32] Burdusel AC, Gherasim O, Grumezescu AM, Mogoanta L, Ficai A,

Andronescu E. Biomedical applications of silver nanoparticles: An up-to-date overview. Nanomaterials 2018; 8: 681.

[33] Hornyak GL, Dutta J, Tibbals HF, Rao A. Introduction to nanoscience. CRC Press, Taylor and Francis LLC, 2008, New York

[34] Braydich-Stolle L, Hussain S, Schlager J, Hofmann MC. In-vitro cytotoxicity of nanoparticles in mammalian germ line stem cells. Toxicol Sci. 2005; 88: 412-419.

[35] Xue M, Li F, Cao T. Fabrication of ultra fine nanostructures using edge transfer printing. Nanoscale 2012; 4: 1939-1947.

[36] Kim HJ, Kim KJ, Kwak DS, A case study on modeling and optimizing photolithography stage of semiconductor fabrication process. Qual. Reliab. Eng. Int., 2010; 26: 765-774.

[37] Zhang XF, Liu ZG, Shen W, Gurunathan S. Silver nanoparticles: Synthesis, characterization, properties, applications, and therapeutic approaches. Int. J. Mol. Sci. 2016; 17: 1534.

[38] Amendola V, Meneghetti M. Laser ablation synthesis in solution and size manipulation of noble metal nanoparticles. Phys. Chem. Chem. Phys. 2009; 11: 3805-3821.

[39] Iravani, S.; Korbekandi, H.; Mirmohammadi, S.V.; Zolfaghari, B. Synthesis of silver nanoparticles: Chemical, physical and biological methods. Res. Pharm. Sci. 2014, 9, 385-406

[40] Kruis FE, Fissan H, Rellinghaus B. Sintering and evaporation characteristics of gas-phase synthesis of size-selected PbS nanoparticles. Mater. Sci. Eng. B 2000; 69: 329-334.

[41] Jung JH, Oh HC, Noh HS, Ji JH, Kim SS. Metal nanoparticle generation using a small ceramic heater with a local heating area. J. Aerosol. Sci. 2006; 37: 1662-1670.

[42] Chen YH, Yeh C.S. Laser ablation method: Use of surfactants to form the dispersed Ag nanoparticles. Colloids Surf. A 2002; 197: 133-139.

[43] Chugh H, Sood D, Chandra I, Tomar V, Dhawan G, Chandra R. Role of gold and silver nanoparticles in cancer nano-medicine. Artif. Cell. Nanomed. Biotechnol. 2018; 46: 1210-1220.

[44] Turkevich J, Kim G. Palladium: Preparation and catalytic properties of particles of uniform size. Science 1970; 169: 873-879.

[45] Turkevich J. Colloidal gold. Part II. Gold. Bull. 1985; 18: 86-91.

[46] Brust M, Walker M, Bethell D, Schiffrin DJ, Whyman R. Synthesis of thiol-derivatised gold nanoparticles in a two-phase liquid-liquid system. J. Chem. Soc. Chem. Commun. 1994: 801-802.

[47] Bai J, Li Y, Du J, Wang S, Zheng J, Yang Q, Chen X. One-pot synthesis of polyacrylamide-gold nanocomposite. Mater. Chem. Phys. 2007; 106: 412-415

[48] Pillai ZS, Kamat PV. What factors control the size and shape of silver nanoparticles in the citrate ion reduction method? J. Phys. Chem. B 2004; 108: 945-951.

[49] Evanoff DD, Chumanov G. Synthesis and optical properties of silver nanoparticles and arrays. Chem. Phys. Chem. 2005; 6: 1221-1231.

[50] Goulet PJG, Lennox RB. New insights into Brust–Schiffrin metal nanoparticle synthesis. J. Am. Chem. Soc. 2010; 132: 9582-9584.

[51] Liang H, Li Z, Wang W, Wu Y, Xu H. Highly surface-roughened "Flower-like" silver nanoparticles for extremely sensitive substrates of surface-enhanced Raman Scattering. Adv Mater. 2009; 21: 4614-4618.

[52] Rycenga M, Cobley CM, Zeng J, Li W, Moran CH, Zhang Q, Qin D, Xia Y. Controlling the synthesis and assembly of silver nanostructures for plasmonic applications. Chem Rev. 2011; 111: 3669-3712

[53] Jia H, Zeng J, Song W, An J, Zhao B. Preparation of silver nanoparticles by photoreduction for surface-enhanced Raman scattering. Thin Solid Films 2006; 2: 281-287

[54] Tsuji T, Okazaki Y, Tsuji M. Photo-induced morphological conversions of silver nanoparticles prepared using laser ablation in water-enhanced morphological conversions using halogen etching. J Photochem Photobiol. A. 2008; 194:247-253

[55] Kshirsagar P, Sangaru SS, Malvindi MA, Martiradonna L, Cingolani R, Pomp PP. Synthesis of highly stable silver nanoparticles by photoreduction and their size fractionation by phase transfer method. Colloids Surf A Physicochem Eng Asp. 2011; 392: 264-270.

[56] Zaarour M, El-Roz M, Dong B, Retoux R, Aad R, Cardin J, Dufour C, Gourbilleau F, Gilson JP, Mintova S. Photochemical preparation of silver nanoparticles supported on zeolite crystals. Langmuir 2014; 30: 6250-6256

[57] Potts JR, Dreyer DR, Bielawski CW, Ruoff RS. Graphene-based polymer nanocomposites. Polym. 2011; 52: 5-25.

[58] Li SM, Jia N, Zhu JF, Ma MG, Xu F, Wang B, Sun RC. Rapid microwave-assisted preparation and characterization of cellulose–silver nanocomposites. Carbohyd Polym. 2011; 83:422-429.

[59] Zhao X, Xia Y, Li Q, Ma X, Quan F, Geng C, Han Z. Microwave-assisted synthesis of silver nanoparticles using sodium alginate and their antibacterial activity. Colloids Surf A, 2014; 444: 180-188.

[60] Singh AK, Raykar VS. Synthesis of silver nanofluids with polyvinylpyrrolidone (PVP) and their transport properties. Colloid Polym Sci. 2008; 286: 1667-1673.

[61] Guo Q, Ghadiri R, Weigel T, Aumann A, Gurevich EL, Esen C, Medenbach O, Cheng W, Chichkov B, Ostendorf A. Comparison of in Situ and ex Situ Methods for Synthesis of Two-Photon Polymerization Polymer Nanocomposites. Polym. 2014; 6: 2037-2050.

[62] Dhibar S, Das CK. Silver nanoparticles decorated polypyrrole/graphene nanocomposite: a potential candidate for next-generation supercapacitor electrode material. J Appl. Polym. Sci. 2017; 134: 44724.

[63] Meeporn K., Thongbai P. Improved dielectric properties of poly (vinylidene fluoride) polymer nanocomposites fileld with Ag nanoparticles and nickelate ceramic particles. Appl. Surf. Sci. 2019; 481: 1160-1166.

[64] Li Q, Lu J, Zeng X, Sun R, Wu J, Xu JB, Wong CP. Enhanced thermal conductivity for Ag-deposited alumina sphere/epoxy resin composites through manipulating interfacial thermal resistance. Compos. Part A: Appl. Sci. Manuf. 2018; 107: 561-569.

[65] Sun JJ, Yao YM, Zeng XL, Pan GR, Hu JT, Huang Y. Preparation of boron nitride nanosheet/nanofibrillated cellulose nanocomposites with ultrahigh thermal conductivity via engineering interfacial thermal resistance. Adv. Mater. Interfaces 2017; 4:1700563-1700572.

[66] Wang Z, Cheng Y, Yang M. Dielectric properties and thermal conductivity of epoxy composites using core/shell structured Si/SiO_2/polydopamine. Compos B Eng 2018; 140: 83-90.

[67] Yang K, Huang X, He J. Strawberry-like core-shell Ag@polydopamine@ $BaTiO_3$ hybrid nanoparticles for high-k polymer nanocomposites with high energy density and low dielectric loss. Adv Mater Interf 2016;2:1500361.

[68] Zhou Z, Wang H, Zhu Z, Yang H, Zhang Q. Enhanced dielectric, electromechanical and hydrophobic behaviors of core-shell AgNWs@SiO_2/PDMS composites. Colloids Surf. A Physicochem Eng Asp. 2019; 563: 59-67.

[69] Gatti D, Haus H, Matysek M, Frohnapfel B, Tropea C, Schlaak HF. The dielectric breakdown limit of silicone dielectric elastomer actuators. Appl. Phys. Lett. 2014; 104: 052905.

[70] Cui Y, Zhang T, Feng Y, Zhang C, Chi Q, Zhang Y, Chen Q, Wang X, Lei Q. Excellent energy storage density and efficiency in blend polymer-based composites by design of core-shell structured inorganic fibers and sandwich structured films. Compos. Part B. Eng. 2019; 177: 107429.

[71] Qian K, Lv XG, Chen S, Luo H, Zhang D. Interfacial engineering tailoring the dielectric behavior and energy density of $BaTiO_3$/P(VDF-TrFE-CTFE) nanocomposites by regulating a liquid-crystalline polymer modifier structure. Dalton Trans 2018; 47:12759-12768.

[72] Chen JW, Yu XM, Yang FQ, Fan Y, Jiang YW, Zhou YX, Duan ZK. Enhanced energy density of polymer composites filled with $BaTiO_3$@Ag nanofibers for pulse power application. J Mater. Sci. Mater. Electron 2017; 28:8043-8050.

[73] Luo S, Yu S, Fang F, Lai M, Sun R, Wong CP. Critical interparticle distance for the remarkably enhanced dielectric constant of $BaTiO_3$-Ag hybrids filled polyvinylidene fluoride composites. Appl. Phys. Lett. 2014; 104: 252903.

[74] Qi L, Lee BI, Samuels WD, Exarhos GJ, Parler SG. Three-Phase Percolative Silver–BaTiO3–Epoxy Nanocomposites with High Dielectric Constants. J. Appl. Polym. Sci. 2006; 102: 967-971.

[75] Liang F, Zhang L, Lu WZ, Wan QX, Fan GF. Dielectric performance of polymer-based composites containing core-shell Ag@TiO_2 nanoparticle fillers. Appl. Phys, Lett. 2016; 108: 072902.

[76] Kuang X, Liu Z, Zhu H. Dielectric Properties of Ag@C/PVDF Composites. J. Appl. Polym. Sci. 2013; 129: 3411-3416.

[77] Sahu G, Das M, Yadav M, Sahoo BP, Tripathy J. Dielectric Relaxation Behavior of Silver Nanoparticles and Graphene Oxide Embedded Poly(vinyl alcohol) Nanocomposite Film: An Effect of Ionic Liquid and Temperature. Polym. 2020; 12: 374.

[78] Xu N, Xiao X, Yang H, Yu E, Zhang Q. Enhanced dielectric constant and suppressed dielectric loss of ternary composites based on Ag-P(VDF-HFP) matrix and TiO_2 nanowires. Ceram. Int. 2016; 42: 12475-12481.

[79] Phromviyo N, Chanlek N, Thongbai P, Maensiri S. Enhanced dielectric permittivity with retaining low loss in poly(vinylidene fluoride) by incorporating with Ag nanoparticles synthesized via hydrothermal method. Appl. Surf. Sci. 2018; 446: 59-65.

[80] Ghosh B, Calderón RMT, Samue EG, Hevia A. Enhanced dielectric properties of PVDF-$CaCu_3Ti_4O_{12}$:Ag composite films. Mater. Chem. Phys. 2017; 196: 302-309.

[81] Wang H, Xie H, Wang S, Gao Z, Li C, Hu gh, Xiong C. Enhanced dielectric property and energy storage density of PVDF-HFP based dielectric composites by incorporation of silver nanoparticles-decorated exfoliated montmorillonite nanoplatelets. Compos. Part A: Appl. Sci. Manuf. 2018;108: 62-68.

[82] Zou YH, Han L, Yuan GL, Liu BL, Zhao XL, Tian BB, Wang JL, Sun S, Sun JL, Meng XJ, Chu JH. Enhanced ferroelectric and dielectric propertyes of the P(VDF-TrFE)/Ag nanoparticles composite thin films. J Mater Sci: Mater Electron. 2014; 25: 3461-3465.

[83] Wang M, Pan X, Qi X, Zhang N, Huang T, Yang JH, Wan Y. Fabrication of Ag@$BaTiO_3$ hybrid nanofibers via coaxial electrospinning toward polymeric composites with highly enhanced dielectric performances. Comos. commun. 2020; 21: 100411.

[84] Zhu P, Weng L, Zhang X, Wang X, Guan L, Liu L. Graphene@ poly(dopamine)-Ag core–shell nanoplatelets as fillers to enhance the dielectric performance of polymer composites. J. Mater. Sci. 2020; 55: 7665-7679.

[85] Xiao X, Yang H, Xu N, Hu L, Zhang Q. High performance of P (VDF-HFP)/Ag@TiO_2 hybrid films with enhanced dielectric permittivity and low dielectric loss. RSC Adv. 2015; 5:79342-79347.

[86] Yang Y, Sun H, Yin D, Lu Z, Wei J, Xiong R, Shi J, Wang Z, Liu Z, Lei Q. High performance of Polyimide/ $CaCu_3Ti_4O_{12}$@Ag hybrid films with enhanced dielectric permittivity and low dielectric loss. J. Mater. Chem. A, 2015; 3: 4916-4921.

[87] Chen X, Liang F, Lu W, Jin Z, Zhao Y, Fu M. High Permittivity Nanocomposites Embedded with Ag/ TiO_2 Core–Shell Nanoparticles Modified by Phosphonic Acid. Polym. 2018; 10: 586-596.

[88] Zhu H, Liu Z, Wang F, Yan K. Influence of shell thickness on dielectric properties of composites filled with Ag@SiO_2 nanoparticles. RSC Adv. 2016; 6: 64634-64639.

[89] Fang F, Yang W, Yu S, Luo S, Sun R. Mechanism of high dielectric performance of polymer composites induced by $BaTiO_3$ supporting Ag hybrid fillers. Appl. Phys. Lett. 2014; 104: 132909.

[90] Wageh S, He L,Al-Ghamdi AA, Al-Turkid YA, Tjong SC. Nano silver-anchored reduced graphene oxide sheets for enhanced dielectric performance of polymer nanocomposites. RSC Adv. 2014; 4: 28426-28431.

[91] Ma PC, Tang BZ, Kim JK. Effect of CNT decoration with silver nanoparticles on electrical conductivity of CNT-polymer composites. Carbon 2008; 46:1497-1505.

[92] Sahu S, Sahoo AP, Shubhadarshinee L, Ramakrishna DS, Barick AK. Effect of Polyaniline-Coated Carbon Nanotube and Nanosilver Hybrid Nanoparticles on the Dielectric Properties of Poly(methyl Methacrylate) Nanocomposites. Polym. Compos., 2018; 39:1294-1305.

[93] Yusof Y, Zaidi M, Johan M. Enhanced Structural, Thermal, and Electrical Properties of Multiwalled Carbon Nanotubes Hybridized with Silver Nanoparticles. J. Nanomater. 2016; 1-9.

[94] He L, Tjong SC. Facile synthesis of silver-decorated reduced graphene oxide as a hybrid filler material for electrically conductive polymer composites. RSC Adv. 2015; 5: 15070-15076.

[95] Hussain T, Jabeen S, Shehzad K, Mujahid A, Ahmad MN, Farooqi ZH,

Raza MH. Polyaniline/Silver Decorated-MWCNT Composites with Enhanced Electrical and Thermal Properties. Polym. Compos. 2018; 39: 1346-1353.

[96] Hsiao ST, Ma CCM, Tien HW, Liao WH, Wang YS, Li SM, Chuang WP. Preparation and characterization of silver nanoparticle-reduced graphene oxide decorated electrospun polyurethane fiber composites with an improved electrical property. Compos. Sci. Technol. 2015; 118: 171-177.

[97] Nayak SK, Mohanty S, Nayak SK. Silver (Ag) nanoparticle-decorated expanded graphite (EG) epoxy composite: evaluating thermal and electrical properties. J. Mater. Sci.: Mater. Electron. 2019; 30: 20574-20587.

Chapter 2

Synthesis, Characterization of Dichlorofluorescein Silver Nanoparticles (DCF-SNPs) and Their Effect on Seed Germination of *Vigna radiata*

Bipin Deochand Lade, Arti Sanjay Shanware and Ruchika M. Barapatre

Abstract

The main objective of this study was to investigate whether dichlorofluorescein (DCF) is adequate for the formulation of stable dichlorofluorescein-induced silver nanoparticles under the boiling method to analyze their effects on the seed germination of Mung seeds *(Vigna radiata)*. Preliminary dichlorofluoresceine nanoparticles (DCF-SNPs) synthesis evidence by noticing the solution color transformed from a light green color to a dark brown color. The 2.5 ml of dichlorofluoresceine (DCF) solution was found sufficient for the formulation of dichlorofluoresceine induced silver nanoparticles at boiling conditions. Purified dichlorofluoresceine nanoparticles (DCF-SNPs) measure an average diameter of 293 nm where the majority of nanoparticles were around 159 nm in size with the surface load of-9.35 mV zeta potential value. The impact of dichlorofluorescein silver nanoparticles (DCF-SNPs) on the germination percentage of *V. radiata* has shown that, the 25% concentration of DCF-SNPs is excellent for the growth of Mung seeds (*V. radiata*). Overall, the dichlorofluorescein silver nanoparticles may be constructive for improving the percentage of seed germination at 25% of its concentration and may also be useful for fluorescent measurement using the confocal microscopy technique. Hence, dichlorofluorescein silver nanoparticles (DCF-SNPs) are proposed as an efficient detection system for nanoparticles in agrochemicals for plants.

Keywords: dichlorofluoresceine (DCF), *Vigna radiata*, dichlorofluoresceine silver nanoparticles (DCF-SNP), Zeta potential, confocal microscopy studies

1. Introduction

Dyes are colored entities that are chemically attached to the matrix fiber and accumulate during the drying period, providing color through the systematic absorption of light and increasing the speed of the fiber dyeing process [1]. There are two types of dyes which are natural or organic dyes and chemically synthesized dyes. In this chapter we are going to deal with organic dyes.

1.1 Organic dye

Natural dyes are derived from natural resources and, are typically categorized as plant, animal, mineral and microbial dyes, though plants are the key sources of natural dyes [2]. Examples of natural and organic coloring dyes are blue dye which derived from plant leaves, while red dye, Madder and Morinda from roots. Brazil Wood an old-world dye comes from wood of plant. Safflower and saffron dye made from their flowers, however, rhizomes of turmeric use to make dye. On the other hand yellow dye derived from roots, leaves, stems and flowers [2]. The organic dyes are colored as they appear in the visible light spectrum (400–700 nm) comprising of one chromosphere, a conjugate framework with a dual bond alternating arrangement with a single bond and an electron resonance that stabilizes in organic compounds [3]. Based on the chemical structure and characteristics properties, dyes are classified as *Azo* dyes, *Anthraquinone* dyes, Nitro dyes, *Diphenlmethane* dyes, *Triphenylmethane* dyes, *Xanthene* dyes, *Phthaleins* dyes, *Indigoid* and *Thionidigoid* dyes [4]. These organic coloring powders are smaller, thicker, finely fragmented crystalline solids, which are water soluble, whereas they are insoluble in application media such as ink or paint for the use of optical media, image sensors [1], photosensitizer and coloration [5]. Improved color functionality has been studied in recent times by adding functional properties such as antimicrobial, UV protection, insect repellent, etc. [6]. In addition, organic coloring is a major supplier for floral, dried flowers, pesticides, ice-melt, deicing, reservoir, lake, water tracing, leak detection, greenhouse, livestock, seed treatment, crop protection, fertilizer staining, food, cosmetic and environmental issues [7]. The multitude application of organic dyes has enable to use them for coloring, detection, biosensor, tracking in the test samples. One of the best examples of tracking dye is dichlorofluorescein (DCF).

1.2 Dichlorofluoresceine (DCF)

Fluorescein based titrations shows effective results at 0.005 N concentrations of chloride and if the formulation has neutral alkaline conditions [8]. 2′,7’-Dichloro- and 2′,7′-difluorofluoresceins are superior alternatives to underivatized fluorescein. Quadrupole ion trap mass spectrometer for the analysis of the gas phase properties of three charge states of fluorescein, indicating that dianions and cations do not emit detectable fluorescence in the gas phase. Monoanions, on the other hand, do fluorescence and are useful for experiments [9]. Dichlorofluorescein (DFC) is a natural, crystalline organic, coloring agent that substitutes for chloride at 2 and 7 positions and originated in the fluorescein family. **Figure 1** illustrates the molecular

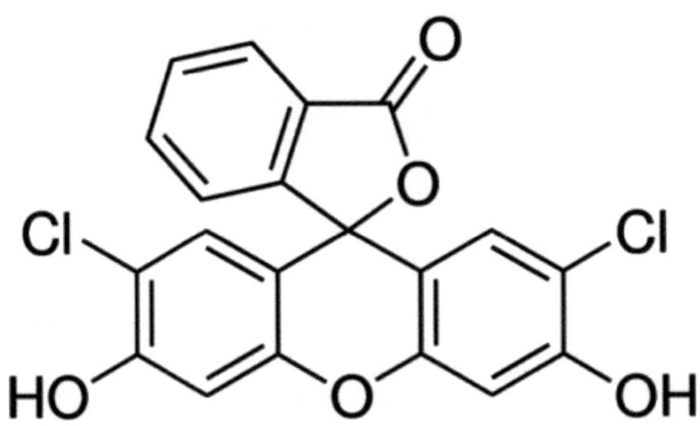

Figure 1.
Demonstrates the molecular structure of Dichlorofluorescein with empirical formula as $C_{20}H_{10}cl_2O_5$ *[10].*

structure of dichlorofluorescein. The molecular weight of dichlorofluorescein (PubChem SID: 24894041) is 401.20 g/mol, a melting point of 280 °C and it is used as an indicator which is not prone to soluble, absorbs, disintegrates or infuses silver or halide ion, but changes color at the end of precipitate due to absorption phenomena [8]. It was therefore used as a quantitative agentometry titrant indicator, consisting of a known concentration of silver nitrate, to estimate the molarity of chloride in the sample as specified in the Fajans method [11].

1.2.1 Dichlorofluorescein (DCF) applications

Organic Dichlorofluorescein (DCF) applications are even less documented, however it has demonstrated application as an indicator for halide titration argentometry, gaseous dianion studies, targeting it as a probe for imaging, heamotherapy and histological applications [8, 10, 12]. Generally, non fluorescent 2′ 7' Dichlorodihydrofluorscein diacetate is used in human hepatocellular carcinoma cells to monitor the oxidation of 2′,7′-dichlorofluorescin-diacetate (DCF-DA) to an extremely fluorescent 2′,7′ dichlorofluorescein (DCF) compound due to the presence of reactive oxygen species using a fluorometric microplate assay [13]. And thereby, 2′,7′ dichlorofluorescein (DCF) may be very significant in the assessment of the human organ culture model. In the same way, organic dyes can be used to research the tracing of in the host body of insects, plants and animals [12]. DCF can be the reducing agent for metallic salt (iron, copper, zinc, silver, gold, etc.) for the formation of metallic nanoparticles for various agriculture applications.

1.3 Silver nitrate (AgNO3)

Silver nitrate (AgNO3) is an inorganic compound that appears to have a colorless, white crystalline composition with molecular weight of 169.873 g/mol. In its solid state, it has a density of 4.35 grams per cubic centimeter and its density in the liquid state at a temperature is 210 °C corresponds to 3.97 g/cm^3. The melting and boiling points of silver nitrate are 414 °F and 824 °F respectively. In 1800s, silver nitrate was used for the treatment of ulcerations and infected wounds and stomach ulcers [14–15]. Silver nitrate is a chemical with a wide range of applications including antiseptic, suturing, eye disease disinfectant, burned wounds, wart and granulation tissue reduction, ulceration, dental cavity retention, etc. [16]. Other noteworthy medical applications of silver include wire or coated suture topical therapy for osteocutaneous fistulae, and foil coverings for burn wounds [14]. Silver nitrate compound is a productive source of creation for many other silver compounds used for the medical, biotechnological, nanotechnological, pharmaceutical as well as several other industries. **Figure 2** shows the silver nitrate molecular structure. Silver nitrate is frequently used chemical in several areas in agriculture including control growth, flowering development, dormancy and by spraying it on the growth tip of plants [17].

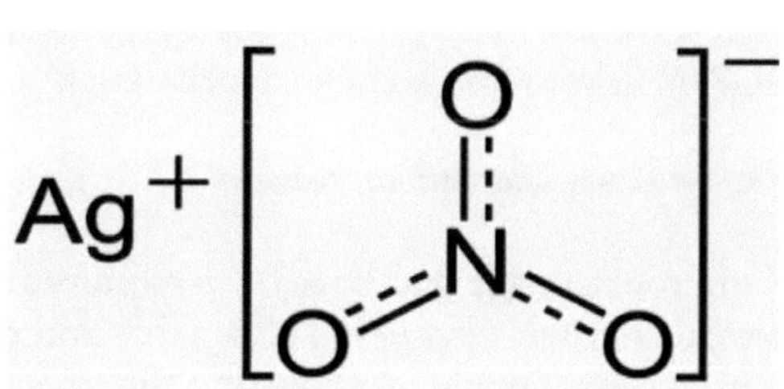

Figure 2.
Shows the molecular structure of silver nitrate [16].

Silver nitrate has a long use in nanotechnology for acting as a protective layer in stabilizing nanoparticles from further agglomeration. During the reduction of metal salt formation narrow size of nanoparticles are obtained. It absorbs on the surface of particle provide stabilization & diffusion barrier in the growth of particle [18]. Therefore, this silver salt are the best source for silver nanoparticles synthesis.

1.4 Silver nanoparticles (AgNPs)

Silver nanoparticles (AgNPs) range from 1 to 100 nm in size, which are fundamentally synthesized by physical, chemical and biological approaches. Silver nanoparticles were being used as antimicrobial agents in a wide variety of applications, which include disinfecting medical instruments and home appliances to water treatment [19]. Even though the other biological properties of silver nanoparticles such as antimicrobial, antifungal, anti-inflammatory, anti-cancer and anti-angiogenesis [20] already have enabled them to be extensively used in the fields of medicine and dentistry, diagnostics, therapeutic, medical care, health and food applications [21]. The other medical applications, including wound repair, bone healing, dental applications, vaccine adjuvant, antidiabetic agent, and biosensing [22]. The techniques of synthesis of metal silver nanoparticles have certain benefits as well as drawbacks. Therefore, depending on the application, the selection of the procedure is presumed and therefore, depending upon the application the selection of the methodology is considered. A well recorded manuscript available in the literature on the physical, chemical and biological preparation of silver nanoparticles. Nanoparticles have proved to be efficient agrochemical agents in order to improve the crop productivity, reducing the pests, increasing the nutrient uptake, inhibiting the pathogens and act as 'magic bullets' serving as herbicides, pesticides and fertilizers etc. [23–24]. The biological activity of AgNPs depends on factors including surface chemistry, size, size distribution, shape, particle morphology, particle composition, coating/capping, agglomeration, and dissolution rate, particle reactivity in solution, efficiency of ion release, and cell type, and the type of reducing agents used for the synthesis of AgNPs are a crucial factor for the determination of cytotoxicity [20]. The physicochemical properties of nanoparticles enhance the bioavailability of therapeutic agents after both systemic and local administration and other hand it can affect cellular uptake, biological distribution, penetration into biological barriers, and resultant therapeutic effects [20]. There are several methods for creating nanoparticles, including co precipitation, hydrothermal synthesis, inert gas condensation, ion sputtering scattering, micro emulsion, microwave, pulse laser ablation, sol–gel, sono chemical, spark discharge, template synthesis, and biological synthesis. We shall now briefly look into the methods for the synthesis of nanoparticles [25].

1.5 Synthesis of silver nanoparticles

There are large number of mrthods for bthe synthesis of silver nanoparticles. While in the present study we will discuss using microorganisms, plant extract and we proposed to use DCF for silver nanoparticles synthesis.

1.5.1 Silver nanoparticles synthesis using microorganism

There are various microorganism that have been explored for the synthesis of silver nanoparticles due to their advantage of reliable and ecofriendly process. The microorganisms used for reducing and capping the silver salts produces the various size, shape and morphology of silver nanoparticles. Novel *Nocardiopsis species, Brevibac teriumfrigoritolerans* strain, *Klebsiella pneumoniae,*

Sr no	Plants	Size (nm)	Plant's part	Shape
1.	*Abutilon indicum*	7–17	Leaves	Spherical
2.	Acalyphaindica	0.5	Leaves	—
3.	Acalyphaindica	20–30	Leaves	Spherical
4.	*Acorus calamus*	31.83	Rhizome	Spherical
5.	*Allium sativum*	4–22	Leaves	Spherical
6.	*Aloe vera*	50–350	Leaves	Spherical,
7.	Alternanthera dentate	50–100	Leaves	Spherical
8.	*Argyreia nervosa*	20–50	Seeds	—
9.	Boerhaaviadiusa	25	Whole plant	Spherical
10.	*Brassica rapa*	16.4	Leaves	—
11.	*Calotropis procera*	19–45	Plant	Spherical
12.	*Carica papaya*	25–50	Leaves	circular,
13.	Centellaasiatica	30–50	Leaves	Spherical
14.	*Citrus sinensis*	10–35	Peel	Spherical
15.	Cocciniaindica	10–20	Leaves	—
16.	Cocous nucifera	22	Inflorescence	Spherical
17.	Cymbopogancitratus	32	Leaves	—
18.	*Datura metel*	16–40	Leaves	Quasilinear
19.	Eclipta prostrate	35–60	Leaves	pentagons,
20.	Eucalyptus hybrid	50–150	Peel	spherical
21.	Ficuscarica	13	Leaves	—
22.	*Garcinia mangostana*	35	Leaves	—
23.	Melia dubia	35	Leaves	Spherical
24.	Memecylonedule	20–50	Leaves	hexagonal
25.	*Moringa oleifera*	57	Leaves	—
26.	Musa paradisiacal	20	Peel	—
27.	*Nelumbo nucifera*	25–80	Leaves	triangular
28.	Nelumbo nucifera	25–80	Leaves	triangular
29.	*Passiflora foetida*	—	Leaves	Coral
30.	Pistaciaatlantica	10–50	Seeds	Spherical
31.	Pogostemonbenghalensis	>80	Leaves	—
32.	*Portulaca oleracea*	<60	Leaves	—
33.	Premnaherbacea	10–30	Leaves	Spherical
34.	Psoraleacorylifolia	100–110	Seeds	—
35.	Swieteniamahogani	50	Leaves	—
36.	Tea extract	20–90	Leaves	Spherical
37.	*Thevetia peruviana*	10–30	Latex	Spherical
38.	Trachyspermumammi	87, 99.8	Seeds	—

Sr no	Plants	Size (nm)	Plant's part	Shape
39.	*Tribulus terrestris*	16–28	Fruit	Spherical
40.	*Vitex negundo*	5 & 10–30	Leaves	Spherical
41.	*Vitis vinifera*	30–40	Fruit	circular,
42.	Ziziphoratenuior	8–40	Leaves	Spherical

Table 1.
Green synthesis of silver nanoparticles by different researchers using plant extracts [31].

Escherichia coli, and Pseudomonas jessinii are some of the examples of micro organisms used for silver nanoparticles synthesis [26]. Besides, the production of silver nanoparticles form *Bacillus clausii* cultured form *Enterogermina* is explored [27]. The aqueous extract of cynobacterial *Oscillatoria limnetica* fresh biomass was used for the green synthesis of AgNPs and it takes about 30–60 hours for the reduction and stabilizing the synthesize of metallic nanoparticles that ranges from 3.30–17.97 nm in size [28].

1.5.2 Silver nanoparticles synthesis using plants

The plants part being organic and eco friendly are extensively used for synthesis of silver nanoparticles (AgNPs). The plant are the hot spots for the phytochemicals and secondary metabolites that are used constantly for various medicinal purposes including antimicrobial, antifungal, anti inflammatory, wound healing, antidiabatic etc. The plant sources such as leaves, stem, roots, flowers possessing the medicinal properties are used for the formulation of silver nanoparticles which carries the specific medicinal compound to reduce and capped the silver salt and present at the outer layer of silver to make them stable [29]. Therefore, the plant based silver nanoparticles possess the duel properties from silver and one from capped compounds from plants. The different plant leaf extracts for examples pine, ginkgo, magnolia, mango, neem, *oscimum scantum*, are used for their extracellular synthesis of silver nanoparticles. The biological silver nanoparticles production has the faster synthesis rates than the chemical methods and potentially be used in various foods, agriculture, chemical and medical application. The aqueous peel extract of *Annona squamosa* has been used successfully for synthesis of silver nanoparticles of irregular spherical in shape with the average particle size of 35 nm, at room temperature [30]. Green synthesis of silver nanoparticles by different plant extracts are described in **Table 1**. There are some of the synthesis methods that are constantly used for nanoparticles synthesis such as high temperature, pressure, sunlight condition, in autoclave [29]. In contrast, there are other publication that performed the synthesis process at room temperature.

Diversity of compounds, polymers, exopolysaccharides, proteins, lipids and other compounds such as natural dyes could be a good source for reducing metal salts to form stable stained nanoparticles which can have various applications such as pesticides, nutrients, hormones delivery for sustainable agriculture. In the light of the above addressed interesting information on dichlorofluorescein, silver nitrate, silver nanoparticles, it is evident that silver nanoparticles have an enormous application in various fields. It was noticed that none of the papers reported the synthesis of silver nanoparticles using any dye. The direct use of dichlorofluorescein (DCF) for the reduction of silver salt can also produce silver nanoparticles and could be used for the study of absorption and biotransformation in seeds and plants. The aim of this research is therefore to conduct the synthesis and characterization of dichlorofluorescein induced dichlorofluorescein silver nanoparticles (DCF-SNPs) under boiling method to evaluate their effects on the seed germination of *Vigna radiata* and to propose that they would be used for real-time transformation in the living plant using confocal microscopy.

2. Materials and methods

The materials and the methodology adopted for the synthesis of dichlorofluorescein silver nanoparticles (DCF-SNP), characterization and application are described below. Material required: Requirement specification for the formulation of silver nanoparticles utilize dichlorofluorescein, NaOH, distilled water, AgNO3 (1 mm), conical flask, beaker, aluminum foil, volumetric flask, etc. Silver nitrate (AgNO3): silver nitrate was used for the synthesis of dichlorofluorescein silver nanoparticles (DCF-SNPs) and was then used with the seed germination assay. Instruments for study: The following instruments such as Fourier transform infrared spectroscopy (FTIR), Zeta Potential (ZP) and Nanoparticles tracking analysis (NTA) were used for characterization of synthesized dichlorofluorescein silver nanoparticles (DCF-SNPs) [32].

2.1 Preparation of dichlorofluorescein (DCF) solution

Dichlorofluorescein (DCF) solution is prepared using 10 mg concentrate in 100 ml of distilled water. The solution is entirely blended until it becomes a greenish liquid. A 2.5 mL of the prepared dichlorofluorescein (DCF) solution was used for the preparation of 100 mL silver nanoparticles.

2.2 Synthesis of silver nitrate ($AgNO_3$) solution

The silver nitrate solution is prepared by using 1.698 g silver nitrate powder in 100 ml of distilled water to form 100 mM concentration. From his prepared 100 ml of silver nitrate solution; 1 ml is used for preparation of 1 mM silver nitrate solution for silver nanoparticles synthesis. The complete process is performed under absence of light.

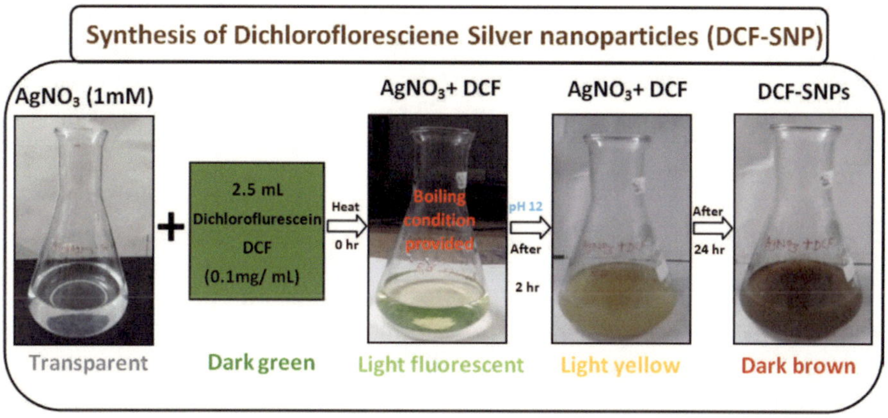

Figure 3.
Displays the synthesis of dichlorofluorescein silver nanoparticles (DCF-SNP) employing AgNO3 and dichlorofluorescein under heating conditions. The change in color of the reaction mixture was noticed from light green to dark brown, demonstrating the formation of silver nanoparticles.

2.3 Synthesis of dichlorofluorescein silver nanoparticles

The dichlorofluorescein silver nanoparticles (DCF-SNPs) is prepared by mixing 97.5 ml of AgNO3 (1 mm) and 2.5 ml of dichlorofluorescein in conical flask at constant stirring. The complete reaction process is carried out under boiling condition for 1–2 min, the pH of the mixture is adjusted to 12 and the reaction completes after 24 hours. The chemical constituents or the functional group that are presents in the Dichlorofluoresceine (DCF) structure reduced the silver salt to form the silver nanoparticles called as dichlorofluorescein silver nanoparticles (DCF-SNPs). The complete process of synthesis of dichlorofluorescein silver nanoparticles (DCF-SNPs) using $AgNO_3$ and dichlorofluorescein at 0 hr., 2 hr. and 24 hr. and observed changes in color from light green to dark brown, confirming the synthesis of silver nanoparticles synthesis is given in **Figure 3**.

3. Characterization of DCF-SNP

Various instruments are used to characterize the synthesized dichlorofluorescein silver nanoparticles. These are some of the techniques which we used in our investigations, such as FTIR analysis, that are used to classify a functional group that has a reducing and stabilizing properties. Finally, zeta potential and nanoparticles tracking analysis are used to track the surface charge and the size of the silver nanoparticles.

3.1 Fourier transform infra red (FTIR)

Dichlorofluorescein silver nanoparticles (DCF-SNPs) are used in the FTIR analysis to recognize the main functional groups involved in the reduction and capping of silver salt for the development of stable dichlorofluorescein silver nanoparticles (DCF-SNPs). In the FTIR instrument, the radiation falls on the sample and causes changes in the vibration and rotational motion of the molecules at a wavelength of 4000–440 cm − 1 consisting of near and far infrared frequencies and the FTIR spectrum of the formulation was recorded.

3.2 Nanoparticles tracking analysis (NTA)

The size of the synthesized dichlorofluorescein silver nanoparticles (DCF-SNPs) is calculated using Nanosight (LM-20, UK) at the Department of Biotechnology, Sant Gadge Baba Amravati University, India. Dichlorofluorescein silver nanoparticles (DCF-SNP) were diluted in 0.5 ml nuclease-free water and injected into the sample chamber. The instrument parameters are calibrated as specified and the device is analyzed nanoparticles samples to calculate the size of the nanoparticles. The data are summarized on the computer screen and are retrieved in PDF format for analysis.

3.3 Zeta-potential analysis (ZP)

The zeta potential characterization of dichlorofluorescein silver nanoparticles (DCF-SNPs) was tested by the Department of Biotechnology, Sant Gadge Baba Amravati University, Maharashtra, India. A zeta potential is used to assess the potential surface charge of Dichlorofluorescein Silver nanoparticles (DCF-SNP) using a zetasizer (nano ZS, malvern instrument Ltd., UK). In particular, the liquid samples of the dichlorofluorescein silver nanoparticles DCF-SNP (5 ml) were diluted with double distilled water (50 ml) using NaCl as an electrolyte suspension solution 2 M NaCl). In addition, the samples are injected into the sample slot of Zeta potential instrument and the good and important results data are reported in PDF formats. In each case, an average of three different measurements made while the values of the zeta potential ranged from +30 mV to-30 mV.

3.4 Effect of dichlorofluorescein silver nanoparticles (DCF-SNPs) on seeds germination of *Vigna radiata*

Dichlorofluorescein silver nanoparticles (DCF-SNP) was used to investigate its effect on the germination of Mung Bean (*Vigna radiata*) seeds. Mung bean seeds were purchased from the market and stored in a dry place under room temperature in the dark. Here, four separate concentrations of 25 per cent, 50 per cent, 75 per cent and 100 per cent (v/v) of dichlorofluorescein silver nanoparticles (DCF-SNPs) dispersion were prepared in distilled water. The Whatman no.3 filter paper is layered on the sterile Petri dishes of 12 cm diameter and germination test analysis was carried out. In this test, the seeds were surface sterilized with 0.1% Hgcl2 solution and rinsed three times with distilled water In total, 10 seeds of Mung Bean (*V. radiata*) were placed in the respective Petri dishes. The solution of each concentration was transferred to each Petri dish and the treatment was administered daily at only enough doses to hydrate the seeds. The petri dish was then placed in a dark seed germinator and held at 25 °C. Seed with root tip 1 mm and above was considered germinated. Percent germination and root and shoot length (in mm) were recorded every 24 hours up to 72 hours. The root length was determined from the region under the hypocotylis to the end of the root cap. The length of the shoot was measured to the nearest millimeter of the root hypocotyl transission zone to the middle of the cotyledon. The length of the shoot and root was calculated with the aid of a string and a scale [33].

4. Results and discussions

Dichlorofluorescein silver nanoparticles (DCF-SNPs) are used by different analytical methods to characterize them at the nanoscale level. The technique used for characterization were fourier transform infrared spectroscopy (FTIR), zeta potential (ZP) analysis and nanoparticles tracking analysis (NTA) using NanoSight LM-20. In addition, the synthesized dichlorofluorescein silver nanoparticles (DCF-SNPs) was used to successfully test its effect on the germination of Mung Bean (*V. radiata*) seeds. Essentially, the characterization of the synthesized dichlorofluorescein silver nanoparticles (DCF-SNP) was primarily carried out by observing the change in color of the reaction mixture from light green to dark brown. In addition different nanoparticles characterization techniques are used to chemically characterize nanoparticles. In the UV–visible spectrum analysis a single, strong, and broad surface plasmon resonance (SPR) peak was observed at 419 nm that confirmed the synthesis of Dichlorofluoresceine Silver nanoparticles. In same way, [34] UV–vis results displayed that the SPRs becomes sharper and shifts towards lower wavelength (*Escherichia hermannii* = 438 nm, *Citrobacter sedlakii* = 441 nm), shows that the particle size of AgNPs decreased. The work of [35] suggested that the green synthesis of gallic acid-coated silver nanoparticles UV–vis absorption spectrum shows typical and narrow absorption peak at approximately 400 nm.

4.1 Visualization of (DCF-SNPs)

Dichlorofluorescein silver nanoparticles (DCF-SNP) were successfully fabricated using AgNO3 (1 mM) and DCF (2.5 per cent) by boiling the reactants at 12 pH. The reaction is observed for color shift and is found to be dark brown in color after 24 hours. The synthesized dichlorofluorescein silver nanoparticles DCF-SNP before and after the reaction and color shift are shown in **Figure 4**.

The silver nanoparticles have been synthesized successfully using DCF has been synthesized by mixing 97.5 ml of $AgNO_3$ (1 mM) and 2.5 ml of dichlorofluoresceine in conical flask. It takes 75 sec to boil and the color changes from orangish to greenish color. However, [34] uses silver nitrate ($AgNO_3$) of 10–3 M concentration to the reaction vessels containing the bacterial isolate supernatants and reaction completes after 24 hours.

4.2 Fourier transforms infrared spectroscopy (FTIR) analysis

The FTIR measurement were carried out in order to identify the involvement of different functional groups present in dichlorofluorescein compound solution responsible for the bioreduction of Ag + and the capping of dichlorofluorescein silver nanoparticles (DCF-SNPs). The observed FTIR intense bands for Dichlorofluorescein compound solution were compared with the standard IR band ranges and this enables to identify the functional group at 1042.809 cm^{-1}, 1074.148 cm^{-1}, 1153.629 cm^{-1}, 1262.412 cm^{-1}, 1373.712 cm^{-1}, 1472.978 cm^{-1}, 1634.676 cm^{-1}, 1997.814 cm^{-1}, 2113.782 cm^{-1}, 2204.477 cm^{-1}, 2261.689 cm^{-1}, 2300.828 cm^{-1} and 3262.457 cm^{-1} for possibly capped the dichlorofluorescein silver nanoparticles. The FTIR spectrum of dichlorofluoresceine silver nanoparticles (DCF-SNPs) is shown in **Figure 5** and the details in terms of wave number, bond and intensity are given in (**Table 2**). **Table 3** illustrates a rather more specific FTIR spectrum of dichlorofluorescein silver nanoparticles (DCF-SNPs) including that of the area covered, peak height from left to right edge and centre.

The fourier transform infrared spectroscopy results confirms that absorption bands at 1042.809 cm^{-1}, 1074.148 cm^{-1}, 1153.629 cm^{-1}, 1262.412 cm^{-1}, 1373.712 cm^{-1}, 1472.978 cm^{-1}, 1634.676 cm^{-1}, 1997.814 cm^{-1}, 2113.782 cm^{-1}, 2204.477 cm^{-1}, 2261.689 cm^{-1}, 2300.828 cm^{-1} and 3262.457 cm^{-1} as the wave numbers for the functional groups amine, tertiary alcohol, aromatic ester, alkane, conjugated alkene, alkyne and isocynate that have taken part reducing the silver salt to form

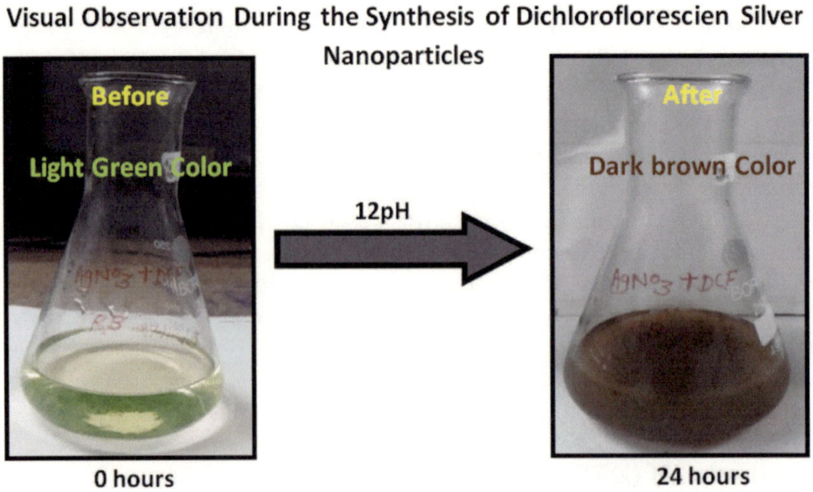

Figure 4.
Visual observation of color change during the dichlorofluorescein silver nanoparticles (DCF-SNPs) nucleation stage.

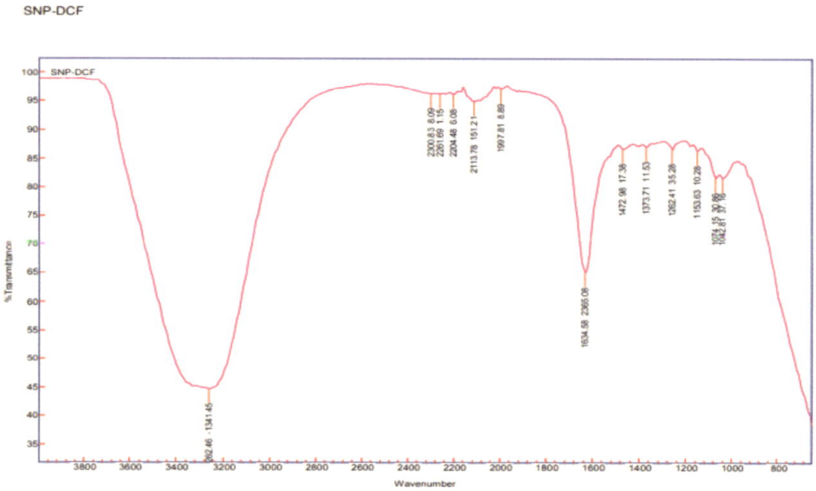

Figure 5.
Shows the FTIR spectrum of dichlorofluorescein silver nanoparticles (DCF-SNPs) which has been reduced via dichlorofluorescein (DCF) solution.

Sr. No.	Wavenumber (cm^{-1})	Bond	Functional group	Intensity
1.	1042.809	C-N stretching	Amine	Medium
2.	1074.148	C-N stretching	Amine	Strong
3.	1153.629	C-O stretching	Tertiary alcohol	Strong
4.	1262.412	C-O stretching	Aromatic easter	Strong
5.	1373.712	O-H bonding	Alcohol	Medium
6.	1472.978	C-H bonding	Alkane	Variable
7.	1634.676	C-C stretching	Conjugated alkene	Medium
8.	1997.814	C=C=C stretching	Allene	Medium
9.	2113.782	C C stretching	Alkyne	Weak
10.	2204.477	C C stretching	Alkyne	Weak
11.	22161.689	N=C=O stretching	Strong, broad	Isocynate
12.	2300.828	—	—	—
13.	3262.457	O-H stretching	Strong, broad	Alcohol

Table 2.
Shows the specifics of the FTIR spectrum of dichlorofluorescein silver nanoparticles (DCF-SNPs) in terms of wave number (cm^{-1}), bond, functional group and bonding strength.

dichlorofluorescein silver nanoparticles (DCF-SNP). Similarly, [36] showed single aldehyde, OH stretching and aldehyde, amide, carbonyl, ethylene, methoxy compounds present in Cannonball Leaves extract and involved in the reduction of silver salt for the creation of silver nanoparticles using the FTIR spectrum.

4.3 Nanoparticles tracking (NTA) analysis

Nanoparticles tracking and analysis (NTA) was carried out using NanoSight LM-20 to determine the dispersion characteristics, i. e. size and distribution of silver nanoparticles of dichlorofluoresceine (DCF-SNPs). The nanoparticles

Peak Name	Area	Height	Left Edge	Right Edge	Center
Peakr1	−37.159	1.234	1052.004	1010.969	1042.809
Peak2	−30.857	1.572	1104.232	1066.927	1074.148
Peak3	−10.283	0.895	1160.189	1141.537	1153.629
Peak4	−35.279	1.485	1279.566	1234.800	1262.412
Peak5	−11.528	0.547	1384.020	1335.523	1373.712
Peak6	−17.379	0.721	1481.013	1451.169	1472.978
Peak7	−2365.076	25.540	1764.532	1563.085	1634.576
Peak8	−8.886	0.502	2014.477	1984.633	1997.814
Peak9	−151.214	2.514	2137.584	2070.434	2113.782
Peak10	−6.076	0.452	2215.924	2197.272	2204.477
Peak11	−1.151	0.104	2268.151	2245.768	2261.689
Peak12	−8.086	0.027	2365.145	2297.996	2300.828
Peak13	1341.449	4.760	3290.312	2932.183	3262.457

Table 3.
Shows the more specific FTIR spectrum of dichlorofluorescein silver nanoparticles (DCF-SNPs) including the area covered, peak height from left to right edge and Centre.

tracking and analysis measure the size of individual nanoparticles in a suspension by their brownian motions from which the intensity of particle size distribution are obtained. The size of dichlorofluorescein silver nanoparticles (DCF-SNPs) from nanoparticles tracking and analysis was found to be less than 293 nm. The size distribution histogram of dichlorofluoresceine silver nanoparticles using nanoparticles tracking and analysis (NTA) is represented in **Figure 6** and the 3-D plot of dichlorofluoresceine silver nanoparticles size distribution intensity can be seen in **Figure 7**.

It could be seen extremely obviously from the histogram that the synthesized dichlorofluorescein silver nanoparticles (DCF-SNPs) ranged from 66 nm to 293 nm in size. The very few dichlorofluoresceine silver nanoparticles (DCF-SNPs) are

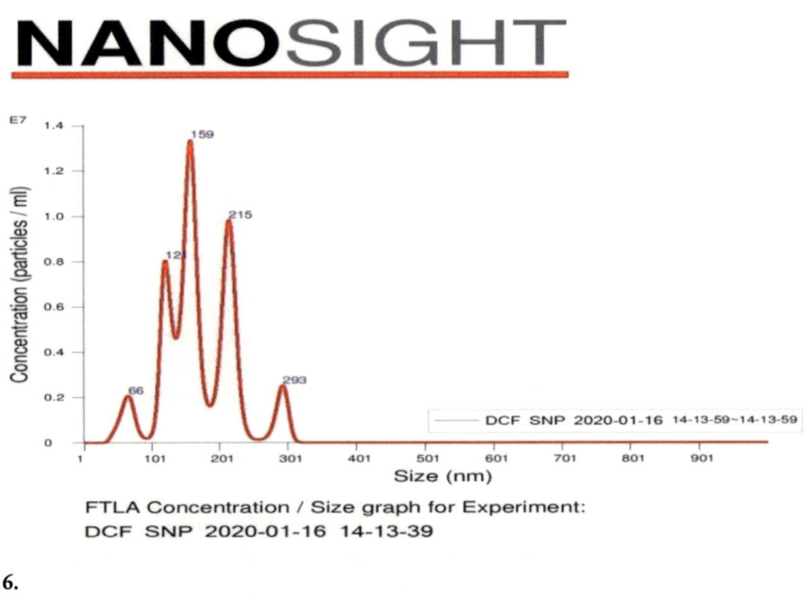

Figure 6.
Shows the size distribution histogram of dichlorofluoresceine silver nanoparticles using nanoparticles tracking and analysis (NTA).

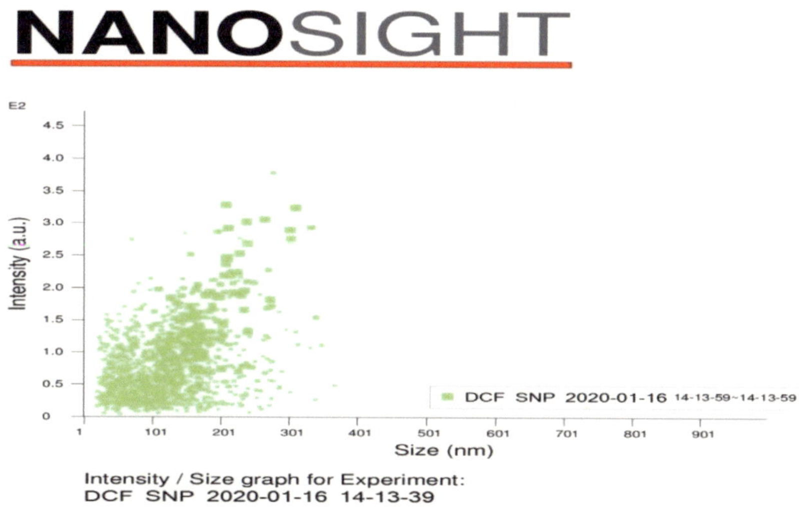

Figure 7.
Shows the 3-D plot for size distribution intensity of dichlorofluorescein silver nanoparticles (DCF-SNPs).

112 nm, 159 nm, 215 nm and 293 nm in size. However the substantial majority of dichlorofluorescein silver nanoparticles (DCF-SNP) are 159 nm wide.

The synthesized dichlorofluorescein silver nanoparticles (DCF-SNPs) intensity of their size distribution in nanoscale clearly demonstrates that most dichlorofluorescein silver nanoparticles (DCF-SNPs) are similar in diameter (159 nm) and few are clustered in scales above 300 nm. The synthesized dichlorofluorescein silver nanoparticles (DCF-SNPs) are therefore stabled formulated which could be used for chemical analysis in agriculture.

4.4 Zeta potential (ZP) analysis

The synthesized dichlorofluorescein silver nanoparticles (DCF-SNPs) has a zeta potential value of-9.35 mV implying that the dichlorofluorescein silver nanoparticles (DCF-SNP) have a high negative surface load. The zeta variance efficiency was found to be 6.10 and 0.0196, reflecting that the zeta potential synthesized dichlorofluorescein silver nanoparticles are significant and could be used to associate other materials or molecules to achieve secondary effects. **Figure 8** displays the zeta potential graph for dichlorofluorescein silver nanoparticles (DCF-SNP).

The dichlorofluoresceine silver nanoparticles (DCFs) are in ranged between 40–293 nm size in diameter confirmed by analyzing nanoparticles tracking and analysis (NTA). The size distribution and zeta potential of dichlorofluoresceine silver nanoparticles were determined by DLS and it is confirmed that the dichlorofluoresceine silver nanoparticles obtained are colloidal in nature, with average diameter approximately 159 nm and the corresponding average zeta potential for dichlorofluoresceine silver nanoparticles as −9.35 mV. In contrast, the work of Saeb et al. (2014) obtained silver nanoparticles using bacterial isolates gave the highest value of zeta potential of −30 mV, which indicates a good stability. However, unexpectedly, this zeta potential value was drastically decreased to −18.3 mV and − 9.5 mV after 30 and 90 days respectively.

4.5 Effect of DCF-SNPs on seeds germination of *Vigna radiata*

The effect of dichlorofluorescein silver nanoparticles (DCF-SNPs) on seed germination of Mung bean (*V. radiata*) was conducted and the root and shoot

			Mean (mV)	Area (%)	Width (mV)
Zeta Potential (mV):	-9.35	Peak 1:	-9.35	100.0	6.10
Zeta Deviation (mV):	6.10	Peak 2:	0.00	0.0	0.00
Conductivity (mS/cm):	0.0196	Peak 3:	0.00	0.0	0.00
Result quality	Good				

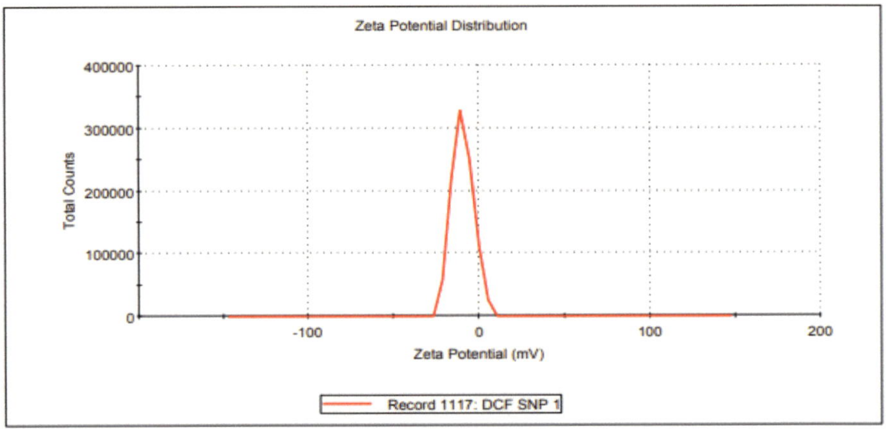

Figure 8.
Shows the zeta potential graph for surface charge on the dichlorofluoresceine silver nanoparticles (DCF-SNPs).

lengths were recorded every 24 hours. In the present experiment, the production of radicals exceeding 1 cm is interpreted to be positive growth otherwise indicated as negative. The effect of dichlorofluorescein silver nanoparticles (DCF-SNPs) on seed germination of Mung bean is shown in the **Figure 9**.

From the **Figure 9** it is noted that as the concentration of dichlorofluoresceine silver nanoparticles increases the root and shoot length decreases as compared to controls once. After 96 hrs, the Mung beans treated with 25% of dichlorofluoresceine silver nanoparticles (DCF-SNPs) shows growth in root and shoot length when compared with the positive control. However, the concentration of DCF-SNPs increases from 50% to 100%, there was inhibition of growth observed when compared to 25% of dichlorofluoresceine silver nanoparticles (**Figure 10**).

The effect of dichlorofluoresceine silver nanoparticles on seed germination of the Mung bean (*V. radiata*) was carried out and the size of root and shoot lengths were recorded after every 24 hour. Percentage of seed germination were substantially influenced by the addition of dichlorofluorescein silver nanoparticles (DCF-SNPs). The 25% concentration of DCF-SNPs treated Mung seeds indicates an excellent growth at 72 hours. In specific, the seed germination rate increases from 75% to 95%. While, as the concentration increases from 50% to 100%, the growth rate get varies or unpredicted. Our results are similar to the work of the [33] found that the as the concentration of ZnO NPs increased there was decrease in germination of seeds. Control showed statistically significant difference and could not improve shoot length. Besides, There was an increase in germination significantly with zinc oxide nano particles treated seeds at different concentrations viz., 20 mg shown 100%, 40 mg-95%, 60 mg-90%, 80 mg-90% and 100 mg of ZnO NPs shown 85% germination in Mung bean seeds [33].

The [37] study the effect of silver nanoparticles on the seed germination and plant growth and found that the highest germination rate for corn seeds, was 6.5 seeds/day, which was observed after exposure to 1.5 mg/ml of silver nanoparticles and the highest germination percentage (73.33%) and highest germination rate (1.59 seeds/day) for watermelon were recorded at 2 mg/ml silver nanoparticles.

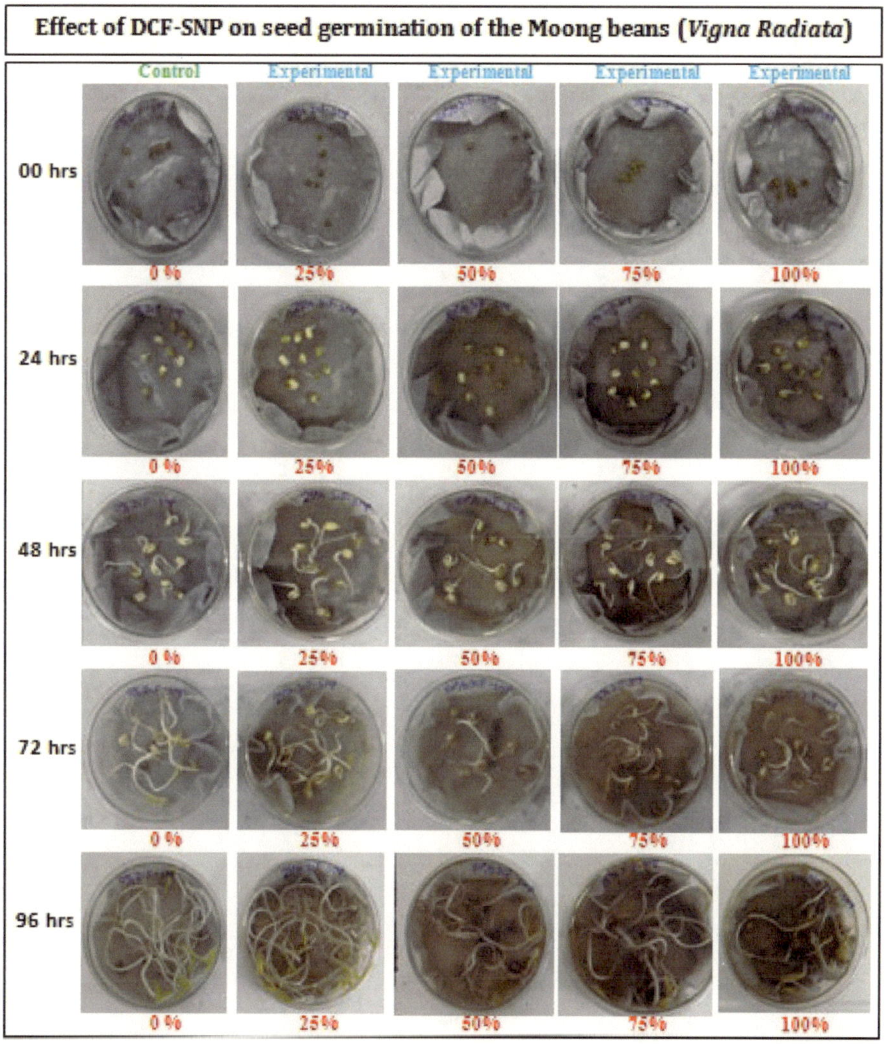

Figure 9.
Shows the time dependent toxicity effect of varying concentrations of dichlorofluoresceine silver nanoparticles (DCF-SNPs) on mung beans.

Due to interactions of dichlorofluorescein silver nanoparticles, the percentage of germination and length of root and shoot has indeed been affected. The average length of root are measure after 24 hrs, 48 hrs and 72 hrs and found to be 6.6 mm, 22.2 mm and 41.7 mm respectively. Similarly, the average length of shoot are measure after 48 hrs and 72 hrs and found to be 2.4 mm and 9 mm respectively. The experiment showed the average length of root and shoot at 72 hours was highest which is 41.7 mm and 9 mm respectively (**Table 4**).

In the present study, the dichlorofluoresceine silver nanoparticles showed unpredicted effects on root and shoot length when treated with the various concentrations of dichlorofluoresceine silver nanoparticles (DCF-SNPs). The higher concentration of nanoparticles may be attributed to toxic level of nanoparticles which has been seen in present experimentation that above certain level of concentration the seedlings respond in different way and causes subsequent declines in growth. The work of [33] evidence the same results their study stating that at low concentrations the ZnO nanoparticles shows good effect on root and shoot was more prominent.

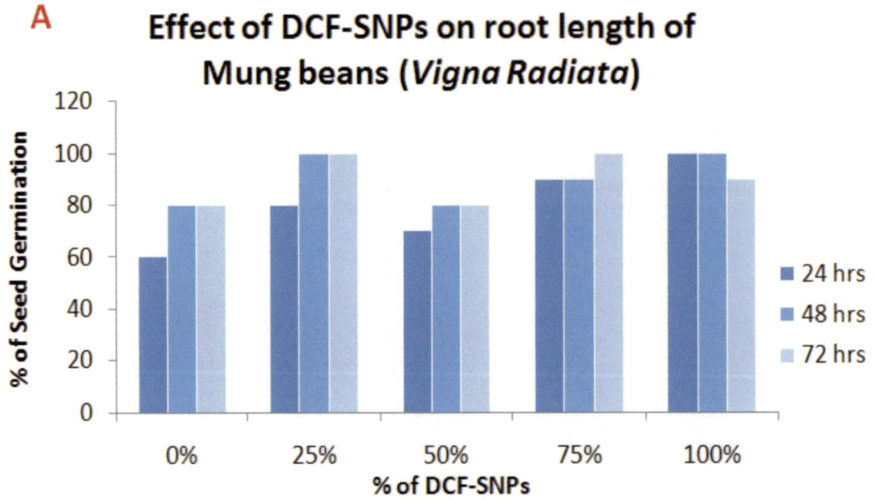

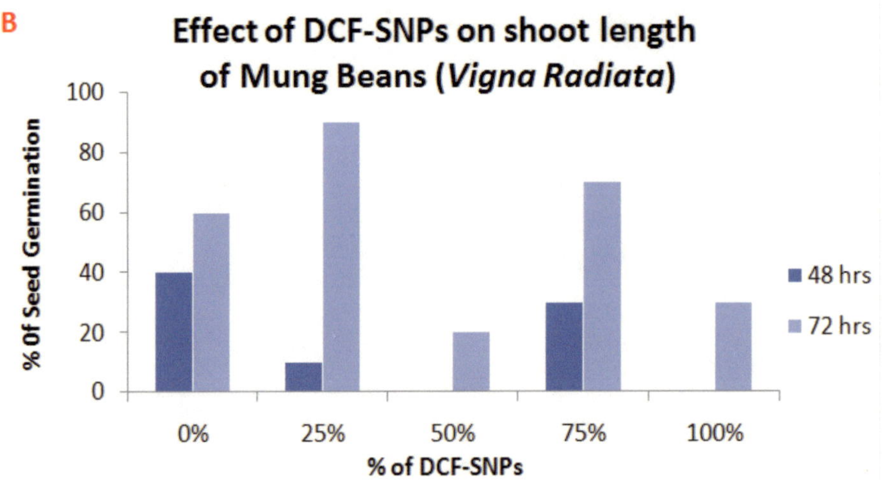

Figure 10.
Shows the dichlorofluorescein silver nanoparticles effect on a: Root length and B: Shoot length of mung beans (V. radiata) seed after 24 hr., 48 hr., 72 hr.

Sr. no	Conc. of DCF-SNPs	Average root length			Average shoot length		
		24 hrs	48 hrs	72 hrs	24 hrs	48 hrs	72 hrs
1.	0%	7 mm	16.1 mm	49.8 mm	—	4 mm	17.5 mm
2.	25%	7.1 mm	24.1 mm	50.2 mm	—	2 mm	16.3 mm
3.	50%	6 mm	22.3 mm	34.5 mm	—	—	2.5 mm
4.	75%	7.1 mm	27.4 mm	42.2 mm	—	1.3 mm	3.7 mm
5.	100%	6 mm	21.4 mm	32 mm	—	—	5.3 mm
6.	Total Avg.	6.6 mm	22.2 mm	41.7 mm		2.4 mm	9 mm

Table 4.
Indicates the average percentage of the observed root and shoot length of germinated mung bean after 24 hrs, 48 hrs, 72 hrs.

Therefore, the 25% concentration of dichlorofluoresceine silver nanoparticles (DCF-SNPs) though show positive effects on the seed germinations of mung beans it could have an advantage of using as the tracking the bio active compound, fertilizers, pesticides, hormone, minerals transfer into the plant system. As the dichlorofluoresceine silver nanoparticles (DCF-SNPs) is a organic dye it could be coupled with the non toxic materials or polymer that could have avoid harmless to plants and at the same time assist to deliver bio active essential compound in plants. This bio uptake, biotransformation, and bioaccumulation of Fluorescent dichlorofluoresceine silver nanoparticles (DCF-SNP) could be studied using the Confocal Laser-Scanning Microscopy. A study done by [38] already used Confocal laser scanning microscopy (CLSM), Leica TCS SP2 microscope (Leica Inc., Buffalo Grove, IL) to visualize the fluorescent Zein nanoparticles translocation in sugar cane.

5. Conclusion

In the current study, the synthesized dichlorofluoresceine silver nanoparticles (DCF-SNPs) were synthesized successfully by boiling method. In addition, dichlorofluoresceine silver nanoparticles (DCF-SNPs) were characterized by different techniques for measuring the particles size, morphology, functional group and surface charge. Moreover, the different concentrations of dichlorofluoresceine silver nanoparticles (DCF-SNPs) effects on germination of mung beans (*V. radiata*) and the length of root and shoot were studied. The synthesized dichlorofluoresceine silver nanoparticles (DCF-SNPs) are less than 159 nm size that interact and activates growth related gene. Therefore, 25% concentration of dichlorofluoresceine silver nanoparticles (DCF-SNP) is tested positive for shoot and root growth as compare to control. The 25% concentration of dichlorofluoresceine silver nanoparticles (DCF-SNPs) is good dye for conjugation with other bioactive compounds and useful for tracking the bio active compounds, bio uptake, biotransformation and bioaccumulation into the plant system using confocal laser scanning microscopy.

Acknowledgements

The authors would like to thank Rajiv Gandhi Biotechnology Centre, Rashtrasant Tukdoji Maharaj Nagpur University, L.I.T. Premises, Nagpur-440033 (M.S.), India, for providing research space and laboratory facility for experimentations.

Conflict of interest

The authors declare no conflict of interest.

Footnotes

The FTIR analysis of dichlorofluorescein silver nanoparticles (DCF-SNPs) were done at the Narsamma Hirayya Arts Commerce & Science college, Amravati, Maharashtra, India.

Thanks

Author would like to thank the Principal and Dr. Khandekar, of Narsamma Hirayya Arts Commerce & Science college, Kiran Nagar Near Farshi Stop,

Amravati, Maharashtra, India, for FTIR analysis. In addition, authors thanks Dr. Aniket Gade, Sant Gadge Baba Amravati University, MS, India for the zeta potential and Nanoparticles tracking and analysis (NTA) characterization of formulated dichlorofluoresceine silver nanoparticles (DCF-SNP).

Author details

Bipin Deochand Lade*, Arti Sanjay Shanware and Ruchika M. Barapatre
Rajiv Gandhi Biotechnology Centre, Rashtrasant Tukadoji Maharaj Nagpur University, L.I.T. Premises, Nagpur, Maharashtra, India

*Address all correspondence to: dbipinlade@gmail.com

References

[1] Gürses A. Chapter 2 Dyes and Pigments: Their Structure and Properties. SpringerBriefs in Green Chemistry for Sustainability. 2016;13-29. DOI : 10.1007/978-3-319-33892-7_2.

[2] Saxena S and Raja ASM. Natural Dyes: Sources, Chemistry, Application and Sustainability Issues. Textile Science and Clothing Technology. 2014; 37-80. DOI: 10.1007/978-981-287-065-0_2.

[3] WHO (2010). IARC Working Group on the Evaluation of Carcinogenic Risks to Humans. Some Aromatic Amines, Organic Dyes, and Related Exposures. Lyon (FR): International Agency for Research on Cancer; 2010. (IARC Monographs on the Evaluation of Carcinogenic Risks to Humans, No. 99.) General Introduction To The Chemistry Of Dyes.

[4] Singh K, Kumar P and Srivastava R. An Overview of Textile Dyes and Their Removal Techniques: Indian Perspective. Poll Res. 2017;36(4): 790-797.

[5] Verma S and Gupta G. Natural dyes and its applications: A brief review. International Journal of Research and Analytical Reviews. 2017;4(4):57-60.

[6] Pargai D, Jahan S and Gahlot M. Functional Properties of Natural Dyed Textiles. Chemistry and Technology of Natural and Synthetic Dyes and Pigments. 2020;1-19. DOI: 10.5772/intechopen.88933.

[7] Organic dye and pigments. ORCO, Colorant solution for all industries [Internet]. 2020. Available from: https://www.organicdye.com/agriculture/.

[8] Kolthoff I M, Lauer W M and Sunde CJ. The use of Dichlorofluorescein as an Adsorption Indicator for the Argentometric Titration of Chlorides. Journal of American Chemical Society. 1929; 51, 11: 3273-3277. DOI:https://pubs.acs.org/doi/abs/10.1021/ja01386a014.

[9] Yao H and Jockusch RA. Fluorescence and electronic action spectroscopy of mass-selected gas-phase fluorescein, 2′,7′-dichlorofluorescein, and 2′,7′-difluorofluorescein ions. Journal of physical chemistry. 2013;117(6):1351-9. DOI: 10.1021/jp309767f. Epub 2013 Jan 23.

[10] Bambach K and Rider T H. Volumetric Determinations of Halides: Use of Dichlorofluorescein as an Adsorption Indicator. Industrial & Engineering Chemistry Analytical Edition. 1935; 7 (3): , 165-166. https://doi.org/10.1021/ac50095a012.

[11] Merck. Molecular structure of 2′,7′-dichlorofluorescein [Internet]. 2020. Available from: https://www.sigmaaldrich.com/catalog/search?term=111843-78-8&interface=CAS%20No.&N=0+&mode=partialmax&lang=en&region=IN&focus=product.

[12] Buccigrossi V, Laudiero G, Russo C, Miele E, Sofia M, Monini M, Maria F R, Guarino A. Chloride secretion induced by rotavirus is oxidative stress-dependent and inhibited by Saccharomyces boulardii in human enterocytes. PloS one. 2014;9(6): e99830. DOI: 10.1371/journal.pone.0099830. eCollection 2014.

[13] Sickinger A N. Cellular DCF-DA assay. Oxidant generating capacity as a metric to allow grouping of nanomaterials and prediction of human health effects. Standard operation procedure (SOP): 2015; Wp4 1. Vol. 1 p.p. 1-6. Availablefrom:https://www.nanopartikel.info/files/projekte/nanOxiMet/SOP/nanOxiMet_SOP_DCF-DA_V1.pdf.

[14] Politano AD, Campbell KT, Rosenberger LH, and Sawyer RG. Use of Silver in the Prevention and Treatment of Infections: Silver Review. Surgical

Infections. 2013;14(1):8-20. DOI: 10.1089/sur.2011.097.

[15] National Library of Medicine, National Centre For Biotechnology Information. Silver nitrate Pubchem: (2020). Available from: https://pubchem.ncbi.nlm.nih.gov/compound/Silver-nitrate.

[16] Gao SS, Zhao IS, Duffin S, Duangthip D, Man Lo EC and Chu CH. Revitalising Silver Nitrate for Caries Management. International Journal of Environmental Research and Public Health. 2018; DOI:15,80; doi:10.3390/ijerph15010080.

[17] Karakaya D and Padem H. Effects of silver nitrate application on cucumbers (*Cucumis sativus* L.) morphology. African Journal of Biotechnology. 2012; 11(72):13664-13669. DOI: https://doi.org/10.5897/AJB11.3230.

[18] Muzamil M, Khalid N, Aziz MD and Abbas SA. Synthesis of silver nanoparticles by silver salt reduction and its characterization. Materials Science and Engineering. 2014; 60:1-9. DOI: 10.1088/1757-899X/60/1/012034.

[19] Kholoud MM, El-Nour A, EftaihaA, Al-WarthanA, Reda A.A. Ammar. Synthesis and applications of silver nanoparticles. Arabian Journal of Chemistry. 2010;3:135-140. DOI:10.1016/j.arabjc.2010.04.008.

[20] Zhang, Xi-Feng et al. "Silver Nanoparticles: Synthesis, Characterization, Properties, Applications, and Therapeutic Approaches." International journal of molecular sciences vol. 17,9 1534. 2016; doi:10.3390/ijms17091534.

[21] Burdușel, A. C., Gherasim, O., Grumezescu, A. M., Mogoantă, L., Ficai, A., & Andronescu, E. Biomedical Applications of Silver Nanoparticles: An Up-to-Date Overview. Nanomaterials (Basel, Switzerland). (2018): 8(9), 681. https://doi.org/10.3390/nano8090681.

[22] Xu L, Yi-Yi Wang, Huang J, Chun-Yuan Chen, Zhen-Xing Wang and Xie H. Silver nanoparticles: Synthesis, medical applications and biosafety. Theranostics. 2020;10(20):8996-9031. DOI:10.7150/thno.45413.

[23] Anand R and Bhagat *M. silver* nanoparticles (AgNPs): as nanopesticides and nanofertilizers. MOJ Biology and Medicine. 2019; 4(1): 19-20. DOI: 10.15406/mojbm.2019.04.00107.

[24] Lade BD and Gogle DP. Nanobiopesticide: synthesis and Applications in Plant Safety, Nano-biotechnology Applications in Plant Protection, Chapter 10, Volume 2 (Eds: Kamel Abd-Elsalam, and Ram Prasad, *Springer Nature, USA* under "*The Nanotechnology in the Life Sciences*" series; 2019. 169-189, volume 2, DOI: 10.1007/978-3-030-13296-5 9.

[25] Rane AV, Kanny K, Abitha VK and Thomas S. Methods for Synthesis of Nanoparticles and Fabrication of Nanocomposites. Synthesis of Inorganic Nanomaterials. 2018;121-139. DOI: https://doi.org/10.1016/b978-0-08-101975-7.00005-1.

[26] Baltazar-Encarnación E, Escárcega-González CE, Vasto-Anzaldo XG,Cantú-Cárdenas ME andMorones-Ramírez JR. Silver Nanoparticles Synthesized through Green Methods Using *Escherichia coli* Top 10 (Ec-Ts) Growth Culture Medium Exhibit Antimicrobial Properties against Nongrowing Bacterial Strains. Journal of Nanomaterials. 2019;1-9. DOI: https://doi.org/10.1155/2019/4637325.

[27] Mukherjee K, Gupta R, Humar G, Kumari S, Biswas S and Padmanabhan P. silver nanoparticle form *Bacillus clausii* and computational profiling of nitrate reductase enzyme involved in production. Journal of Genetic Engineering and Biotechnology. 2018. 16 (2); 527-536. https://www.sciencedirect.com/science/article/pii/S1687157X18300313.

[28] Hamouda, R.A., Hussein, M.H., Abo-elmagd, R.A. *et al.* Synthesis and biological characterization of silver nanoparticles derived from the cyanobacterium *Oscillatoria limnetica*. *Science Report.* 9, 13071 (2019). https://doi.org/10.1038/s41598-019-49444-y.

[29] Bipin D. Lade and Arti S. Shanware (January 21st 2020). Phytonanofabrication: Methodology and Factors Affecting Biosynthesis of Nanoparticles, Smart Nanosystems for Biomedicine, Optoelectronics and Catalysis, Tatyana Shabatina and Vladimir Bochenkov, IntechOpen, DOI: 10.5772/intechopen.90918.

[30] Kumar R, Roopan SM, Prabhakarn A, Khanna VG, Chakroborty S. Agricultural waste *Annona squamosa* peel extract: biosynthesis of silver nanoparticles. Spectrochim Acta A Mol Biomol Spectrosc. 2012 May;90:173-6. doi: 10.1016/j.saa.2012.01.029. Epub 2012 Jan 28. PMID: 22336049.

[31] Ahmed S, Ahmad M, Swami BL, Ikram S. A review on plants extract mediated synthesis of silver nanoparticles for antimicrobial applications: A green expertise. Journal of Advanced Research. 2016;7(1):17-28. DOI: https://doi.org/10.1016/j.jare.2015.02.007.

[32] Nano Composix. Zeta Potential Analysis Of Nanoparticles, version 1.1. Nanocomposix, San Diego, 2012; CA, pp 1-6.

[33] Jayarambabu N, Kumari B S, Rao K V and Prabhu YT. Germination and Growth Characteristics of Mungbean Seeds (*Vigna radiata* L.) affected by Synthesized Zinc Oxide Nanoparticles. International Journal of Current Engineering and Technology. 2014; 4 (5); http://inpressco.com/wp-content/uploads/2014/09/Paper593411-3416.pdf.

[34] Saeb AT, Alshammari AS, Al-BrahimH, Al-RubeaanKA. Production of silver nanoparticles with strong and stable antimicrobial activity against highly pathogenic and multidrug resistant bacteria. The scientific world journal. 2014 ;2014:704708. DOI: 10.1155/2014/704708.

[35] Li D, Liu Z, Yaun Y and Niu F. Green synthesis of gallic acid-coated silver nanoparticles with high antimicrobial activity and low cytotoxicity to normal cells. Process Biochemistry. 2015; 50(3): DOI: 10.1016/j.procbio.2015.01.002.

[36] Devraj P, Kumari P, Chirom A and Renganathan A. Synthesis and Characterization of Silver Nanoparticles Using Cannonball Leaves and Their Cytotoxic Activity against MCF-7 Cell Line. Journal of Nanocomposites. 2013;1-5. DOI: https://doi.org/10.1155/2013/598328.

[37] Almutairi Z and Alharbi A. Effect of Silver Nanoparticles on Seed Germination of Crop Plants. International Journal of Nuclear and Quantum Engineering. 2016;9:594-598.

[38] Alisha Prasad, Carlos E. Astete, Andreea E. Bodoki, McKenzie Windham, Ede Bodoki, and Cristina M. Sabliov. Journal of Agricultural and Food Chemistry. 2018 66 (26), 6544-6551.DOI: 10.1021/acs.jafc.7b02487.

Chapter 3

Multifunctional Silver Nanoparticles: Synthesis and Applications

Nguyen Hoang Nam

Abstract

Multifunctional silver nanoparticles have attracted widely due to their potential applications. Based on the properties of individual silver nanoparticles, such as plasmonic and antibacterial properties, silver nanoparticles can become multifunctional by surface modifications with various surfactants or they can be combined in core-shell and composite structures with the magnetic nanoparticles to form bifunctional nanoparticles. After reviewing the methods of synthesis and applications of silver nanoparticles, the chapter describes the synthesis and the properties of the new types of multifunctional silver nanomaterials based on the plasmonic behaviors of silver nanoparticles and the iron oxide Fe_3O_4 superparamagnetic nanoparticles. One type is a simple combination of silver nanoparticles and iron oxide nanoparticles in a silica matrix Fe_3O_4/Ag-4ATP@SiO_2. Other types are the core-shell structured nanoparticles, where Fe_3O_4 nanoparticles play as the core and silver nanoparticles are the outer shell, so-called Fe_3O_4@SiO_2-Ag and Fe_3O_4-Ag. In the Fe_3O_4@SiO_2-Ag, silver nanoparticles are reduced on the surface of silica-coated magnetic core, while in Fe_3O_4-Ag, silver nanoparticles are directly reduced on the amino groups functionalized on the surface of magnetic nanoparticles without coating with silica. Both of types of the multifunctional silver nanoparticles show the plasmonic and magnetic properties similar as the individual silver and iron oxide nanoparticles. Finally, some applications of those multifunctional silver nanoparticles will be discussed.

Keywords: Silver nanoparticles, magnetic nanoparticles, plasmonic-magnetic bifunctional, core-shell nanoparticles, multifunctional nanoparticles, combination synthesis, bioapplications

1. Introduction: Silver nanoparticles and their properties

Silver was used for long time in human life as jewelry and houseware. It is a transition metal which is soft and has the highest reflectivity. Among its many useful applications, silver also has known as an infection controller from ancient civilizations [1]. In the Han Dynasty in China *circa* 1500 BC, the earliest use of silver for therapeutic purposes was recorded [2]. Silver vessels and plates were commonly used during the Phoenician, Macedonian, and Persian empires [3]. Families of the higher socioeconomic classes during the middle-ages were so acquainted with the usage of silver. The result is that they developed bluish skin discolorations known as argyria, an affliction which may have led to the term 'blue blood' to describe members of the aristocracy [4]. In recent years, silver has gained attention as an antimicrobial agent due to the increase in bacterial antibiotic resistance, along with the development of nanosilver studies opening up the huge potential applications of silver in biomedicine. Today, modern medicine uses medical silver forms, such as silver sulfadiazine, silver nitrate, colloidal silver and/or silver in the nanoparticle form. Applications of the silver nanoparticles can be found in personal care products, home products, medical applications, and agricultural/industrial applications. Silver nanoparticles can be defined as silver particles with the size ranged from 1 to 100 nm. When the size of particles decreases to the nano scale, the surface-volume ratio dramatically increases, then bring the nanoparticles become highly applicable in applications requiring large surface contact, such as catalysis or antibacterial therapeutic. Silver in aqueous solution, or in its monoatomic state (Ag^+), can be biologically active [5]. It is similar to the form of silver sulfadiazine or silver nitrate, which usually be used in wound healing treatments. In nanocrystalline form, silver metallic (Ag^0) also can be biologically active when it releases ions or either in its native form. To date, the mechanisms by which silver acts on bacterial can be listed as follows:

- Silver ions react with the peptidoglycan component and form pores and puncture the bacterial cell wall [6];

- Silver cations enter into the bacterial cell, disrupting metabolic pathways resulting in generation of reactive oxygen species and inhibiting cellular respiration [7];

- In the cell, silver disrupt DNA replication cycle and DNA itself [8, 9].

Using those mechanisms, the wide and ever-expanding applications of silver nanoparticles in medicine, health care, and other daily life activities were developed [10].

In addition to antibacterial applications above, silver nanoparticles are being integrated into products from photovoltaics to chemical sensors due to their unique optical, electrical, and thermal properties. The most interesting property of silver nanoparticles is plasmonic absorption as shown in **Figure 1a**. The plasmons are the oscillations of free electrons in metallic nanoparticles that are excited by incident light as can be illustrated in **Figure 1d**.

The plasmons are the consequence of the formation of a dipole in the nanoparticles due to incident electromagnetic waves, which lead the electrons migrate to restore initial state. The incident electromagnetic waves oscillate forcing the electrons to oscillate at the same frequency as the light, which called resonant frequency. These coupling are called surface plasmon resonance because the free electrons are distributed on the surface of silver nanoparticles and, therefore, the resonant frequency strongly depends on the size and shape of the silver nanoparticles. In experiments, the surface plasmonic absorptions are detected by monitoring the resonant frequency as a function of incident wavelength. **Figure 1a** exhibits the absorption spectra of silver nanoparticles prepared by the sonoelectrochemical method with various concentration of surfactant to control the size of nanoparticles. In **Figure 1a**, two absorption peaks related to two size distributions of silver nanoparticles can be observed. Nanocrystalline silver has the face-centered cubic structure with the X-ray diffraction XRD peaks of (111), (200) and (220) as can be seen in **Figure 1b**. The prepared silver nanoparticles have the transmission electron microscope (TEM) images as shown in **Figure 1c** with size distribution varies when the current changes accordingly. In **Figure 1d**, the schematic structure

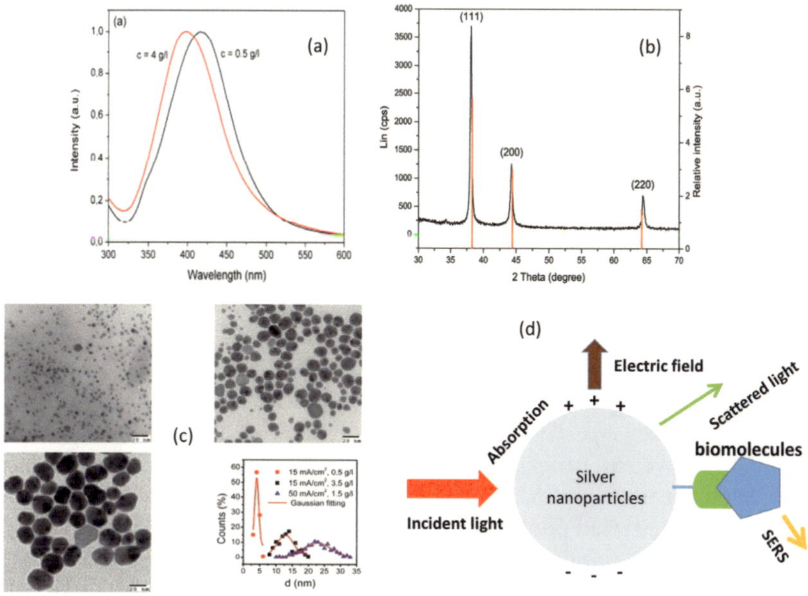

Figure 1.
Silver nanoparticles prepared by sonoelectrochemical method. (a): Absorption spectra of silver nanoparticles (b) XRD pattern of silver nanoparticles (c) TEM images and size distribution of silver nanoparticles (d) schematic illustration of properties of silver nanoparticles.

of functionalized silver nanoparticles is illustrated. As described above, the absorption or the scattering cross-sections describe the intensity of a given frequency to be absorbed or scattered, respectively. The absorptions are due to the collective electrons oscillations to form surface plasmon resonances, which exhibit enhanced near-field amplitude at the resonance wavelength. This near-field is highly localized at the nanoparticles and decays rapidly away from the nanoparticles, though far-field scattering by the nanoparticles is also enhanced by the resonance. Under the effect of incident laser, these resonance increases significantly the intensity of the Raman scattering signals of the vibration of molecules/biomolecules functionalized on the surface of nanoparticles. Observing these enhanced Raman signals under the incident wavelength, the Surface Enhance Raman Scattering (SERS) technique can be used to recognize the vibration of molecules conjugated on the surface of the nanoparticles, then can be applied in bio-labeling. Under a certain incident wavelength, the enhanced Raman signals are sensitive with the shape and size distribution of silver nanoparticles, which can be varied with different method of synthesis.

Silver nanoparticles can be produced from chemical, physical, physicochemical methods, as well as biological or green methods [11–13]. Chemical reduction is the most common approach for synthesis of silver nanoparticles. The reducing agents, such as sodium borohydride $NaBH_4$, sodium citrate ... reduce Ag^+ to form metallic silver Ag^0, which aggregate into oligometric clusters, then form metallic silver particles [14–16]. During the production process, the dispersive nanoparticles need to be stabilized and protected from the agglomeration by using surfactants as protective agents. Those functional comprising surfactants, such as thiols, amines, alcohols, etc. interact with particle surfaces and stabilize particles growth, as well as protect nanoparticles from agglomeration, sedimentation, or losing their surface properties. One of the effective protective agents to stabilize nanoparticles is polymeric compound, such as polyvinyl alcohol, polyethylene glycol, polymethylmethacrylate, etc. Silver nanoparticles were prepared by injecting them with various rates at high temperature up to 100 °C [17].

Microemulsion techniques are effective to synthesize uniform and size controllable silver nanoparticles [18]. The metal precursor and the reducing agent are in two immiscible phases aqueous organic-systems. The interface between the two liquids and the intensity of inter-phase transports affect the rate of the reactions between metal precursors and reducing agents, leading to the metal clusters formation at the interface. The metal clusters are stabilized by coating with stabilizer molecules occurring in the non-polar aqueous medium.

Physical approaches can be used to synthesize silver nanoparticles with advantages of the absence of solvent contamination. Laser ablation is a method to synthesize silver nanoparticles by laser ablation of metallic bulk materials in solution [19]. Evaporation-condensation and arc discharge are also effective physical methods to synthesize silver nanoparticles [20, 21].

Physicochemical approaches, which use a physical method to support chemical reduction, are very effective methods to synthesize silver nanoparticles. Photoreduction is a clean process to fabricate the silver nanoparticles in various mediums, such as cells, emulsion, polymer films, etc. One effective and simple method is the UV-initiated photoreduction, which use the UV light to synthesize silver nanoparticles in stabilizing agents for prevention of the aggregation [22]. Besides, the electrochemical and sonoelectrochemical methods can be used to synthesize silver nanoparticles with controlling the particles size by adjusting electrolysis and ultrasonic parameters [23, 24]. A modified sonoelectrodeposition method was used to synthesize silver nanoparticles [16, 25]. In this modification the cathode is a silver plate, which allows producing silver nanoparticles with the size range of 4–30 nm dispersed in a non-toxic medium. Furthermore, microwave-assisted synthesis provides silver nanoparticles with smaller sizes, narrower size distributions, and a higher degree of crystallization [26].

Biological or green methods need to develop as environmentally and economically friendly processes, which do not use toxic chemicals in the protocol. These methods use organisms, such as prokaryotic bacterial cells, eukaryotic fungi and plants to synthesize silver nanoparticles [27]. Biological methods could be used for synthesis of highly stable and well-characterized nanoparticles. Sizes and morphologies of synthesized silver nanoparticles can be controlled by altering some critical conditions, including pH, mixing speed, light, exposure time, temperature, substrate concentration, buffer strength, biomass and substrate concentration. In these methods, the critical aspects, such as types of organisms, genetical and inheritable properties of organisms, election of the biocatalyst states, enzyme activity, optimal conditions for cell growth and optimal reaction conditions have been considered.

2. Multifunctional silver nanoparticles and their applications

Multifunctional nanoparticles have attracted much interest because of their ability to combine various properties, such as electronic, optical, catalytic, surface enhance, and magnetic properties, by using the combination of individual nanoparticles and/or functionalize the core nanoparticles with appropriate agents [28–31, 15]. Combining of individual nanoparticles enhances functionality or creates new properties compared to those of individual nanoparticles. For instance, nanoparticles with the combination of plasmonic and magnetic properties can be a good candidate for bio imaging, diagnosis and therapy using their plasmonic properties. In parallel, they can be used for targeted drug delivery and magnetic resonance imaging under an external field, as well as allow bio tissues to be targeted and separated with their superparamagnetic properties.

Individual silver nanoparticles are special plasmonic nanoparticles which can be used as an antibacterial agent and/or as the plasmonic agent in biomedicine applications. They can be multifunctional if they are functionalized with appropriate surfactants allowing to apply silver nanoparticles in more than one purpose, such as in labeling together with in therapy. They also can be multifunctional if they can be combined with other individual nanoparticles, such as magnetic or semiconductor nanoparticles, in order to have both the properties/functionalities of individual nanoparticles in the combined one. In this section, some simple combinations of silver nanoparticles and magnetic nanoparticles, which so-called multifunctional silver nanoparticles, will be described.

One type of the magnetic nanoparticles, which most often used for bio applications due to their good superparamagnetic feature and simple in production, is iron oxide Fe_3O_4 nanoparticles [28]. During the production, they can be stabilized and more biocompatible by coating with polymers, chitosan, starch and silica, in which silica has shown much higher stability than the other coatings [32]. The amorphous silica shells can also behave as carriers for the immobilization of functional nanoparticles and avoid their aggregation, as well as provide the ability of further functionalizing. In order to combine magnetic nanoparticles Fe_3O_4 and silver nanoparticles to form multifunctional silver nanoparticles, silica shell plays an important role, which keeps the stability of bifunctional plasmonic-magnetic nanocomposites. It can be the outer shell of the combination of 4-Aminothiophenol (4-ATP) functionalized silver nanoparticles and magnetic Fe_3O_4 nanoparticles as shown in **Figure 2a**, which protect and keep both functionalities of individual nanoparticles [15]. It also can be act as a very thin protection layer of the magnetic core, and then silver ion was observed and reduced on the surface of the silica layer to form plasmonic-magnetic nanocomposites as can be seen in **Figure 2b** [29]. In comparison, Fe_3O_4 magnetic nanoparticles are functionalized with 3-aminopropyl-triethoxysilance (APTES) and silver nanoparticles are reduced on the amino ($-NH_2$) groups of this functionalized layer to form multifunctional silver nanocomposites, which have a one-atom thin layer of Si-O on the surface of Fe_3O_4 core as can be seen in **Figure 2c** [30, 31].

The magnetic nanoparticles in all listed composites were synthesized by co-precipitation method [28], where Fe^{2+}/Fe^{3+} with 1:2 molar ratio was diluted to 0.01/0.02 M concentration from $FeCl_2$ and $FeCl_3$. Polyvinylpyrolidone (PVP) was used as stabilizer to restrain the nanoparticles size. The mixture was stirred and kept warm at 60–70 °C. NH_4OH, then was added to the warmed mixture to have

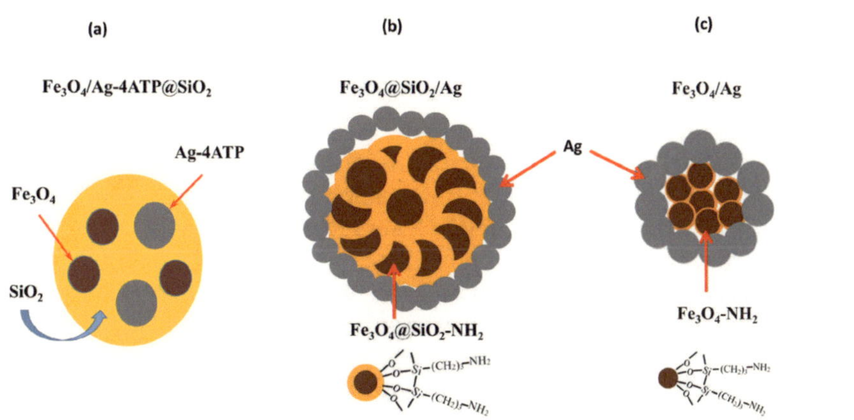

Figure 2.
Schematic structure of silver multifunctional nanoparticles.

the black color precipitation, which was collected by purification with magnetic separation and washing with ethanol and distilled water. This is the most common and simplest way to produce Fe_3O_4 magnetic nanoparticles.

In order to combine plasmonic and magnetic properties of individual nanoparticles, the simplest thinking is to cover both of them in a suitable shell, such as silica. Biocompatible silica layer can be used as stabilizer, as well as the dock for further targeted functionalizing. Inside the silica shell, the silver nanoparticles need to exhibit all the plasmonic behaviors, which are properties of multifunctional silver nanoparticles. To confirm that, the silver nanoparticles were prepared with functionalized 4-ATP, which will be indicator for the surface plasmon resonance, on the surface to form Ag-4ATP nanoparticles [15]. They were produced by reducing silver acetate $AgCH_3COO$ solution at 0.01 M concentration using $NaBH_4$ as reductant with the present of PVP as surface activator. The solution was vigorously stirred before 4-ATP was added and the covalent S-Ag linkages were formed between the silver nanoparticles surface and the 4-ATP molecules. The prepared silver nanoparticles functionalized with 4-ATP (Ag-4ATP) was combined with the produced Fe_3O_4 magnetic nanoparticles by using inverse micro emulsion method, which was created from the mixing of hydrophobic phase of toluene and hydrophilic phase that was made from the Ag-4ATP and Fe_3O_4 aqueous solutions. Under sonic bath, the mixture was moderated for 2 h before adding tetraethyl orthosilicate (TEOS), which react with water in solution to form silica shell. The combination of magnetic and plasmonic nanoparticles in silica shell is symbolled by Fe_3O_4/Ag-4ATP@SiO_2.

In order to produce plasmonic-magnetic nanocomposites in "core-shell" form, Fe_3O_4 magnetic nanoparticles were first coated by silica layer, and then were functionalized with APTES. The magnetic nanoparticles were dispersed in water mixed with ammonia solution. TEOS was added into the mixture to form a silica layer on the magnetic nanoparticles. After washing, the collected nanoparticles were functionalized with the amino group by a silanol reaction of APTES in a mixture containing silica-coated magnetic nanoparticles and ammonia Fe_3O_4@SiO_2-NH_2. Finally, the plasmonic-magnetic nanocomposites denoted by Fe_3O_4@SiO_2-Ag were prepared by the reduction of $AgNO_3$ onto the surface of amino functionalized magnetic nanoparticles. The reduction of silver ions was carried out in pH of 11 by using sodium borohydride $NaBH_4$ as a reducing agent, under the influence of an ultrasonic wave. This method is called ultrasound-assisted chemical method [29].

Magnetic nanoparticles can also be directly functionalized with APTES without coating with silica layer [30, 31]. They were dispersed in ethanol before adding the APTES and NH_4OH with assistance of ultrasound. The next step is the reduction of $AgNO_3$ onto the surface of amino functionalized magnetic nanoparticles and the final product denoted by Fe_3O_4-Ag has the structure as expected in **Figure 2c**.

The prepared multifunctional nanoparticles/composites need to be confirmed the structure and morphology by using X-ray diffractometer (XRD) and high resolution TEM (HRTEM). An example can be seen in **Figure 3**, where the XRD patterns in **Figure 3(i)** of Ag-4ATP, Fe_3O_4 nanoparticles and the combination of them in silica shell Fe_3O_4/Ag-4ATP@SiO_2 are shown together [15]. They demonstrate that the combination contains both the individual nanoparticles, where the XRD pattern of the combination shows both the characteristic peaks of silver and magnetic nanoparticles. The XRD results of Fe_3O_4@SiO_2-Ag and Fe_3O_4-Ag composites also lead to the similar conclusions [29, 31]. The TEM images in **Figure 3(ii)** conclude that the combination contains both the individual nanoparticles due to the existence of bigger nanoparticles of silver and smaller one of magnetic nanoparticles, in consistence with the results observed from XRD patterns. The pattern D in **Figure 3(ii)** shows the schematic model of the combination in pattern C indicating

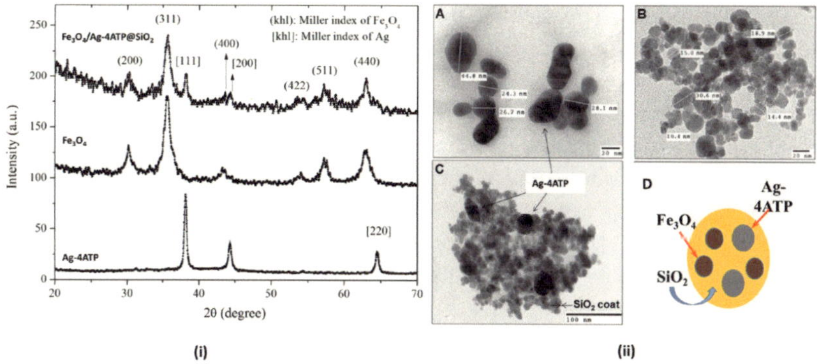

Figure 3.
XRD pattern (i) and TEM images (ii) of the combination of Fe_3O_4 and Ag-4ATP nanoparticles in silica matrix.

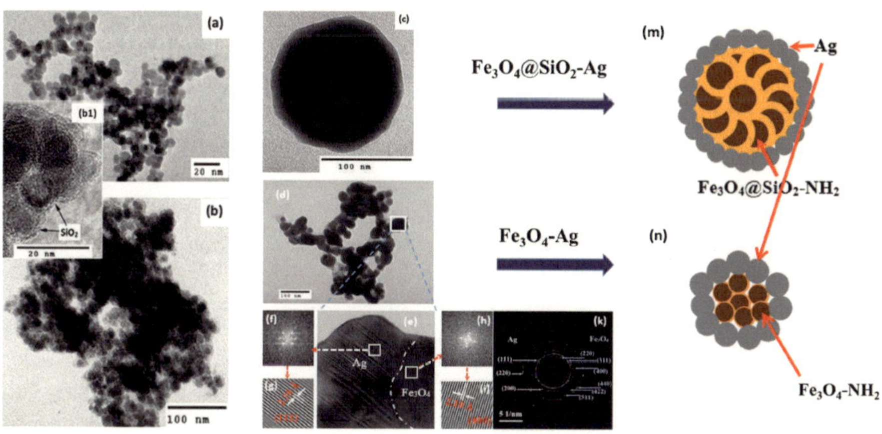

Figure 4.
TEM and HRTEM images and the schematic structure of core-shell silver multifunctional nanoparticles.

that the combination is covered by a thin layer of silica and have the size of around 150–200 nm, which can be applicable for biomedicine, despite that they are not so homogenous.

The HRTEM images of plasmonic-magnetic Fe_3O_4@SiO_2-Ag and Fe_3O_4-Ag composites are shown in **Figure 4** [29–31], where the **Figure 4a** shows the image of magnetic core. **Figures 4b** and **4b1** show the TEM and HRTEM image of the magnetic nanoparticles coated with silica Fe_3O_4@SiO_2, respectively, while the **Figure 4c** shows the final composite image after Ag nanoparticles were reduced on the amino groups NH_2- around the magnetic core. In this TEM image, two layers can be seen: the inner core is high contrast, which can be the Fe_3O_4@SiO_2 nanoparticles, while the outer layer with low contrast can be the silver nanoparticles layer. The outer layer is of thickness of around 10 nm, in consistence with that conducted from XRD results. The possible structure of the nanocomposite Fe_3O_4@SiO_2-Ag with the size of around 150 nm have schematically shown in **Figure 4m**. This possible structure was also strengthened by the XRD results, which also show both the peaks of Fe_3O_4 and Ag nanocrystalines [29]. Similarly, the **Figure 4d** shows the TEM images of Fe_3O_4-Ag nanocomposites and the enlarged HRTEM image in **Figure 4e** shows the composites structure, where the crystalline patterns of silver part and Fe_3O_4 parts can be clearly seen. The patterns in **Figure 4f** and **g** show the lattice and diffraction images of silver part in composites, while **Figure 4h** and **i** show those of magnetic crystalline, respectively. The diffraction of the composites can be seen in **Figure 4k** and then the schematic structure of Fe_3O_4-Ag is illustrated in **Figure 4n**. It is noted that Fe_3O_4-Ag, which have the magnetic core without the coating of silica shell, have more simple structure than that of Fe_3O_4@SiO_2-Ag. In addition, they have the size of around 50 nm which is much smaller than that of Fe_3O_4@SiO_2-Ag.

In this type of "core-shell" nanocomposites of Fe_3O_4@SiO_2-Ag, the steps of forming nanocomposites can be checked by using Fourier-Transform Infrared (FTIR) spectroscopy. In the **Figure 5**, one can see the FTIR spectra of (a) Fe_3O_4, (b) Fe_3O_4@SiO_2 and (c) Fe_3O_4@SiO_2 functionalized with amino group Fe_3O_4@SiO_2-NH_2. All the spectra show the absorption peaks at 426 cm^{-1} and 530 cm^{-1} of F-O vibration corresponded to the magnetic core, where peak at 530 cm^{-1} is broaden and shifted to 537 cm^{-1} in coated nanoparticles. The peaks at 1045 cm^{-1} in (b) and (c) can be assigned to stretching vibration of Si-O-Si. The peaks at 882 cm^{-1}, 1087 cm^{-1}, 1271 cm^{-1}, and 2979 cm^{-1} are only observed in Fe_3O_4@SiO_2-NH_2, which can be assigned to a twisting vibration of the -CH_2 group [33, 34], a stretching vibration of C-N [35], a wag vibration of -CH_2 [33] or a C-N binding vibration [36] and a stretching vibration of -CH_2 [33, 37, 38], respectively. These indicated that the Fe_3O_4 nanoparticles were coated by silica layers, then amino groups were successfully functionalized onto the surface of Fe_3O_4@SiO_2 nanoparticles, as schematically shown in right side of the **Figure 5**. In Fe_3O_4-Ag, the X-ray photoelectron spectroscopy (XPS) results also lead to the same conclusion that the Fe_3O_4 magnetic

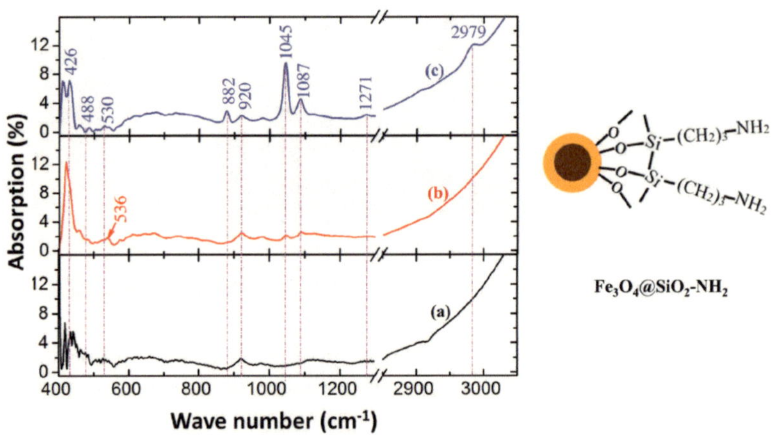

Figure 5.
The FTIR spectra of the Fe_3O_4, Fe_3O_4@SiO_2 and Fe_3O_4@SiO_2-NH_2 nanoparticles.

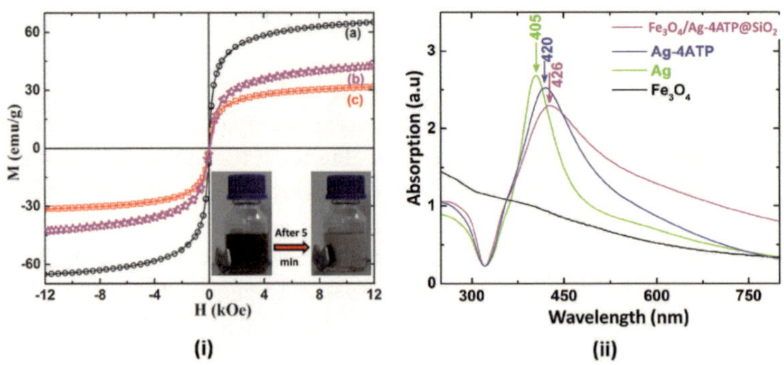

Figure 6.
Magnetic (i) and optical (ii) properties of Fe_3O_4/Ag-4ATP@SiO_2.

nanoparticles were successfully functionalized by NH_2 amino groups, then the Ag reduced on those to form multifunctional silver nanoparticles [30].

In order to confirm the multifunctionalities of nanocomposites, the magnetic and the optical properties of the final product were measured in comparing with those of the individual magnetic and silver nanoparticles. **Figure 6(i)** shows the magnetization curves of Fe_3O_4 nanoparticles in (a), Fe_3O_4/Ag-4ATP/SiO_2 with different ratio of Fe/Ag in (b) and (c), respectively. Both of them show the superparamagnetic properties without the coercivity and applicable in biomedical applications. The magnetization of the Fe_3O_4 magnetic nanoparticles has highest value of around 64 emu/g under the external field of 12 kOe. The magnetization decreases when the magnetic nanoparticles were combined with non-magnetic Ag-4ATP nanoparticles. When the ratio of Fe/Ag decreases from (b) to (c), the magnetization also decreases, indicating that the combination nanoparticles have magnetic properties as expected from the combined structure and can be used in biomedical applications. The inset shows the ability to be applied in magnetic separation of the multifunctional nanoparticles after 5 min. In parallel, the absorption spectra of Fe_3O_4, Ag, Ag-4ATP and Fe_3O_4/Ag-4ATP/SiO_2 nanoparticles were shown in **Figure 6(ii)**. The absorption peak of silver nanoparticles was broadened and shifted from 405 nm to 420 nm and then 426 nm, when the silver nanoparticles were functionalized with 4ATP then put into the combination of Fe_3O_4 and Ag-4ATP nanoparticles inside the silica matrix.

The Fe_3O_4 nanoparticles do not show the similar absorption peaks in the spectrum. The combination nanoparticles then have the plasmonic properties same as the individual Ag nanoparticles themself, therefore, they can be used in labeling or diagnostic applications based on the plasmonic properties.

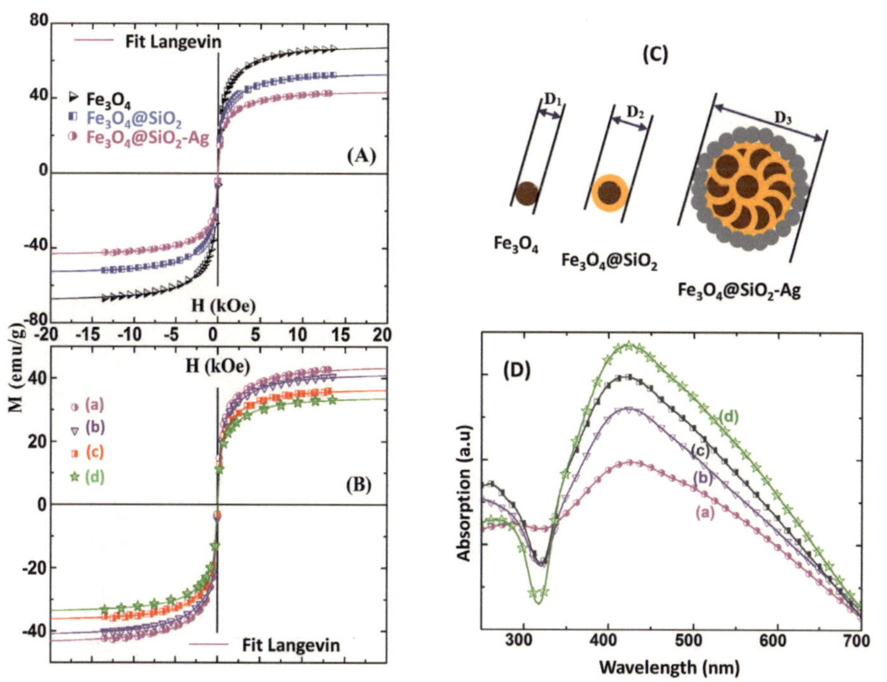

Figure 7.
(A) Magnetization curves of Fe_3O_4, Fe_3O_4@SiO_2 and Fe_3O_4@SiO_2-Ag nanoparticles (B) magnetization curves of Fe_3O_4@SiO_2-Ag nanoparticles with various ratios of Fe/Ag (c) schematic structure of from the core to the shell of the Fe_3O_4@SiO_2-Ag nanoparticles (D) absorption spectra of Fe_3O_4@SiO_2-Ag nanoparticles with various rations of Fe/Ag.

The $Fe_3O_4@SiO_2$-Ag "core-shell" multifunctional silver nanoparticles also show both the superparamagnetic and the plasmonic properties, which change when the Fe/Ag ratio changes [29]. **Figure 7A** shows the magnetization curves of Fe_3O_4, $Fe_3O_4@SiO_2$, and $Fe_3O_4@SiO_2$-Ag nanoparticles under the external field. The magnetization of those decreases when the core Fe_3O_4 magnetic nanoparticles were coated with silica, then silver nanoparticles was reduced onto the surface of coated nanoparticles, as schematically shows in **Figure 7C**. When the amount of reduced silver nanoparticles on the surface of coated nanoparticles increases, or, in another words, the ratio of Fe/Ag of the core-shell nanoparticles decreases, the magnetization of those decreases accordingly as shown in **Figure 7B**. In this figure, (a), (b), (c) and (d) denote the magnetization curves of the nanoparticles with the decrease of the ratio Fe/Ag. In consistence, the absorption spectra of those exhibit the increase in the absorption peak of silver nanoparticles as can be seen in **Figure 7D**. Those results also imply that the core-shell nanoparticles combine the plasmonic and magnetic properties, which can be used together in the applications of the multifunctional silver nanoparticles. The Fe_3O_4-Ag composites, as expected, show the same picture [30, 31].

The plasmonic-magnetic nanoparticles can be applied in the applications which can use both the magnetic and plasmonic properties of them. In order to check the applicability in biomedical applications, the multifunctional silver nanoparticles were or surface modified with aptamer then were checked the modification by SERS using their surface plasmon resonance behavior. The aptamer can help the nanoparticles to conjugate with the cells, such as A549 lung cancer cells. Using the magnetic separator and the superparamagnetic properties of the multifunctional nanoparticles, the cells which conjugated with the nanoparticles can be separated from original sample then can be seen under microscope. This procedure is illustrated in **Figure 8(i)**. The results of using $Fe_3O_4@SiO_2$-Ag "core-shell" multifunctional nanoparticles to label the lung cancer cells A549 are shown in **Figure 8(ii)**. In this Figure, patterns B and D are bright-field and dark-field microscopy images of only A549 cancer cells. Patterns B1, D2 and B2, D2 show the bright-field and dark-field images of the A549 cells conjugated with $Fe_3O_4@SiO_2$-Ag after annealing in 1 h and 2 h, respectively. It can be seen that the cells were observed much more clearly.

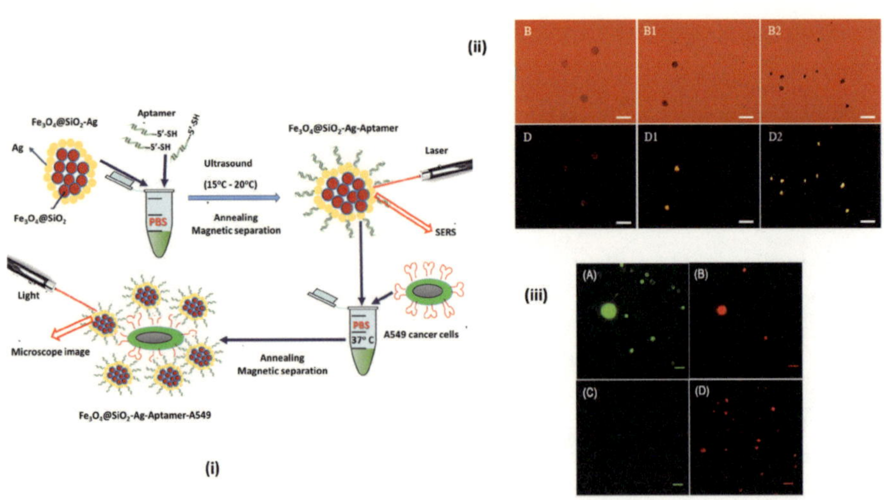

Figure 8.
(i) Schematic procedure of the application of $Fe_3O_4@SiO_2$-Ag nanoparticles (ii) bright (B) and dark (D) field images of A549 cancer cells with and without conjugation with $Fe_3O_4@SiO_2$-Ag nanoparticles (iii) isolation and detection CD34+ stem cells using Fe_3O_4-Ag nanoparticles.

The multifunctional silver nanoparticles can also be functionalized with amino groups then connect to antibody of cells through the bridge of EDC catalyst to form Fe_3O_4-Ag-antiCD34 nanoparticles [31]. **Figure 8(iii)** shows the microscopic images of the FITC labeled CD45+ cells (A), the EDC labeled CD34+ cells in the bone marrow samples before being conjugated with the Fe_3O_4-Ag-antiCD34 nanoparticles. Bone marrow is a source of blood stem cells, which contain around 10 million cells/mL, wherein includes 65% CD45+ leukemia cells and 5–7% CD34+ stem cells. After being conjugated with Fe_3O_4-Ag-antiCD34, (C) shows that no green luminescent CD45+ cells were observed, and (D) shows that only the red luminescent CD34+ cells were isolated from the bone marrow samples. The EDC catalyst enhanced the reaction of the amino group on the surface of Fe_3O_4-Ag with the carboxyl group of the ECD-antiCD34 antibody molecules [39]. The conjugated product, Fe_3O_4-Ag-antiCD34 nanoparticles, emits red light under the excitation of 494 nm incident light. The fluorescence basically originated from the ECD-fluorophore of the ECD-antiCD34 molecules. By using similar methods, multifunctional silver nanoparticles can be used in many applications, which need the plasmonic and magnetic properties together.

3. Conclusion

The chapter told a story about the synthesis and applying the silver multifunctional nanoparticles in the combination of individual silver nanoparticles and Fe_3O_4 magnetic nanoparticles. They can be combined in a silica matrix, or the core magnetic nanoparticles were functionalized with various methods and creating the outer shell with silver nanoparticles in order to have the combination of plasmonic and magnetic properties both in once. They can be used similar as individual silver nanoparticles, or individual magnetic nanoparticles with those applications. But the most important of the new type of these materials is that they can use both the properties together in order to enhance the number of possible applications, especially in biomedicine. It is expected that the multifunctionality of nanoparticles will lead them to be used for nano-biosensors and manufacturing nano-devices.

Acknowledgements

The author would like to thank Dr. Luu Manh Quynh and Dr. Chu Tien Dung for their contributions. The author acknowledges the financial support of the project code DTDL.CN-02/18.

Author details

Nguyen Hoang Nam
University of Science, Vietnam National University, Hanoi, Vietnam

*Address all correspondence to: namnh@hus.edu.vn

References

[1] Melaiye A, Youngs WJ. Silver and its application as an antimicrobial agent. Expert Opinion Therapeutic Patents. 2005;15:125-130. DOI: 10.1517/13543776.15.2.125

[2] Yamada K. The origins of acupuncture and moxibustion: The two phases of the formation of ancient medicine : The origins of acupuncture and moxibustion, the origins of decoction. International Research Center for Japanese Studies: Kyoto, Japan; 1998.154p.

[3] Alexander JW. History of the medical use of silver. Surgical Infections (Larchmt). 2009;10(3):289-292. DOI: 10.1089/sur.2008.9941

[4] Davies O. They Didn't Listen, They Didn't Know How. Author House; Bloomington, IN, USA; 2013. p.805.

[5] Hoffman RK, Surkiewicz BF, Chambers LA, Phillips CR. Bactericidal action of movidyn. Industrial& Engineering Chemistry. 1953;45:2571-2573. DOI: 10.1021/ie50527a054

[6] Jung WK, Koo HC, Kim KW, Shin S, Kim SH, Park YH. Antibacterial activity and mechanism of action of the silver ion in Staphylococcus aureus and Escherichia coli. Applied and Environmental Microbiology. 2008;74(7):2171-2178. DOI: 10.1128/AEM.02001-07

[7] Morones-Ramirez JR, Winkler JA, Spina CS, Collins JJ. Silver enhances antibiotic activity against gram-negative bacteria. Science Translational Medicine. 2013;5(190):190ra81. DOI: 10.1126/scitranslmed.3006276

[8] Yakabe Y, Sano T, Ushio H, Yasunaga T. Kinetic studies of the interaction between silver ion and deoxyribonucleic acid. Chemistry Letter. 1980;9:373-376. DOI: 10.1246/cl.1980.373.

[9] Asmaa AS, Prabhat K, Jack F, Ravi KG, Ryan HM, Mason R, Venkata RK, Suresh KT, Jingyi C, Yong W. Silver Ions Caused Faster Diffusive Dynamics of Histone-Like Nucleoid-Structuring Proteins in Live Bacteria. Applied and Environmental Microbiology. 2020;86(6):e02479–e02419. DOI: 10.1128/AEM.02479-19

[10] Wilson S, Ross TB, Blaskovich MAT, Zyta MZ. Antimicrobiral Silver in Medicinal and Consumer Applications: A Patent Review of the Past Decade (2007-2017). Antibiotic (Basel). 2018;7(4):93. DOI: 10.3390/antibiotics7040093

[11] Iravani S, Korbekandi H, Mirmohammadi SV, Zolfaghari B. Synthesis of silver nanoparticles: chemical, physical and biological methods. Research in Pharmaceutical Sciences. 2014 Nov-Dec;9(6):385-406.

[12] Natsuki J, Natsuki T, Hashimoto Y. A review of silver nanoparticles: Synthesis Methods, Properties and Applications. International Journal of Materials Science and Applications. 2015;4(5):325. DOI: 10.11648/j.ijmsa.20150405.17

[13] Asim AY, Khalid U, Mohamad NMI. Silver nanoparticles: various methods of synthesis, size affecting factors and their potential applications - a review. Applied Nanoscience. 2020;10:1369-1378. DOI:10.1007/s13204-020-01318-w

[14] Wiley B, Sun Y, Mayers B, Xia Y. Shape-controlled synthesis of metal nanostructures: the case of silver. Chemistry. 2005;11(2):454-463. DOI: 10.1002/chem.200400927

[15] Dung CT, Loc NQ, Huong PT, Duong DTT, Hong TT, Quynh LM,

Nam NH. Combination of 4-ATP coated silver nanoparticles and magnetic Fe_3O_4 nanoparticles by inverse emulsion method. VNU Journal of Science: Mathematics – Physics. 2014;30(2):1-9.

[16] Nguyen HL, Nguyen HN, Nguyen HH, Luu MQ, Nguyen MH. Nanoparticles: synthesis and applications in life science and environmental technology. Advances in Natural Sciences: Nanoscience and Nanotechnology. 2015;6:015008. DOI:10.1088/2043-6262/6/1/015008

[17] Zhang Y, Peng H, Huang W, Zhou Y, Yan D. Facile preparation and characterization of highly antimicrobial colloid Ag or Au nanoparticles. Journal of Colloid and Interface Science. 2008;325(2):371-376. DOI: 10.1016/j.jcis.2008.05.063

[18] Krutyakov Y, Olenin A, Kudrinskii A, Dzhurik P, Lisichkin G. Aggregative stability and polydispersity of silver nanoparticles prepared using two-phase aqueous organic systems. Nanotechnology in Russia. 2008;3:303-310. DOI: 10.1134/S1995078008050054

[19] Mafune F, Kohno J, Takeda Y, Kondow T, Sawabe H. Structure and stability of silver nanoparticles in aqueous solution produced by laser ablation. Journal of Physical Chemistry B. 2000;104:8333-8337. DOI: 10.1021/jp001803b

[20] Kruis F, Fissan H, Rellinghaus B. Sintering and evaporation characteristics of gas-phase synthesis of size-selected PbS nanoparticles. Materials Science and Engineering B. 2000;69:329-334. DOI: 10.1016/S0921-5107(99)00298-6

[21] Tien DC, Tseng KH, Liao CY, Huang JC, Tsung TT. Discovery of ionic silver in silver nanoparticle suspension fabricated by arc discharge method. Journal of Alloys and Compounds. 2008;463:408-411. DOI: 10.1016/j.jallcom.2007.09.048

[22] Huang H, Yang Y. Preparation of silver nanoparticles in inorganic clay suspensions. Composites Science and Technology. 2008;68:2948-2953. DOI: 10.1016/j.compscitech.2007.10.003

[23] Johans C, Clohessy J, Fantini S, Kontturi K, Cunnane VJ. Electrosynthesis of polyphenylpyrrole coated silver particles at a liquid-liquid interface. Electrochemistry Communication. 2002;4:227-230. DOI: 10.1021/la048277q

[24] Socol Y, Abramson O, Gedanken A, Meshorer Y, Berenstein L, Zaban A. Suspensive electrode formation in pulsed sonoelectrochemical synthesis of silver nanoparticles. Langmuir. 2002;18:4736-4740. DOI: 10.1021/la015689f

[25] Tran QT, Nguyen VS, Hoang TKD, Nguyen HL, Bui TT, Nguyen TVA, Nguyen DH, Nguyen HH. Preparation and properties of silver nanoparticles loaded in activated carbon for biological and environmental applications. Journal of Hazardous Materials 2011;192(3):1321-1329. DOI:10.1016/j.jhazmat.2011.06.044

[26] Nadagouda MN, Speth TF, Varma RS. Microwave-assisted green synthesis of silver nanostructures. Account of Chemical Research. 2011; 44(7):469-478. DOI: 10.1021/ar1001457

[27] Mohanpuria P, Rana NK, Yadav SK. Biosynthesis of nanoparticles: technological concepts and future applications. Journal of Nanoparticle Research. 2008;10:507-517. DOI: 10.1007/s11051-007-9275-x

[28] Nam NH, Luong NH. Nanoparticles: synthesis and applications. In: Grumezescu V, Grumezescu AM. Materials for biomedical engineering:

Inorganic micro- and nanostructures. Elsevier Inc; 2019. p. 211-240. DOI: 10.1016/B978-0-08-102814-8.00008-1

[29] Dung CT, Doanh SC, Quynh LM, Hong TT, Quach TD, Kim HD, Nam NH. Synthesis of bifunctional Fe_3O_4@SiO_2-Ag magnetic-plasmonic nanoparticles by an ultrasound assisted chemical method. Journal of Electronic Materials 2017;46(6):3646-3653. DOI: 10.1007/s11664-017-5421-0

[30] Dung CT, Huong PT, Nam NH. Synthesis of Fe_3O_4-Ag magnetic-plasmonic hybrid nanocomposites. To be submitted.

[31] Quynh LM, Dung CT, Mai BT, Huy HV, Loc NQ, Hoa NQ, Thach PT, Anh BV, Thao CT, Nam NH, Nhung HTM, Long NN, Vu LV. Development of Fe_3O_4/Ag core/shell-based multifunctional immunomagnetic nanoparticles for isolation and detection of CD34+ stem cells. Journal of immunoassay and immunochemistry 2018;39(3):308-322. DOI: 10.1080/15321819.2018.1488725

[32] Wang J, Shah ZH, Zhang S, Lu R. Silica-based nanocomposites via reverse microemulsions: classification, preparations, and applications. Nanoscale. 2014;6:4418-4437. DOI: 10.1039/C3NR06025J

[33] Li YS, Church J, Woodhead AL, Moussa FS. Preparation and characterization of silica coated iron oxide magnetic nano-particles. Spectrochimica Acta Part A: Molecular and Biomolecular Spectroscopy. 2010;76(5):484-489. DOI: 10.1016/j.saa.2010.04.004

[34] Etgar L, Schuchardt G, Costenaro D, Carniato F, Bisio C, Zakeeruddin SM, Nazeeruddin MK, Marchese L, Graetzel M. Enhanceing the open circuit voltage of dye sensitized solar cells by surface engineering of silica particles in a gel electrolyte. Journal of Materiral Chemistry A. 2013;1:10142. DOI: 10.1039/c3ta11436h

[35] Lien YH, Wu TM, Preparation and characterization of thermosensitive polymers grafted onto silica-coated iron oxide nanoparticles. Journal of Colloid and Interface Science. 2008;326:517-521. DOI: 10.1016/j.jcis.2008.06.020

[36] Hassannejad Z, Khosroshahi ME, Firouzi M, Fabrication and characterization of Magnetoplasmonic liposome carriers. Nanoscience and Technology Open Access 2014;1(2):1-9. DOI: 10.15226/2374-8141/1/2/00109

[37] Mahdavi M, Ahmad MB, Haron MJ, Gharayebi Y, Shameli K, Nadi B. Fabrication and characterization of SiO_2/(3-Aminopropyl) triethoxylane-coated magnetite nanoparticles for Lead(II) removal from aqueous solution. Journal of Inorganic and Organometallic Polymers. 2013:23:599-607. DOI 10.1007/s10904-013-9820-2

[38] Li Y, Chen CX, Li JT, Yang Y, Lin ZM, Surface charges and optical characteristic of colloidal cubic SiC nanocrystals. Nanoscale Research Letters. 2011;6:454. DOI: 10.1186/1556-276X-6-454

[39] Quynh LM, Nam NH, Kong K, Nhung NT, Notingher I, Henini M, Luong NH. Surface-Enhanced Raman Spectroscopy Study of 4-ATP on Gold Nanoparticles for Basal Cell Carcinoma Fingerprint Detection. Journal of Electronic Materials. 2016;45(5):2563-2568. DOI: 10.1007/s11664-016-4421-9

Chapter 4

Silver Nanoparticles: Properties, Synthesis, Characterization, Applications and Future Trends

Sunil T. Galatage, Aditya S. Hebalkar, Shradhey V. Dhobale, Omkar R. Mali, Pranav S. Kumbhar, Supriya V. Nikade and Suresh G. Killedar

Abstract

Nanotechnology is an expanding area of research where we use to deal with the materials in Nano-dimension. The conventional procedures for synthesizing metal nanoparticles need to sophisticated and costly instruments or high-priced chemicals. Moreover, the techniques may not be environmentally safe. Therefore "green" technologies for synthesis of nanoparticles are always preferred which is simple, convenient, eco-friendly and cost effective. Green synthesis of nanoparticle is a novel way to synthesis nanoparticles by using biological sources. It is gaining attention due to its cost effective, ecofriendly and large scale production possibilities. Silver nanoparticles (AgNPs) are one of the most vital and fascinating nanomaterials among several metallic nanoparticles that are involved in biomedical applications. It has vital importance in nanoscience and naomedicines to treat and prevent vital disease in human beings especially in cancer treatment. In current work we discussed different methods for synthesis of AgNPs like biological, chemical and physical along with its characterization. We have also discussed vital importance of AgNPs to cure life threatnign diseases like cancer along with antidiabetic, antifungal, antiviral and antimicrobial alog with its molecular mode of action etc. Finally we conclude by discussing future prospects and possible applications of silver nano particles.

Keywords: green synthesis, silver nanoparticles, nonmaterials, anticancer and antidiabetic

1. Introduction

Currently, improving and protecting our environment using green chemistry have become important issues in many fields of research. The most promising approach for generating new fields in biomedical sciences is the pharmaceutical application of nanoparticles (NPs) [1]. Due to ascension of industrial era and explosion of world population large amount of hazardous chemicals and gases released in environment in which adversely affecting our nature. Due avoid this and to protect our nature currently we world is focusing on development of natural products nanoparaticles. Biomolecules are highly compatible with nanotechnology which makes unique assembly for development of metal nanoparticles of biological molecules which are authentic and coast effective [2]. From the ancient era noval medicinal potential of silver has been known and proven for its antimicrobial potential [3]. Silver nano particles (AgNPs) and its related products were tremendously venomous and showed broad spectrum antibacterial potential against sixteen bacterial species [4, 5]. Nanotechnology is future era in material science which develops and upgrades qulaites of particles such as size, and morphology which provide entry of nonmaterial in future quality material building in almost every field [6]. Nanotechnologies have been used to develop nanoparticles-based targeted drug carriers [7]. Metal nanoparticles have a high specific surface area and a high fraction of surface atoms because of the unique physicochemical characteristics of nanoparticles [8, 9]. In that they include catalytic activity, optical properties and electronic properties, antibacterial properties, and magnetic properties [10, 11]. The nanoscale materials have emerged as novel "antimicrobial agents" due to their high surface area to volume ratio and their unique chemical and physical properties [12, 13]. In recent years development of metallic nanoparicles is an emerging field of research in material science. Crystalline nanosilver gained prime importance and has superior applicability in detection of biomolecules, antibacterial, electronics, diagnostic applications in health care system etc. Apart from novel applicability of AgNPs researchers still in search of advance methods to synthesize eco-friendly and coast effective tools to develop AgNPs [14, 15]. As silver posse's broad spectrum potential against bacterial and microbial species which specially utilized in industries it has key role in healthcare systems [16]. Nitrate group of silver potentially responsible for its broad spectrum antibacterial potential and as it convert in to AgNPs surface area is drastically increased which improve microbial exposure time and area [17–19]. Different techniques are available to synthesize AgNPs such as physical, chemical and biological. Though chemical method is rapid it utilizes capping agents for synthesis which is costly and produces adverse and toxic effects. This demands development of safe, ecofriendly, coast effective tool for synthesis of AgNPs and focused on biological methods such as green synthesis which is non toxic and developed using plant origin materials and overcomes disadvantages of earlier approaches. Moreover, use of plant extracts also reduces the cost of microorganism's isolation and culture media enhancing the cost competitive feasibility over nanoparticles synthesis by microorganisms [20]. Applicability of AgNPs is primly due to its nanoscale size and shape as compared to bulk. Due to these unique properties researchers are hunting of novel methods to synthesize AgNPs with prissily controllable size and shape [21–24]. Apart from excellent inhibitory potential of AgNPs in recent years most of the pathogenic bacteria developed resistant against it which is major concern of health care system. Chemical and physical approaches consumes ample of time, energy, money and generate toxic side effects. Nowadays green synthesis utilize microbes, fungi and medicinal plants which are easily available, convenient to handle and wide source of metabolites to synthesize AgNPs gained prime importance due to its nontoxic and ecofriendly properties [25]. Currently AgNPs are synthesized from natural herbs having medicinal potential such as synthesis of various metal nanoparticles using fungi like *Aspergillus terreus*, *Paecilomyces lilacinus* and *Fusarium* [26]. *Penicillium* sp. [27] *Fusarium oxysporum* [28] and *Euphorbia hirata*, green tea, neem, starch aloevera, lemon etc. [29–32]. AgNPs mainly binds to cell wall and penetrate deep inside the cell wall which produces cellular damage by interacting with DNA, proteins inside the cell which leads to cell death [33–37].

2. Need for green synthesis and silver nanoparticles

Silver is a basic element which is non-toxic belonging thermal and electrical potential [38]. Silver demand will likely to rise as silver find new uses, particularly in textiles, plastics and medical industries, surgical, dental resigns, coated water filters, sanitizers, detergents, soap and wound dressings. Applicability in healthcare for treatment of mental illness, convulsions, de addiction of narcotic products along with sexually transmitted diseases like syphilis and gonorrhea leads to changing the pattern of silver emission as these technologies and products diffuse through the global economy [39–41]. Green synthesis is an emerging approach which overcomes demerits of physiochemical approaches by utilization of natural herbs which are nontoxic [42, 43]. Green synthesized nanosilver offer many advantages like utilization of phytochemicals, antioxidants acts as naturally occurring reducing agents, coast efficient, large scale manufacturing highly beneficial and usage of toxic chemicals, high pressure, energy are avoided. Nanosilver can be engineered by different techniques such as irradiation, reduction, electrochemical and chryochemical synthesis. Nanosilver can be molded in to desired shapes and bear unique properties like permeability by pH and dissolved ions as compare to routine metals [44, 45]. As AgNPs generate larger surface area per unit mass which improves contact time nanosilver customer market an demand drastically raised in wide verity of industries along with healthcare, food packing, textiles, cosmetics etc. [46].

3. Silver nano particles

Generally AgNPs are nanoparticles of silver having size range between 1 and 100 nm in size having unique properties such as electrical, optical and magnetic having wide range of applicability [47]. Green chemistry is and encouraging approach mainly utilize nanosilver along with natural biomolecules such as polysaccharides, tollens which overcomes drawbacks of conventional methods and produce AgNPs which are ecofriedly, nontoxic and coast effective [48, 49]. Metallic silver ions are inactive but once it come contact with reducing agent ionization occurs and it get converted in its active form. Ionic silver is active form of silver which binds to cell wall of bacteria leading to major structural changes in cell morphology. AgNPs causes de-naturation of RNA and DNA replication which further leads to cell death [50]. Silver is also called as oligodynamic due to its bactericidal potential at minimum concentration. That's why it has been largely used in medical products [51, 52].

4. Methods for synthesis of silver nanoparticle

4.1 Physical approaches

In physical approach of synthesis of AgNPs evaporation and condensation has major importance. Temperature gradient play important role in cooling of vapors at desired rate. A chance of contamination by solvent has been removed by physical approach as no solvent has been used in physical method and uniform distribution of particle size precisely obtained [53, 54]. Minimum inhibitory concentration in toxicity studies can be easily achieved by production of nano scale nanoparticles in high concentration [55]. AgNPs also synthesized by laser ablation of metallic particles [56]. One important advantage of laser ablation technique compared to other methods for production of metal colloids is the absence of chemical reagents in solutions. Therefore, pure and uncontaminated metal colloids for further applications can be prepared by this technique [57]. Wide range of material can be synthesized in nanoparticels by physical method such as Au, Au and PbS etc. Synthesis of AgNPs by tube furnace has ample of disadvantages such as require larger space, high power, rapid rise of environmental temperature etc. AgNPs synthesized by laser ablation strongly depend on laser wavelength, time of laser pulse, laser fluence, the ablation time duration and the effective liquid medium. Ejection of AgNPs synthesized by laser ablation requires little power and particle size is precisely depends on laser fluence. However morphology, size and shape of AgNPs mainly depend on contact of laser light passing. Also, the formation of nanoparticles by laser ablation is terminated by the surfactant coating. The nanoparticles formed in a solution of high surfactant concentration are smaller than those formed in a solution of low surfactant concentration. One advantage of laser ablation compared to other conventional method for preparing metal colloids is the absence of chemical reagents in solutions. Therefore, pure colloids, which will be useful for further applications, can be produced by this method [58, 59].

4.2 Chemical approaches

Chemical reduction is the most frequently applied method for the preparation of AgNPs as stable, colloidal dispersions in water or organic solvents. Most commonly used reductant is citrate. In aqueous solution reduction of silver occurs and nanosize colloidal silver ions are generated. Stability of any colloidal dispersion has prime importance and which could be achieved by stabilizing agent (dodecanethiol) which adsorbed on surface and produce protective sheath. It can avoid agglomeration and crystal growth of the system. During the synthesis of AgNPs minute changes in parameters (Polymers) makes drastic changes in size, shape, morphology, polydispersibility index, self assembling and zeta potential (Stability). Frequently used ingredients in synthesis of AgNPs and AuNPs are glycol derivatives Polyvinyl pyrrolidone (PVP) and Polyethylene glycol (PEG). Polyacrylamide play dual function such as reducing and stabilizing agent in synthesis of AuNPs [59, 60]. Surfactants containing functional groups such as amines, thoils and acids play important role in stability of colloidal dispersion which protects the system from crystal growth, coaleseces and agglomeration. Currently AuNPs developed by modified tollens method utilize saccharides and silver hydrosols and reducing agent which yield AgNPs in the range of 50–200 nm and 20–50 nm respectively [61].

4.3 Biological approaches

Biotechnology is an emerging tool to develop biological synthesis of AgNPs. Besides this magnetic nanoparticles has great antibacterial potential due to improved surface area to treat raised microbial resistant against many antibiotics and medicines [62]. Currently green chemistry is rapidly growing technique utilized for synthesis of AgNPs with naturally occurring stabilizing, reducing and capping agents to synthesize AgNPs without toxic adverse effects [63]. Reduction of metal ions by combined efforts of herbs and certain enzymes, proteins, microorganisms, bacteria and fungi etc. in biological synthesis has been successfully reported [64].

4.4 Synthesis of silver nanoparticles by fungi

High production yield AgNPs synthesized by fungi obtained when compared to bacteria due to fungi secret higher amount of proteins that directly responsible for increased production [65]. Higher production rate is mainly due to silver ions entered in to fungal cell wall which leads to reduction of silver ions by fungal enzymes such as naphthoquinones and anthraquinones [66]. Slower rate and process is only disadvantage associated with fungal synthesis of AgNPs hence green synthesis approach is more preferred over the other techniques [67].

4.5 Synthesis of silver nanoparticles by bacteria

Pseudomonas stutzeri which is the first strain of bacteria form which AgNPs were synthesized and isolated form Ag amine [68]. Many of the bacterial strains and microorganism developing resistance to metal at lower concentration. Resistance mainly produced due to efflux, change in solubility, toxicity via oxidation/reduction and precipitation of metals [69]. There are evidences that at lower conc. Microorganisms are alive but once exposed to high conc. Metal ions leads to microbial death. In biosynthesis of silver enzyme nitrate reductase convert nitrate to nitrite [70].

4.6 Synthesis of silver nanoparticles by plants

Green synthesis is an excellent tool that can be utilized for synthesis of AgNPs as it uses natural origin medicinal herbs and its extracts which contain wide range of metabolites specifically water soluble flavones, quiones causes rapid rapid and quick reduction of silver when compared to fungi and microbes. Green chemistry approach is safe, cosat efficient, easily scalable to mass productions, easily availability of raw materials at cheaper coast. Phytochemicals directly take part in reduction process of the silver ions a during synthesis of AgNPs (**Figure 1**) [71].

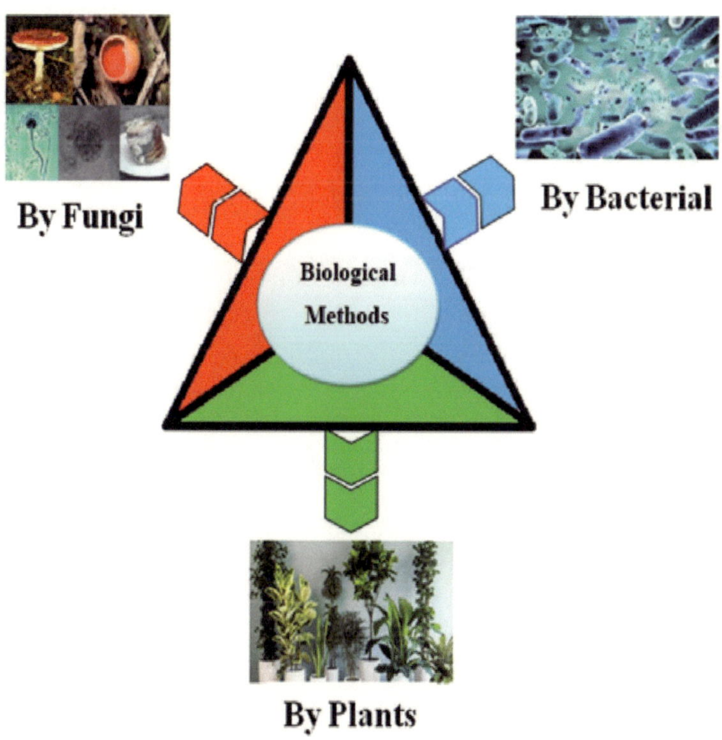

Figure 1.
Biological methods of silver nanoparticles.

5. Mechanism of action of silver nanoparticles

5.1 AgNP's antimicrobial MOA

When AgNP reaches toward cell they release Ag+ ions. These released ion then interact with sulfur and phosphorus containing compound present in cell wall. This lead to disarranged cell wall formation and small pits forms in the cell wall. Formed pit gives access to entry of ions and other foreign material to entry into cell. This increase intracellular osmotic pressure. As pressure built up in the cell, it begins to swell. Finally all these event lead to bursting of cell wall and cell lysis take place. This type of antimicrobial activity is more in gram –ve cell than gram +ve cell. As gram +ve cell have more cross linked peptidoglycan layer and teichic acid in their cell wall. The gram –ve cell have less or no peptidoglycane layer and have more lipopolysaccharide in their cell wall. So the AgNP's easily interact with gram –ve cell due less barrier [72].

5.2 AgNP's anticancer MOA

As described in above when pit formation takes place in the cell wall, the Ag+ ions released by AgNP's get entered into cell. Then they reaches to mitochondria where they interact with thiolgoups and bind to NADPH dehydrogenase enzyme and liberates ROS. These formed ROS in mitochondria interacted with respiratory enzymes damage ATP formation and respiratory cycle of cell. Formed ROS also interact with protein, sulfur and phosphorus containing cell constituent. Also these formed ROS also bind to phosphorus elements of DNA and RNA which lead to inhibit cell replication and protein synthesis. Due to binding with DNA aggregation of damage protein sysnthesis which lead to cell death. Another possible action is by autophagy. AgNP's have ability to induce autophagy by accumulation of autophagolysosmes in human ovarian cancer cell. This autophagy work by mainly 2 ways; at lower level they increases cell life i.e. surviving rate, but when its level increase it lead to cell death (**Figure 2**) [73].

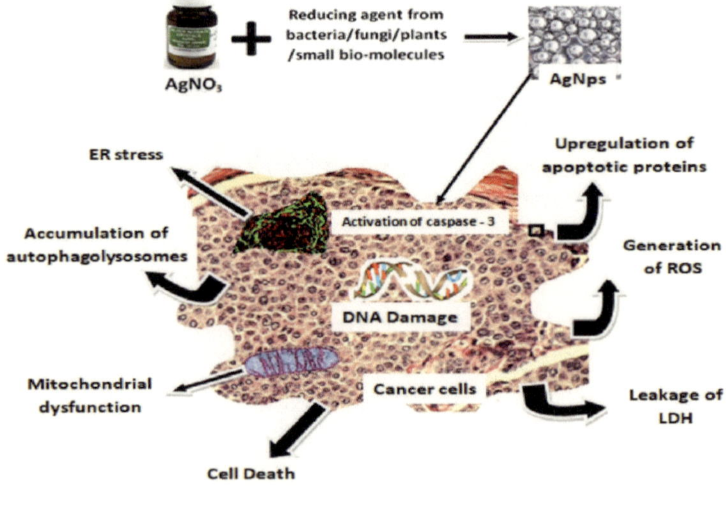

Figure 2.
Anticancer mechanism of action of silver nano particles.

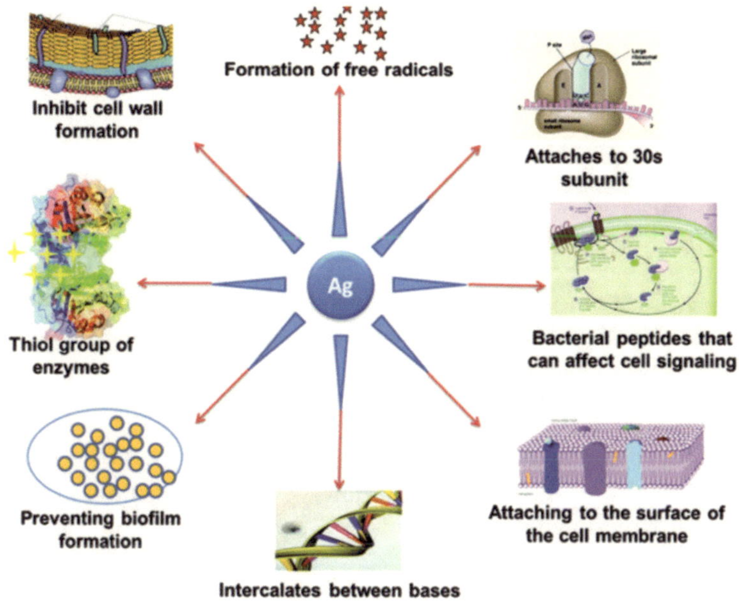

Figure 3.
Factors affecting to the bactericidal effect of silver nanoparticles.

6. Factors affecting bactericidal potential of AgNP's

Primarily morphology i.e. size and shape along with reactivity of AgNP's were responsible for bactericidal potential of AgNP's. Size and surface are inversely proportional to each other as size decreases area increases leads of rapid rise in surface-area to volume ratio. Bactericidal potential inhibit cell wall and free radicals Ag-thiol groups of enzymes Preventing biofilm formation Intercalates between bases Attaching to the surface of the cell membrane Bacterial peptides that can affect cell signaling Attaches to 30 s subunit (**Figure 3**). Silver nanoparticles showing multiple bactericidal actions [74].

7. Charactrisation of AgNP's

7.1 Visual and UV: Visible study

To ascertain either AgNPs are developed or not visual and calorimetric appearance of samples checked by UV–Visible spectrophotometer before and after formulation of AgNPs at different time intervals. Before synthesis of AgNPs silver nitrate is colorless and herbal extract has definite color. Once AgNPs synthesized silver nitrate solution develop yellowish brown color after interacting with herbal extract which is confirmed by surface Plasmon resonance SPR and UV visible absorption in the specific range of 400–475 nm [75].

7.2 FTIR analysis

FTIR spectroscopy is an investigational tool to determine/conform functional groups priesnt in the moiety which is characteristic of that compound. Major functional moieties present in AgNPs and herbal extract were identified by scanning the samples in the range of 4000 to 400 cm−1 [76].

7.3 SEM/TEM analysis

Scanning electron microscopy/Transmission electron microscopy mainly used to study surface morphology of synthesized AgNPs. SEM/TEM plates were prepared by addition of silver nitrate to develop smear of solution on slides. Conductivity was incorporated in system by making thin film of platinum which was coated on slides. Once the slides were ready they were scanned at 20 KV accelerating voltage and high quality images were captured [77].

7.4 X-ray diffraction (XRD) analysis

X-ray diffraction is a modern technique mainly utilized to identify state of matter either it is crystalline or amorphous in nature at different radiation angles. X-ray diffraction determines phases either crystalline/amorphous and cell dimension [78].

8. Application of silver nanoparticles

8.1 Antimicrobial activity

Products prepared with silver nanoparticles have been permitted by no. of accredited bodies including USFDA, USEPA, Korea's testing body and SIAA of japan institute of research. Antimicrobial and antimicrobial potential potential of AgNPs containing silver sulfadiazine is incorporated in to medicines and used in burns to avoid infections. Nowadays AgNPs involved in extending field of nanotechnology and appears in many consumer products that include acne vulgaris cream and for deodorizing sprays. The antimicrobial properties of silver nanoparticles depend on size, environmental conditions (size, Ph, tonic strength) and capping agent. Recently an improvement in antimicrobial activity synergistic effect has been reported when silver naqnoparticles combined with ampicillin, amoxicillin and chloramphenicol on the contrary reports showed antagonistic interaction between silver nanoparticles and amoxicillin or oxacillin antibiotic combined with silver nanoparticles have suggested improve therapeutic activity (**Figure 4**) [79, 80].

8.2 Antiviral activity

Antiviral activity of silver nanoparticles have proven to exert antiviral activity against HIV-1 at non cytotoxic concentration but the mechanism underlying their HIV inhibitory activity has not been fully elucidated. The study from intranasal silver nano particles administration in mice increased survical, lower lung viral titer levels, minor pathologic lesions in lung disease, and remarkable survival benefit after infection with the H3N2 influenza virus, suggesting that AgNPs had significant role in mice survival. Biologically prepared silver nano particles inhibited the viability in herpes simplex virus (HSV) types 1 and 2 and human para influenza virus type 3 based on size and zeta potential. The treatment of vero cells with non-cytotoxic concentrations of silver nanoparticles significantly inhibited by the replication of paste des petits ruminants virus (PPRV). The mechanisms of viral replication are due to the interaction of silver

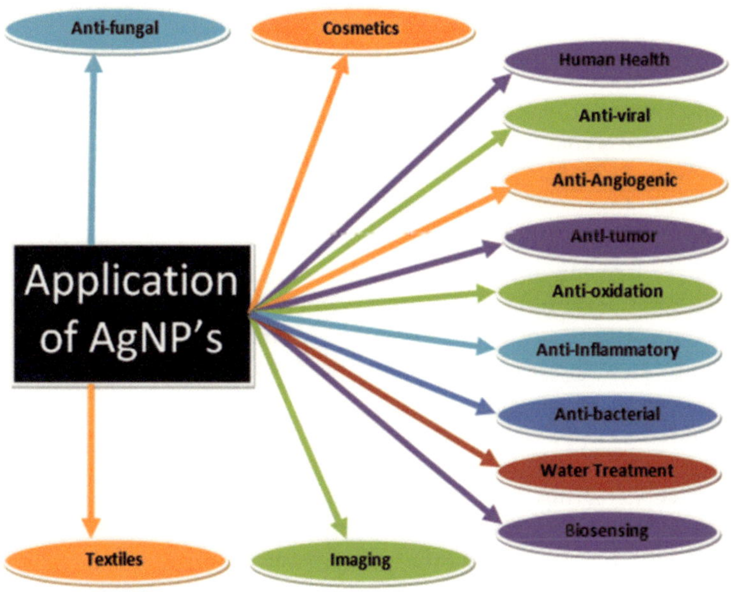

Figure 4.
Applications of silver nanoparticles.

nanoparticles with the virion core. Tannic acid mediated synthesis of various various sizes of silver nanoparticles capable of reducing HSV-2 infectivity both in in-vitro and in-vivo through direct interaction, blocked virus attachment, penetration and further spread [81, 82].

8.3 Antibacterial activity

Silver nano particles are one of the most attractive nonmaterial's for commercialization applications. As antibacterial agents silver nanoparticles were used for wide range of applications from disinfecting medical devices and home appliances to water treatment. AgNPs promisingly used in drastic fields such as healthcare products, food storage, textile and medicinal devices. In antibacterial potential AgNPs free silver ions are released at slower rate along with higher surface area which produces noxious environment and this is the main reason for broad spectrum antibacterial potential of AgNPs [83].

8.4 AgNP's in cancer control

AgNPs has prominent anticancer potential as it discourage mitochondrial respiratory chain, increase reactive oxygen species (ROS) rate of synthesis which finally leads to DNA damage and cancerous cell death. Yu-Guo Yuan in 2018 revealed that the combination of camptothecin and silver nanoparticles treatment significantly increases the levels of cancer cells. It increases oxidative stress markers and decrease ant oxidative stress markers compared to single treatment. Overall these results suggested that camptothecin and silver nanoparticles cause cell death by inducing the mitochondrial membrane permeability change and activation of caspase. The synergistic cytotoxicity and apoptosis effect seems to be associated with enhanced ROS formation and depletion of antioxidant. Certainly a combination of CPT and silver nano particles provide advantageous effect in treatment of cervical cancer compared to immunotherapy [84].

8.5 Antidiabetic activity of AgNP's

Tephrosiatinctoria stem extracts mediated silver nano particle synthesis was evaluated for control of blood suger levels. AgNP's scavenged free radicals, reduced the levels of enzymes that bring about hydrolysis of complex carbohydrates ($\alpha - glucosidase\alpha - amylase$) and as a result of which there is an increase in consumption ratr of glucose. The promising antidiabetic activity of shown by *Ananascomosus* (L.) silver nanoparticles. In dose dependent manner. AC-AgNP's inhibit α-glucosidase enzyme in stomach. Which is helpful in non-insulin diabetic patient. Also the silvernanoparticles synthesized with *Argyreia nervosa*leaf extract shown great antidiabetic activity. They inhibit mainly enzymes that digest the carbohydrates into monosaccharide and reduce blood glucose level [85, 86].

8.6 Different field application of AgNp's

Studies can contracting on the therapeutic applications of AgNP's in the gastrointestinal tract have displayed that gastric cells can be sensitized to radiation by the use of AgNP's and they may bypass the stomach and instead release the drug in small intestine. Apart from the health related applications; Silver Nanoparticles are act as a brilliant heterogeneous catalyst used for reduction of halogenated organic pollutants. Also it increases the bleaching power of organic dyes. The tubular shaped silver Nanoparticles have a very potent catalytic activity so they can used as a catalyst. In case of water treatment when the biosynthesized Silver Nanoparticles which are biologically synthesized on nitrocellulose membrane filters, can used for the promising inhibition and inactivation of microbes like *E. coli* and *Enterococcus faecalis*, etc. Rather as the silver Nanoparticles are the very good antimicrobial agents so they are used as the preservatives in various food and agricultural products [87].

8.7 Antifungal activity of AgNPs

AgNP's play important role as antifungal agents against various diseases caused by fungi. Biologically synthesized AgNP's shows enhanced antifungal activity with fluconazole against *phomaglomerata*, *Candida albicans* species. AgNP's stabilized by sodium dodecyl sulphate showed greater antifungal activity against *Candida albicans* compared to conventional antifungal agents. The AgNP's synthesized by bacillus species exhibit strong antifungal activity against the plant pathogenic fungus*fusariumoxysporum* at concentration of 8 μg/ml. AgNP's shown promising antifungal activity on *T. asahii* with MIC of 0.5ug/ml by damaging cell wall and components of cell. Due to size of nanoparticles they easily penetrate into cell. Where it binds to different cell components and inhibits cell functions. In combination with antimicrobial agents like ketoconazole shown great antifungal activity with MIC less than 0.5 mg/ml against the Malassezia where they give synergistic effect with ketoconazole it form pores in cell to show antifungal activity [88, 89].

8.8 Anti angiogenic activity of AgNP's

Antiangiogenic potential of green synthesized AgNP's in retinal endothelial cells model mainly produced by inhibition, proliferation and migration of BRECs at 500 nM concentration. In CAM model (chicken embryo chorioallantoic membrane) the silver nanoparticles inhibit angiogenesis approximately up to 73%. In comparison to other antiangiogenic molecules. They give dose dependent cytotoxic action on endothelial cell present in blood vessels to inhibit formation of new blood vessel in tumor region. Also the by using the same model i.e. by CAM assay the silver nanoparticle synthesized by *Rubinatinctorum* shown antigiogenic activity. Ru-AgNP's shown inhibitory action on blood vessels. In CAM model, there is decrease in length of embryo resulted out due to the antiangiogenic action of Ru-AgNP's [90, 91].

8.9 Diagnostic, biosensor and gene therapy applications of AgNP's

Nanoparticles have advantage over today's therapies because they can be engineered to have certain properties or to in certain way. They are helpful in cellular imaging. Silver plays an important role in imaging systems due to its stronger and sharper Plasmon resonance. Currently biosensor made with silver used as powerful tool to detect cytochrome P53 of squamous cell cancer of head and neck. Due to the colorimetric sensing property the silver Nanoparticles are applicable to detect the heavy metal ions of nickel, cobalt and mercury along with the sulfide traces. Among all the types of silver Nanoparticles, especially the triangular shaped silver Nanoparticles have higher anisotropy and lightening rod effect which leads to its wide use in manufacturing of Plasmon sensors or Plasmon detectors which are used to detect the mercurial ions in the solution. Also the silver Nanoparticles are used to develop the electrochemical sensor which is used to detect common herbicide atrazine. On the other hand the in situ growth and development of silver Nanoparticles on polydopamine traced filter paper is responsible for the quick collection and detection of green colored residue of malachite [92].

8.10 Anti-inflammatory activity of AgNP's

AgNPs have been known for its antimicrobial but the anti-inflammatory response is still limited. Rats treated intra colonic ally with 4 mg/kg or orally with 40 mg/kg of nanocrystalline silver (NP32101) showed significantly reduced colonic inflammation. AgNPs showed rapid healing and improved cosmetic appreance occurring in dose dependent manner. Silver Nanoparticles made by using the extraction method with petroleum ether and some small amount of ethyl acetate are having potent cyclooxigenase-2 inhibition property. So, as one can add the natural extract of anti-inflammatory activity to this silver Nanoparticles extracted with petroleum ether, the anti-inflammatory activity of the resulted silver Nanoparticles get increased. Recently some scientists were done the extraction of soft coral named nephthea sp. Which already possessing the anti-inflammatory activity and extracted the silver Nanoparticles with petroleum ethers then the produced silver Nanoparticles of nephthea sp. having very potent anti-inflammatory activity which were estimated by analysis and molecular docking methods [93].

9. Future prospects

AgNPs has potential applications in healthcare system and treating infectious diseases and it is emerging as remedies for large no of resistant bacteria infections along with it is known for its anti-inflammatory potential. Apart from it has numerous application in biological and research fields such as electrochemistry, biochemistry, nanoprism synthesis, garments, detergents and soap industry, involved in devising water purification system, and surgical instrument. Nowadays Ag-NPs opened new era as it has used in artificial implants which decreeing dependency on antibiotics. Studies have been revealed that Ag-NPs have novel potential in development of new pharmaceutical dosage forms and AgNPs cures inflammation of bladder which tremendous application in healthcare systems. AgNPs useful in animal models for detection of biosensors [94]. A reliable mechanism responsible for the impressive biological activity of AgNPs is considered to be a key factor in future research. Wide scope to aware control the release of silver and improving the stability of AgNPs used in medical and mechanical devices.

10. Concslusion

Over the past few decades, nanoparticles of noble metals such as silver exhibited significantly distinct physical, chemical and biological properties from their bulk counterparts. Current chapter specifically encounters synthesis, characterization, and bio-applications of silver nanoparticles, with special emphasis on anticancer, antimicrobial activity and its mechanisms. Green chemistry is being exploited for developing silver nanoparticles. Several methods utilized to create silver nanoparticles utilizing plant extracts as reducing or capping agents. Current chapter represents different methods of preparation silver nanoparticles and application of these nanoparticles in different fields.

Acknowledgements

The authors are thankful to Department of Pharmaceutics Sant Gajanan Maharaj College of Pharmacy Mahagaon and Trustees of Sant Gajanan Maharaj College of Pharmacy Mahagaon for providing required guidance and support for completion of this work.

Conflict of interest

The authors declare no conflict of interest.

Notes/thanks/other declarations

Special thanks to Shivtej for continuous support throughout the work.

Abbreviations

AgNP's	Silver Nano Particles
NPs	Nanoparticles
nm	nanometer
$AgNO_3$	Silver Nitrate
PVP	polyvinyl pyrrolidone
PEG	poly ethylene glycol
PMAA	poly methacrylic acid
SPR	surface plasmon resonance
ROS	Reactive oxygen species

Author details

Sunil T. Galatage*, Aditya S. Hebalkar, Shradhey V. Dhobale, Omkar R. Mali, Pranav S. Kumbhar, Supriya V. Nikade and Suresh G. Killedar
Sant Gajanan Maharaj College of Pharmacy, Mahagaon, Maharashtra, India

*Address all correspondence to: gsunil201288@gmail.com

References

[1] Salata OV. "Applications of nanoparticles in biology and medicine". Journal of Nanobiotechnology. 2004; 6:1-6.

[2] Harekrishna B, Bhui D K, Gobinda S P, Sarkar P, Sankar P D., "Green synthesis of silver nanoparticles using latex of Jatropha curca." Colliods and surface A: Physicochem Eng. Aspect. 2009; 39(3): 134-139.

[3] Won Hyuk Suh a, Kenneth S. Suslick b, Galen D. Stucky a, Yoo-Hun Suh c. "Nanotechnology, nanotoxicology, and neuroscience". Progress in Neurobiology. 2009; (87):133-170.

[4] Slawson, RM, Trevors, JT, Lee, H. "Silver accumulation and resistance in pseudomonas stutzeri". Arch. Microbiol. 1992; 158:398-404.

[5] Zhao, GJ, Stevens, SE. "Multiple parameters for the comprehensive evaluation of the susceptibility of Escherichia coli to the silver ion". Biometals. 1998; 11: 27-32.

[6] Kaviya S, Viswanathan B. "Green synthesis of silver nanoparticles using Polyalthia longifolia leaf extract along with D-sorbitol". Journal of nanotechnology. 2011;1-5.

[7] Falanga, A, Vitiello M.T., Cantisani M, Tarallo R, Guarnieri D, Mignogna E, Netti, P, Pedone C, Galdiero M, Galdiero S. "A peptide derived from herpes simplex virus type 1 glycoprotein H: Membrane translocation and applications to the delivery of quantum dots". Nanomedicine. 2011; 10(1016)4-9.

[8] Catauro M, R.M., De Gaaetano FD, Marotta A. "sol–gel processing of drug delivery materials and release kinetics". J Mater Sci Mater Med. 2005; 16(3):261-265.

[9] Crabtree JH, B.R, Siddiqi RA, Huen IT, Handott LL, Fishman A. "The efficacy of silver-ion implanted catheters in reducing peritoneal dialysis-related infections". Perit Dial Int. 2003; 23(4):368-374.

[10] Krolikowska A, Michota A, Bukowska J, "SERS studies on the structure of thioglycolic acid monolayers on silver and gold". Surf Sci. 2003; 532:227-232.

[11] Zhao G, S.J., "Multiple parameters for the comprehensive evaluation of the susceptibility of *Escherichia coli* to the silver ion". Biometals. 1998; 11:27.

[12] Morones J R, Elechiguerra J L, Camacho A, Holt K, Kouri J B, Ramírez J T, Yacaman M.J. "The bactericidal effect of silver nanoparticles". Nanotechnology. 2005; 16:2346-2353.

[13] Helenius, A. "Fields 'virology' fifth edition: Virus entry and Uncoating; LWW: London, UK. 2007; 99-118.

[14] Krolikowska A, K.A., Michota A, Bukowska J, "SERS studies on the structure of thioglycolic acid monolayers on silver and gold". Surf Sci. 2003; 532:227-232.

[15] Zhao G, S.J., "Multiple parameters for the comprehensive evaluation of the susceptibility of *Escherichia coli* to the silver ion". Biometals. 1998; 11:27.

[16] Jiang H, M.S., Wong ACL, Denes FS, "Plasma enhanced deposition of silver nanoparticles onto polymer and metal surfaces for the generation of antimicrobial characteristics". J Appl Polym Sci. 2004; 93:1411-1422.

[17] Duran N, M.P., Alves OL, De Souza GIH, Esposito E, "Mechanistic aspects of biosynthesis of silver nanoparticles by several Fusarium oxysporum strains". J Nanobiotechnol. 2005; 3:8-14.

[18] Becher RO. "Silver ions in the treatment of local infections". Met Based Drugs. 1999; 6(11):297-300.

[19] Klaus T, J.R, Olsson E, Granqvist C-G, "Silverbased crystalline nanoparticles, microbially fabricated". Proc Natl Acad Sci USA. 1999; 96:13611-13614.

[20] Garima Singhal, R.B., Kunal Kasariya, Ashish Ranjan Sharma, Singh PL, "biosynthesis of silver nanoparticles using Ocimum sanctum (Tulsi) leaf". Bioresources and Bioprocessing. 2014; 1:3.

[21] Zhang H, Wu M, Sen A. "In Nano-Antimicrobials: Silver Nanoparticle Antimicrobials and Related Materials". Edited by Cioffi N, Rai M. New York: Springer. 2012; 3-45.

[22] Niraimathi KL, Sudha V, Lavanya R, Brindha P. "Biosynthesis of silver nanoparticles using Alternanthera sessilis (Linn.) extract and their antimicrobial, antioxidant activities". Colloids Surf B Biointerfaces. 2013; 102:288-291.

[23] Jiang, Z.J, Liu, C.Y, Sun, L.W. "Catalytic properties of silver nanoparticles supported on silica spheres". J. Am. Chem. Soc. 2004; 71:2341-2343.

[24] Wijnhoven, S.W.P, Peijnenburg, W.J.G.M, Herberts, C.A, Hagens, W.I, Oomen, A.G, Heugens, E.H.W, Roszek, B, Bisschops, J, Gosens, I, van De Meent, D. "Nano-silver-A review of available data and knowledge gaps in human and environmental risk assessment". Nanotoxicology. 2009; 3: 109-138.

[25] Okafor F, Janen A, Kukhtareva T, Edwards V and Curley M. "Green synthesis of silver nanoparticles, their characterization, application and antibacterial activity". Int. J. Environ. Res. Public Health. 2013; 10:5221-5238.

[26] Devi L S and Joshi S R. "Antimicrobial and Synergistic Effects of Silver Nanoparticles Synthesized Using Soil Fungi of High Altitudes of Eastern Himalaya Mycobiology". 2012; 40(1): 27-34.

[27] Hemanth NKS, K.G., Karthik L, Bhaskara RKV. "Extracellular biosynthesis of silver nanoparticles using the filamentous fungus Penicillium sp". Arch Appl Sci Res. 2010; 2(6):161-167.

[28] Nelson D, P.D, Oswaldo LA, Gabriel IHDS, Elisa E. "Mechanical aspects of biosynthesis of silver nanoparticles by several Fusarium oxysporum strains". Journal of Nanobiotechnology. 2005; 3:8.

[29] Elumalai EK, P.T, Hemachandran J, Vivivan Therasa S, Thirumalai T, David E. "Extracellular synthesis of silver nanoparticles using leaves of Euphorbia hirta and their antibacterial activities". J Phram Sci. 2010; 2(9):549-554.

[30] Ankamwar B, D.C., Ahmad A, Sastry M. "biosynthesis of gold and silver nanoparticles using Emblics Officinalis fruit extract and their phase transfer and Transmetallation in an organic solution". J nanosci nanotechnol. 2005; 5(10):1665-1671.

[31] Kasthuri J, K.K., Rajendran N. "Phyllanthin assisted biosynthesis of silver and gold nanoarticles a novel biological approach". J Nanopart Res. 2009; 11:1075-1085.

[32] Vijayaraghavan K, K.N.S., Udaya Prakash N, Madhankumar D. "Biomimetic synthesis of silver nanoparticles by aqueous extract of *Syzygium aromaticum*". Colloids Surf B Biointerfaces. 2012; 75:33-35.

[33] Amarendra DD, K.G., "Biosynthesis of silver and gold nonoparticles using Chenopodium album leaf extract". Colloid Surface A. 2010; 369(3):27-33.

[34] Pal S L, Jana U, Manna P K, Mohanta G P, Manavalan R. "Nanoparticle: An overview of preparation and characterization". Journal of Applied Pharmaceutical Science. 2011; 1(06):228-234

[35] VJ Mohanraj1 and Y Chen. "Nanoparticles – A Review Tropical Journal of Pharmaceutical Research". 2006; 5 (1):561-573

[36] Horikoshi S And Serpone N. Introduction to Nanoparticles

[37] Xu Z P, Z.Q.P., Lu G Q and Yu A B, "Inorganic Nanoparticles as Carriers for Efficient Cellular Delivery". Chemical Engineering Science. 2006; 61:1027-1040.

[38] Chen X, S.H.J. "Nanosilver: A Nanoproduct in Medical Applications". Toxicol Lett. 2008; 176:1-12.

[39] M Ramya, M.S.S., "Green Synthesis of Silver Nanoparticles". Int. J. Pharm. Med. & Bio. Sc. 2012; 1.

[40] Drake P L. "Exposure-related health effects of silver and silver compounds: A review". Ann Occup Hyg, 2005; 49:575-585.

[41] Prabhu S, Poulose E K. "Silver nanoparticles: Mechanism of antimicrobial action, synthesis, medical applications, and toxicity effects". Prabhu and Poulose International Nano Letters. 2012; 2:32.

[42] Parasharu K. "Bioinspired synthesis of silver nanoparticles". Digest Journal of Nanomaterials and Biostructures. 2009; 4:159-166.

[43] Begum N A, "Mondal colloids and surface B". Biointerfaces. 2009. 71:113-118.

[44] Chun nam l, X.a.S.H. "Nanosilver: A Nanoproduct in Medical Applications". Toxicol Lett. 2008; 176:1-12.

[45] C, L., "Proteomic analysis of the mode of antibacterial action of silver nanoparticles". J Proteome Res, 2006. 5: p. 916-924.

[46] M Ramya, M.S.S. "Green Synthesis of Silver Nanoparticles". Int. J. Pharm. Med. & Bio. Sc. 2012; 1.

[47] Klaus, T.J. R.; Olsson, E. & Granqvist, C.gr., "silver-based crystalline nanoparticles, microbially fabricated.". Proc Natl Acad Sci USA. 1999; 96:13611-13614.

[48] Senapati, S. "Biosynthesis and immobilization of nanoparticles and their applications.". University of pune, India. 2005.

[49] Kruis, F.F., H. & Rellinghaus, B. "Sintering and evaporation characteristics of gas-phase synthesis of size-selected PbS nanoparticles". Mater Sci Eng B. 2000; 69:329-324.

[50] Rai M, Yadav A, Gade A. "Silver nanoparticles as a new generation of antimicrobials". Biotechnology Advances. 2009; 27:76-83.

[51] N. Stobie, B. Duffy, J. Colreavy, P. McHale, S. J. Hinder, D. E. McCormack. "Journal of Colloid and Interface Science". 2010; 345:286-292.

[52] T. Pradeep, Anshup. "Thin Solid Films". 2009; 517:6441-6478

[53] Kruis, F H. & Rellinghaus B. "Sintering and evaporation characteristics of gas-phase synthesis of size-selected PbS nanoparticles". Mater Sci Eng B. 2000; 69:329-324.

[54] Magnusson, M.D., K.; Malm, J.; Bovin, J. & Samuelson, L., "gold nanoparticles: Production, reshaping, and thermal charging". J Nanoparticle Res. 1999; 1:243-251.

[55] Jung, J.O., H.; Noh, H.; Ji, J. & Kim, S., "metal nanoparticle generation using

a small ceramic heater with a local heating area". J Aerosol Sci. 2006; 37:1662-1670.

[56] Mafune, F.K., J.; Takeda, Y.; Kondow, T. & Sawabe, H., "formation of gold nanoparticles by laser ablation in aqueous solution of surfactant". J Phys Chem B. 2001; 105:5114-5120.

[57] Wiley, B.S, Y. Mayers, B. & Xi, Y. "Shape-controlled synthesis of metal nanostructures: the case of silver". Chem Eur. 2005; 11:454-463.

[58] Merga, G.W., R.; Lynn, G.; Milosavljevic, B. & Meisel, D. "redox catalysis on "naked" silver nanoparticles". J Phys Chem C. 2007; 111:12220-12226.

[59] Kholoud M.M, Abou El-Nour, Ala'a Eftaiha, Abdulrhman Al-Warthan, Reda A.A. Ammar. "Synthesis and applications of silver nanoparticles Arabian Journal of Chemistry". 2010; 3:135-140.

[60] Oliveira, M.U., D. Zanchet, D. & Zarbin, A. "Influence of synthetic parameters on the size, structure, and stability of dodecanethiol-stabilized silver nanoparticles". J Colloid Interface Sci. 2005; 292:429-435.

[61] Yin, Y.L., Z-Y.; Zhong, Z.; gates, B. & Venkateswaran, S. "synthesis and characterization of stable aqueous dispersions of silver nanoparticles through the Tollens process". J Mater Chem. 2002; 12:522-527.

[62] Sahayaraj K and Rajesh S. "Bionanoparticles: synthesis and antimicrobial applications".

[63] Ankamwar, B.D, Ahmad, A. & Sastry, M. "Biosynthesis of gold and silver nanoparticles using Emblica officinalis fruit extract, their phase transfer and transmetallation in an organic solution". J Nanosci Nanotechnol. 2005; 5:1665-1671.

[64] Iravani, S., "Green synthesis of metal nanoparticles using plants". Green Chem. 2011; 13:2638-2650.

[65] Mohanpuria, P., Rana, KN, Yadav, S K. "biosynthesis of nanoparticles: Technological concepts and future applications". J. Nanopart. Res. 2008; 10:507-517.

[66] Mukherjee, P., Ahmad, A, Mandal, D, Senapati, S, Sainkar, SR, Khan, MI, Parischa, R, Ajaykumar, PV, Alam, M, Kumar, R, Sastry, M. "Fungus mediated synthesis of silver nanoparticles and their immobilization in the mycelial matrix: A novel biological approach to nanoparticle synthesis". Nano Lett. 2001; 1:515-519.

[67] Ahmad, A., Mukherjee, P, Senapati, S, Mandal, D, Khan, MI, Kumar, R, Sastry, M, "extracellular biosynthesis of silver nanoparticles using the fungus Fusarium oxysporum". Colloids Surf. B Biointerfaces. 2003; 28:313-318.

[68] Haefeli, C., Franklin, C, hardy, K, "plasmid-determined silver resistance in pseudomonas stutzeri isolated from silver mine". J. Bacteriol. 1984; 158:389-392.

[69] Husseiny, M., Aziz, MAE, Badr, Y, Mahmoud, MA. "Biosynthesis of gold nanoparticles using Pseudomonas aeruginosa". Spectrochimica. Acta. Part A. 2006; 67:1003-1006.

[70] Vaidyanathan, R., Gopalram, S, Kalishwaralal, K, Deepak, V, Pandian, SR, Gurunathan, S. "Enhanced silver nanoparticle synthesis by optimization of nitrate reductase activity". Colloids Surf. B Biointerfaces. 2010; 75:335-341.

[71] C, F.A.a.S.M., "Inactivity of Two Noble Metals as Carcinogens". J. Environ Patho. Toxicol. 1978; 1:51-57.

[72] Xi-Feng Zhang 1, Zhi-Guo Liu. Silver Nanoparticles: Synthesis, Characterization, Properties,

Applications, and Therapeutic Approaches. Int. J. Mol. Sci. 2016, 17, 1534.

[73] Hayelom Dargo Beyene, Adhena Ayaliew Werkneh. Synthesis paradigm and applications of silver nanoparticles (AgNPs).Review Sustainable Materials and Technologies 13 (2017) 18-23.

[74] Prateek Mathur, Swati Jha, Suman Ramteke & N. K. Jain.Pharmaceutical aspects of silver nanoparticles, Artificial Cells, Nanomedicine, and Biotechnology 2018 46(1);115-126.

[75] Kumar B, Kumari S, Cumbal L, Debut A. "Synthesis of silver nanoparticles using Sacha inchi (Plukenetia volubilis L.) leaf extracts" Saudi Journal of Biological Science. 2014.

[76] Singh S, Saikia J P, Buragohain A K. "A novel 'green' synthesis of colloidal silver nanoparticles (SNP) using Dillenia indica fruit extract", Colloids and Surfaces B: Biointerfaces 2013 102; 83-85.

[77] Kathireswari P, Gomathi S and Saminathan K. "Green synthesis of silver nanoparticles using Vitex negundo and its antimicrobial activity against human pathogens", Int. J. Curr. Microbiol. App.Sci 2014 3(8); 614-621.

[78] Shoib A. B, Shahid A. M. "Evaluation of antioxidant and antibacterial activity of methanolic extracts of Gentiana kurroo royle" Saudi Journal of Biological Science.2014;1-13

[79] Ponarulselvam S, Panneerselvam C, Murugan K, Aarthi N, Kalimuthu K, Thangamani S. "Synthesis of silver nanoparticles using leaves of Catharanthus roseus Linn. G. Don and their antiplasmodial activities" Asian Pacific Journal of Tropical Biomedicine 2012; 574-580.

[80] Vazquez-Murioz R, A. Meza-Villezcas, P.G.J. Fournier, E. Soria-Castro, K. Juarez-Moreno, A.L. Gallego-Hernandez "Enhancement of antibiotics antimicrobial activity due to silver Nanoparticles impact on the cell membranes." PLOS ONE, 2019; 1-3.

[81] Kholoud M.M. Abou El-nour. Synthesis and applications of silver nanoparticles. Arabian Journal of Chemistry 2010 3; 135-140.

[82] Xi-Feng Zhang, Zhi-Guo Liu, Wei Shen, Sangiliyandi Gurunathan. "Silver nanoparticles – Synthesis, characterization, properties, application and therapeutic approaches" International Journal of Molecular Science. 2016; 6-7.

[83] Zhi Zhang, Wenfei Shen, Jing Xue. "Recent advance in synthetic methods and applications of silver nanoparticles" Nanoscale Research Letters. 2018 (54); 6-7.

[84] Yu-Guo Yuan. "Silver nanoparticles potentiates cytotoxicity and apoptotic potential of camptothecin in human cervical cancer cells" Oxidative Medicine and Cellular Longevity. 2018;1-22.

[85] Gitishree Das, Jayanta Kumar Patra, Trishna Debnath, Abuzar Ansari, Han-Seung Shin "Investigation of antioxidant, antibacterial, antidiabetic, and cytotoxicity potential of silvernanoparticles synthesized using the outerpeel extract of *Ananas comosus* (L.)" Plos one journal. 2019 12;1-19.

[86] Ganesh Dattatraya Saratale, Rijuta Ganesh Saratale, Giovanni Benelli Gopalakrishnan Kumar, Arivalagan Pugazhendhi, Dong-Su Kim, Han-Seung Shin. "Anti-diabetic Potential of Silver Nanoparticles Synthesized with *Argyreia nervosa* Leaf Extract High Synergistic Antibacterial Activity with Standard Antibiotics Against Foodborne Bacteria" Journal of Cluster Science.2017 28;1709-1727.

[87] Sonika Dawadi, Saurav Katuwal, Akash Gupta and members. "Current Research on silver Nanoparticles – synthesis, characterization, and applications" Journal of Nanomaterials. 2021;9-12.

[88] Zhi-Kuan Xia, Qiu-Hua Ma, Shu-Yi Li, De-Quan Zhang, Lin Cong, Yan-Li Tian, Rong-Ya Yang. "The antifungal effect of silver nanoparticles on Trichosporon asahii". Journal of microbiology, immunology and infection. 2014; 1-7.

[89] Javier Esteban Mussin1, María Virginia Roldán, Florencia Rojas, María de los Ángeles Sosa, Nora Pellegri and Gustavo Giusiano. "Antifungal activity of silver nanoparticles in combination with ketoconazole against Malassezia furfur" AMB Expr. 2019 9 (131);1-9.

[90] Brhaish Ali Saeed, Vuanghao Lim, Nor Adlin Yusof, Kang Zi Khor, Heshu Sulaiman Rahman, Nozlena Abdul Samad. "Antiangiogenic properties of nanoparticles: a systematic review". Internatinal Journal of Nanomedicine, 2019; 5135-5146.

[91] Sara Ghandehari, Masoud Homayouni Tabrizi, Pouran Ardalan. "Evaluation of Anti-angiogenic Activity of Silver Nanoparticle Synthesis by Rubina tinctorum L (Ru-AgNPs) Using Chicken Chorioallantoic Membrane (CAM) Assay". Arak Medical University Journal. 2018; 82-90.

[92] Sonika Dawadi, Saurav Katuwal, Akash Gupta. "Current research on silver nanoparticles – Synthesis, characterization, and applications". Journal of Nanomaterials. 2021; 9-12.

[93] Abdelhafez OH, Ali TFS, Fahim JR, Desoukey SY, Ahmed S, Behery FA, Kamel MS, Gulder TAM. "Anti-Inflammatory Potential of Green synthesized Silver Nanoparticles of the Soft Coral nephthea Sp. Supported by Metabolomics analysis and Docking studies". International Journal of nanomedicines. 2020; 5358.

[94] Adnan Haider and Inn-Kyu Kang. Preparation of silver nanoparticles and their industrial and biomedical applications: A comprehensive review. Advances in Materials Science and Engineering 2015; 1-16.

Section 2

Biomedical Uses of Silver Nanostructures

Chapter 5

Biomedical Applications of Silver Nanoparticles

Manikandan Dhayalan, Priadharsini Karikalan, Mohammed Riyaz Savaas Umar and Nalini Srinivasan

Abstract

Nanotechnology is a branch of science and engineering dedicated to materials, having dimensions in the order of nanometer scale and it has been widely used for the development of more efficient technology. Nanoparticles offer many benefits to bulk particles such as increased surface-to-volume ratio, and increased magnetic properties. In recent years, nanotechnology has been embraced by industrial sectors due to its applications in the field of electronic storage systems, biotechnology, magnetic separation and pre concentration of target analytes, targeted drug delivery, and vehicles for gene and drug delivery. Over the year's nanomaterials such as nanoparticles, nanoclusters, nanoreods, nanoshells, and nanocages have been continuously used and modified to enable their use as a diagnostic and therapeutic agent in biomedical applications. Thus, In this chapter, introduction to metal nanoparticles, synthesis (Chemical and green synthesis) and biomedical application silver nanoparticles are presented.

Keywords: Green Synthesis, Metal Nanoparticles , Silver nanoparticles , Biomedical Applications

1. Introduction

The trend of miniaturization combined with technological modernization requirements has led to the substantial rise in exploring nanoparticles. The discoveries of new antibiotics, conventional drugs and chemically modified drugs can not only resolve the microbial resistant issues but also necessitates a prolonged effective metallic nanotechnology in diverse applications. Nanomaterials such as nanoparticles, nanoclusters, nanorods, nanoshells, and nanocages are modified constantly to enable their use as a diagnostic and therapeutic agents applications. Theefficacy of the nanoparticle can be determined by its size, structure, concentration, dimensions and ionic strength accompanied with surface coating can support additional strength and durability as a carrier for a wide range of therapeutic components in several biomedical applications. Nanoparticles offer various benefits to bulk particles with increased surface-to-volume ratio, magnetictarget [1, 2], wound healing properties [3], biocomposite preparation, gene and drug delivery vehicles [1, 2, 4].

Nanoparticles synthesis and characterization have flourished due to their wide-ranging applicability particularly as catalysts in biomedical, optics, and energy fields [5]. Among the classified nanoparticles, metalnanoparticles have fascinated, due to their distinctive physical and chemical properties, selectivity, highly active, and reproducibility. Among different metal nanoparticles, silver nanoparticles (AgNPs) have attracted considerable researcher's attention because of its high electrical

IntechOpen

and thermal conductivity, surface-enhanced Raman scattering, chemical stability, catalytic activity and antimicrobial activity [6, 7]. Silver nanoparticles are increasingly being applied in biomedicines for their respective broad antimicrobial behavior becoming more attractive for use in drug delivery and targeting especially for their tunable hydrophilic - hydrophobic balance and target specific localization surface features as versatile opportunities in drug delivery and modification systems [6, 7]. The above mentioned properties have enabled silver nanoparticles to serve as a material in the development of new generation electronic, optical and sensor devices.

2. Synthesis of metal nanoparticles

In the synthesis of nano materials, particularly metallic nanoparticles, has raised greatest attention over the past decade due to their exclusiveproperty that make them suitable in various fields of science and technology. There is a scarcity of effective methods to synthesis a homogeneous size and shape nanoparticles with limited or no toxicity to the human health and the environment. There are two methods for the synthesis of metallic nanoparticles- top-down and bottom-up approaches [8]. In bottom-up approach, reduction of materials components with further self-assembly process which leads to the formation of nanostructures. Representative examples include Quantum dot and formation of nanoparticles from colloidal dispersion. In Top down approach [9] includes the macroscopic structures which can be externally controlled in the processing of nanostructures, such as ball milling, application of severe plastic deformation [10].

3. Chemical synthesis/green synthesis of metal nanoparticles

Even though nanoparticles can be made using various physicochemical methods their synthesis using nontoxic and environmentally kind biological methods is attractive specially. The biological method (green synthesis) is comparatively easy, economical, and environmentally affable method than the conventional chemical method of synthesis and thus accomplish an upper hand. Numerous studies have shown that characteristics of metallic nanoparticles such as size, stability, physical, chemical properties and morphology are strongly influenced by the experimental conditions. Several routes have been developed for biological or biogenic synthesis of nanoparticles from salts of the corresponding metals [11–14]. Microorganisms, whole plants, plant tissue and fruits, plant extracts and marine algae [15] have been used to synthesis nanoparticles.

Plants are regarded as a highly desirable system for nanoparticle synthesis due to their tremendous capability to produce a broad range of bioactive secondary metabolites with profound reducing potential. As compared to bacteria and algae and, plants are less vulnerable to metal toxicity, thus offering a green substitute for the biosynthesis of metal nanoparticles [16].

4. Green synthesis of silver nanoparticles using leaf, seed, fruit, bark and their potential

Among all metal nanoparticles, silver nanoparticles are of great significance in the field of nanotechnology [17]. Nanoparticles are synthesized by physical, chemical, and biological or green methods. Various chemical and physical methods are proved to be quite expensive and the use of various toxic chemicals that are responsible for various biological risks. This may be the reason for choosing biosynthesis of nanoparticles via green routethat does not employ toxic chemicals and proved to be eco-friendly [18]. Silver nanoparticles are the most prototypical target of green methods [19–23] and can be synthesized using plant extracts.Gold and silver metal nanoparticles weregreen synthesized using the Root Extract of Coleous forskohlii as capping and reducing agent for biomedical applications [24].

5. Biomedical application of silver nanoparticles

Silver nanoparticles have diverted the attention of the scientific community and industrialist itself due to their wide range of applications in industry for the preparation of consumer products and highly accepted application in biomedical fields. Silver has function in antimicrobial, catalytic and biological systems and the unique physical and chemical properties of silver nanoparticles only increase the efficacy of silver. Though there are many mechanisms ascribed to the antimicrobial activity shown by silver nanoparticles, the actual and most reliable mechanism is not fully understood or cannot be generalized as the nanoparticles are found to act on different organisms in different ways.

During the past few years, silver nanoparticles became one of the most examined and explored nanotechnology-derived nanostructures, given the fact that silver nanoparticles proved to have interesting, challenging, and promising characteristics suitable for various biomedical applications [25]. Even though there is limited information regarding the toxicity and in vivo biological behavior, these nanostructures were used for a long time as antibacterial agents in the health industry cosmetics, food storage, textile coatings and some environmental applications [26].

The exclusiveproperty of silver nanoparticles areparticularly advantageous for cancer therapeutics since they led to an improved chemotherapeutic efficiency together with minimal systemic toxicity [27]. AgNPs attracted special attention for this particular domain, and were successfully evaluated as effective anti-tumor drug-delivery systems [28], acting either as passive [29] or active [30] nanocarriers for anticancer drugs. Recent studies evidenced the potential use of AgNPs as vaccine and drug carriers for specific and selective cell or tissue targeting [31]. In addition tothe great optical properties of AgNPs [32–34] the recent improvements in AgNP biocompatibility and stability viasurface modification strongly recommend nanostructured systems based on silver as specific, selective,and versatile candidates for drug-delivery applications [35].

6. Silver nanoparticles as antimicrobial agents

Researches on the synthesis of silver nanoparticles using microbes and plant extract has become active due to its easy accessibility, non-toxicity, wide ranged applications, flexibility and essentially for its biodegradability, sustainability and cost effectivity. Various plants are being effectively used for the synthesis of metal nanoparticles. Various plant parts including fruit peels, leaves, barks, flowers, roots are used in synthesizing silver and other metal nanoparticles. Silver nanoparticles can serve as a medium for the delivery of antibiotics and disinfecting materials. Silver ions (Ag^+) and their respective compounds are highly toxic to broad spectral microorganisms. As a biological approach, different plants.

7. Antibacterial

Silver has been found in our traditional medicines and culinary for a long time. Silver is known to cause bacteriostatic (growth inhibition) and bactericidal (eradicate) properties, hence described as oligodynamic it is metals enclose ions that devastate living cells, like fungi, bacteria, and viruses. Silver in ionic forms strongly interacts with thiol groups of vital enzymes in bacteria and inactivates them and thereby making it lose their ability to replicate their DNA. Silver compounds such as silver nitrate and silver sulfadiazine are being used to prevent bacterial growth in sterilization process of drinking water and also in burn care activities. Silver nanoparticles has been widely used as antibacterial agents for centuries that exerts bactericidal activity even at minimal concentration which has lead its use against antibiotic resistant bacteria and prevents against a broad range of pathogenic microorganisms. Silver nanoparticles have been found to destabilize the membrane potential and deplete the intracellular ATP (Adenosine tri-phosphate) levels by target resulting in death of the bacteria.

8. Antifungal

Silver displays multiple mode of inhibitory mechanisms against microorganisms. Silver nanoparticles can be actively applied in the field of plant protection following the emergence of various resistant fungal pathogens leading to the reduction in agricultural production.The antifungal potential of silver nanoparticles was tested against various human pathogens, plant pathogens, wood degrading fungi including *Aspergillusochraceus*, *Candidaalbicans*. *Macrophominaphaseolina*, *Fusariumoxysporum*, *Fusariumsolani*, *Trichoderma* sp., and *Alternariaalternate* [36], *Raffaelea* sp., *Alternariabrassicicola*, *Botrytiscinerea*, *Cladosporiumcucumerinum*, *Corynesporacassiicola*, *Cylindrocarpondestructans*, *Didymellabryoniae*, *Glomeirellacingulata*, *Monosporascuscannonballus*, *Pythiumaphanidermatum*, *Pythiumspinosum*, *Stemphyliumlycopersici* [37] commercially important fungal pathogens were tested to check the fungicidal properties of silver nanoparticles. The findings suggest that silver nanoparticles are capable of inhibiting the above mentioned pathogens with slight variations according to the silver nanoparticles applied. Most of the fungi showcased higher inhibition rate at low concentrations of silver nanoparticles. Though very little is known about the effects of silver nanoparticles on phytopathogenic fungi, certain studies carried out proved the efficiency of silver nanoparticle on inhibition of mycelial growth and conidial germination.

9. Anticancer

Cancer cells have abnormal metabolic behaviors and genomic expressions by causing various pathological and metabolic alterations in cellular surroundings developed by cell signaling, rapid proliferation, angiogenesis and metastasis. Many studies reported depicts that the use of silver nanoparticles enhances the chemotherapeutic efficacy against multidrug resistant cancer cells emphasized with specifications and combinations. Nanoparticles coated with specific binders can recognize particular surface receptors and targets only the cancerous cells or the anomalous cells. Many platinum nanoparticles and platinum based compounds were approved as anticancer agents. Though many cancer types are susceptible to platinum based drugs accompanied with toxic side effects. Consequently other metal nanoparticles are explored in search of a better anticancer agents, while silver with advantageous antimicrobial activity arose into interest as an effective anticancer agent. Cancer cells such as HepG2 (human liver cancer cells) [24], HCT (Human colon cancer cells), HeLa (Human cervical adenocarcinoma cells), MCF 7 (Human breast adenocarcinoma cells) [24] and various other cancer cells were used to study the cytotoxicity effect of silver nanoparticles. Silver nanoparticles synthesized using different plant extracts showed potentially high cytotoxicity and less cell viability against various cancer cells. Moreover, nanoparticles of 5-35 nm sizes effectively induced cell death through mitochondrial structure targeting [38].

10. Silver nanoparticles for drug-delivery systems

Metallic nanoparticles had emerge as probable antimicrobial agents due to their ultra-small size, high surface to volume ratio, novel physiochemical properties rooted from interaction with microbes including cellular uptake and aggregation leading to toxicity and death of the microbe [39]. Ligand dependent silver release with drug may offer potent synergistic antimicrobial activities not only for drug but also for AgNPs due to their short carbon chain and weak binding atom of oxygen. Therefore, the optimization of the surface ligands such as coordination atoms, carbon chain lengths and terminal groups is very important to prepare nanoparticles for commercial applications against infectious diseases [40]. Research evidences shows that modification of silver nanoparticles could be exploited for drug delivery and are used to modulate the toxic actions of drugs. It also accompanies that as the concentration increases, non-significant reduction in the cytotoxic actions for the silver nanoparticle conjugates were relative to the cytotoxicity of the cells.

11. Silver nanoparticles for catheter modification

In generalmicrobes adhere on the surface of the catheters and growsrapidly forming biofilmsin such environmental conditions leading to bloodstream infections, even worse. Silver impregnated catheters have already been used in clinical fields and silver nanoparticles are applied in number of biomedical devices. Methods like solvent casting, electrospinning, electrospraying, and silver iontophoretic technology were being used for the synthesis of silver impregnated catheters. The nature of silver nanoparticles and the coating incorporation will determine the efficacy and durability of the medical devices.

12. Silver nanoparticles for dental applications

Silver has been proven to be less toxic and a good biocompatible with human cells [41]. Silver nanoparticles are used as endodontics, several areas of dentistry such as implantology, restorative dentistry and dental prostheses. Use of silver nanoparticles in dentistry is mainly to inhibit or decrease the growth of microbial colonization over the dental materials to improve and maintain oral health. Other advantage being the penetration possibility of silver nanoparticles through cell membranes resulting in higher antimicrobial activity especially for biofilm forming microbes. Silver nanoparticles incorporated in dental materials through distinct methods depending on the type of materials. For dental implants titanium samples are immersed in silver nitrate solution and irradiated with UV(Ultraviolet) light after wash and dried [42]. Whereas, for adhesive/resin composite a monomer preferably 2-tert-butylamino-ethyl-methacrylate is added to improve the silver solubility [43]. In order to improve quality and durability of polymeric restorative materials many studies are being performed. Rather than notable advancements, restoration composite materials accumulate more biofilms. Actually an imperfect sealing between the restoration composite material and the cavity wall leads to the colonization of oral microbes resulting in secondary caries leading to replacements. To avoid such complications, restorative materials with antimicrobial property has to be incorporated.

13. Silver nanoparticles for wound healing

The disruption of skin integrity defines the formation of a wound, which can be classified into acute (burns), chronic (diabetic foot) and pressure ulcers [44]. Due to the development of antibiotic resistant and outbreak of infectious diseases, the scientists are eyeing for better replacements. Indeed, there is rise in interest in silver nanoparticles integrated biopolymers in wound healing applications. A review conducted by Sim *et al.*, on the silver based patented products revealed that over 5000 new applications were registered during 2007–2017 [45]. Silver based products are patented and commercialized for their improvised designs and efficacies than the standard dressing materials. The antimicrobial effect of silver significantly reduces the hazardous natureof microorganisms to develop resistance and increases the efficiency against multi-drug-resistant microbes.An active role of silver nanoparticles attributed in wound healing mechanisms along with its distinctive role in preventing infections, which in turn promotes faster healing rates, stimulated proliferation, relocation of keratinocytes and wound contraction [3]. The antimicrobial therapy which mainly supervise the process such as Colonization, proliferation of pathogens along with multidrug confrontation which serves as foremost and imperative aspects of skin and wound care. Enchantingly, silver and silver ions havebeen engaged for thousands of years since their bactericidal activity that include.Antibacterial effects at the multilevel approach that aims at reducing chances of enlargement of confrontation, and this purpose is served via blocking the respiratory enzymes. Efficiency against multi drug resistant organisms. Low systemic toxicity. A significant volume of research data has provided confirms the beneficial effects of silver nanoparticles as biocompatible, however, the interaction mechanism between silver nanoparticles and the microbial flora, along with clinical toxicity studies are still requires deep investigations. Medical products such as bandages, gauzes, sutures, plasters, textile materials, creams and ointments can be functionalizedwith silver nanoparticles for wound healing property. A synergistic effect between silver and silk protein sericin improved tissue regeneration and antimicrobial properties, a natural wound dressing biomaterial approved [46].

14. Silver nanoparticles for bone healing

Human bones are composed of crystallized hydroxyapatite, a form of calcium and phosphate. It is a widely accepted and used body implant material. As a suitable choice for the fabrication of antimicrobial and bioactive bone implants biocompatible hydroxyapatiteintegrated either with metallic or ionic silver forms are used as superficial implant materials. Such hydroxyapatite coatings embedded with silver nanoparticles found to be an effective inhibitors of both Gram positive and Gram negative bacteria [47]. In addition to antimicrobial property, the additive should provide additional optical, mechanical, chemical properties to achieve enhanced biomaterials Kora *et al*, [48].

15. Toxicity of silver nanoparticles

Over the years, silver nanoparticles have been subjected to numerous in vitro and in vivo tests to provide information about their toxic behavior towards living tissues and organisms. The biosafety of silver nanoparticles has gained much attention for its interaction with blood and tissues. Considering their unique physical and chemical properties, it is likely that these silver nanoparticles besides possess distinctive toxicity mechanisms, a better understanding of silver nanoparticles safety is essential, in order to escalate their clinical use [49]. It was proven that nanosilver can cause the formation and intracellular accumulation of ROS (Reactive Oxygen Species), modification of mitochondrial membrane permeability and DNA (Deoxyribo Nucleic Acid) damage. Variousscientific research proved that silver nanoparticles disclosure can induce a decrease in cell viability through different cellular mechanisms. One of these mechanisms is represented by the induction of apoptosis-related genes and the activation of apoptosis mechanism [50].

16. Conclusions

Silver nanoparticles are intensively explored nanostructures for exceptional and enhanced biomedical applications, thanks to their attractive size-related physico-chemical properties and biological functionality, including their high antimicrobial efficiency and non-toxic nature. Silver nanoparticles-based nanomaterials and nano-systemsare appropriate substitutions for drug delivery, wound dressing, tissue scaffold and protective coating applications. Various physicochemical parameters were related to the intrinsic antimicrobial effects exhibited by silver nanoparticles, such as size, shape, concentration, surface charge and colloidal state. Moreover, the available surface of nanosilver allows the coordination of many ligands, thus enabling remarkableoptions with respect to the surface functionalization of silver nanoparticles.

Author details

Manikandan Dhayalan[1*], Priadharsini Karikalan[2], Mohammed Riyaz Savaas Umar[3] and Nalini Srinivasan[4]

1 Department of Chemistry, Dr. Mahalingam College of Engineering and Technology, Pollachi, Tamil Nadu, India

2 CAS in Botany, University of Madras, Chennai, Tamil Nadu, India

3 Department of Biotechnology, Islamiah College (Autonomous), Vaniyambadi, Tamil Nadu, India

4 Centre for Ocean Research, Sathyabama University, Chennai, Tamil Nadu, India

*Address all correspondence to: manikandandhayalan88@gmail.com

References

[1] Dobson J. Gene therapy progress and prospects: magnetic nanoparticle-based gene delivery. Gene Ther. 2006;13:283-7. [PubMed] [Google Scholar]

[2] Rudge S, Peterson C, Vessely C, Koda J, Stevens S, Catterall L. Adsorption and desorption of chemotherapeutic drugs from a magnetically targeted carrier (MTC) J Control Release. 2001;74:335-40. [PubMed] [Google Scholar]

[3] Vijayakumar V, Samal S K, Mohanty S and Nayak S K (2019). Recent advancements in biopolymer and metal nanoparticle-based materials in diabetic wound healing management. International Journal of Biological Macromolecules, 122:137-148.

[4] Appenzeller T. The man who dared to think small. Science. 1991;254:1300. [PubMed] [Google Scholar

[5] J.Y. Song, B.S. Kim, Rapid biological synthesis of silver nanoparticles using plant leaf extracts, Bioprocess Biosyst. Eng. 32 (2009) 79-848.

[6] M. Jeyaraj, M. Rajesh, R. Arun, D. MubarakAli, G. Sathishkumar, G. Sivanandhan, G. Kapil Dev, M. Manickavasagam, K. Premkumar, N. Thajuddin, A. Ganapathi. An investigation on the cytotoxicity and caspase-mediated apoptotic effect of biologically synthesized silver nanoparticles using Podo phyllumhexandrum on human cervical carcinoma cells. Colloids and Surf B: Biointerfaces 102(2013) 708-717.

[7] F. Okafor, A. Janen, T. Kukhtareva, V. Edwards, M. Curley, Green synthesis of silver nanoparticles, their characterization, application and antibacterial activity, Int. J. Environ. Res. Public Health 10 (2013) 5221-5238.

[8] Rampino LD, Nord FF (1941) Preparation of palladium and platinum synthetic high polymer catalysts and the relationship between particle size and rate of hydrogenation. Journal of the American Chemical Society 63: 2745-2749.

[9] Yuliang Wang and Younan Xia (2004)Bottom-Up and Top-Down Approaches to the Synthesis of Monodispersed Spherical Colloids of Low Melting-Point Metals,Nano Letters 2004, 4, 10, 2047-2050.

[10] Jacques Hout ,Nataliya ye. Skryabina and Daniel Fruchart (2012), Application of Severe Plastic Deformation Techniques to Magnesium for Enhanced Hydrogen Sorption Properties,Metals 2012, 2(3), 329-343.

[11] Bar H, Bhui DK, Sahoo GP, Sarkar P, De SP, Misra A. Green synthesis of silver nanoparticles using latex of Jatropha curcas. Colloids Surf A 2009;339: 134-139.

[12] Dhillon GS, Brar SK, Kaur S, Verma M. Green approach for nanoparticle biosynthesis by fungi: current trends and applications. Crit Rev Biotechnol 2012;32:49-73.

[13] Duran N, Marcato PD, De Souza GIH, Alves OL, Esposito E. Antibacterial effect of silver nanoparticles produced by fungal process on textile fabrics and their effluent treat-ment. J Biomed Nanotechnol 2007;3:203-208.

[14] Thakkar KN, Mhatre SS, Parikh RY. Biological synthesis of metallic nanoparticles. Nanomed NanotechnolBiol Med 2010;6: 257-262.

[15] Luangpipat T, Beattie IR, Chisti Y, Haverkamp RG. Gold nanoparticles produced in a microalga. J Nanopart Res 2011;13:6439-6445.

[16] S. Pandey, A.Mewada, M. Thakur, R. Shah, G. Oza,M. Sharon, Biogenic gold nanoparticles as fotillas to fire berberine hydrochloride using folic acid as molecular road map, Mater. Sci. Eng., C 33 (2013) 3716-3722

[17] Ahmad A, Mukherjee P, Senapati S, Mandal D, Khan MI, Kumar R,et al. Extracellular biosynthesis of silver nanoparticles using the fungus Fusarium oxysporum. Colloids Surf B Biointerfaces 2003;28:313-8.3.

[18] Reddy GA, Joy JM, Mitra T, Shabnam S, Shilpa T. Nano silver - A review. Int J Adv Pharm 2012;1(2):9-15.4.

[19] Hu B. Microwave assisted rapid facile green synthesis of uniform silver nanoparticles: Self-assembly into multi-layered films and their optical properties. J Phys Chem C 2008;112(30):11169-174.5.

[20] Sankar S, ValliNachiyar C. Microbial synthesis and characterization of silver nanoparticles using endophytic bacterium Bacillus cereus: A novel source in benign synthesis. Glob J Med Res 2012;12:43-9.6.

[21] Darroudi M. Green synthesis of colloidal silver nanoparticle by sonochemical method. Adv Mater Lett 2012;66:117-20.7.

[22] Subba Rao Y, Kotakadi VS, Prasad TN, Reddy AV, Sai Gopal DV. Green synthesis and spectral characterization of silver nanoparticles from Lakshmi Tulasi (Ocimum sanctum) leaf extract. Spectrochim Acta A Mol Biomol Spectrosc 2013;103:156-159

[23] Elizabath Antony, Shylaja Gunasekaran, MythiliSathiavelu, SathiaveluArunachalam, A Review On Use Of Plant Extracts For Gold And Silver Nanoparticle Synthesis And Their Potential Activities Against Food Pathogens, Asian J Pharm Clin Res, Vol 9, Issue 4, 2016, 18-23.

[24] Manikandan Dhayalan, Michael Immanuel Jesse Denison, A. Manikandan , Nagendra Gandhi N, Kathiravan Krishnan , Abdulhadi Baykal, (2018)Biogenic Synthesis, Characterization of Gold and Silver Nanoparticles from Coleus forskohlii and their Clinical Importance", Journal of Photochemistry & Photobiology, B: Biology, 183,, Pages 251-257.

[25] Alexandra-Cristina Burdușel, Oana Gherasim, Alexandru Mihai Grumezescu,2 Laurențiu Mogoantă,4Anton Ficai, and Ecaterina Andronescu , (2015)Biomedical applications of silver nanoparticles:An Up-to-Date Overview,Nanomaterials (Basel). 2018 Sep; 8(9): 681

[26] Shakeel Ahmed, Mudasir Ahmad, BabuLalSwamiSaiqaIkram (2016),A review on plants extract mediated synthesis of silver nanoparticles for antimicrobial applications: A green expertiseJournal of Advanced Research, 7(1), 17-28.

[27] Jyotsna Thayath , Keechilat Pavithran Shantikumar V Nair ManzoorKoyakutty (2021)Cancer nanomedicine developed from total human serum: a novel approach for making personalized nanomedicine , Nanomedicine (Lond) 2021 May;16(12):997-1015

[28] Benyettou, F.; Rezgui, R.; Ravaux, F.; Jaber, T.; Blumer, K.; Jouiad, M.; Motte, L.; Olsen, J.C.; Platas-Iglesias, C.; Magzoub, M.; et al.Synthesis of silver nanoparticles for the dual delivery of doxorubicin and alendronate to cancer cells. J. Mater. Chem. B 2015, 3, 7237-7245.

[29] Barbinta-Patrascu, M.E.; Badea, N.; Pirvu, C.; Bacalum, M.; Ungureanu, C.; Nadejde, P.L.; Ion, C.; Rau, I. Multifunctional soft hybrid

bio-platforms based on nano-silver and natural compounds. Mater. Sci. Eng. C 2016, 69, 922-932.

[30] Poudel, B.K.; Soe, Z.C.; Ruttala, H.B.; Gupta, B.; Ramasamy, T.; Thapa, R.K.; Gautam, M.; Ou, W.; Nguyen, H.T.; Jeong, J.-H.; et al. In situ fabrication of mesoporous silica-coated silver-gold hollow nanoshell for remotely controllable chemo-photothermal therapy via phase-change molecule as gatekeepers. Int. J. Pharm. 2018, 548, 92-103.

[31] Rai, M.; Ingle, A.P.; Gupta, I.; Brandelli, A. Bioactivity of noble metal nanoparticles decorated with biopolymers and their application in drug delivery. Int. J. Pharm. 2015, 496, 159-172.

[32] Sarkar, S.; Das, R. Shape effect on the optical properties of anisotropic silver nanocrystals. J. Lumin. 2018, 198, 464-470.

[33] Delgado-Beleño, Y.; Martinez-Nuñez, C.E.; Cortez-Valadez, M.; Flores-López, N.S.; Flores-Acosta, M. Optical properties of silver, silver sulfide and silver selenide nanoparticles and antibacterial applications. Mater. Res. Bull. 2018, 99, 385-392.

[34] dos Santos Courrol D.; Regina Borges Lopes C.; da Silva Cordeiro T.; Regina Franzolin, M.; Dias Vieira Junior, N.; ElgulSamad, R.; CoronatoCourrol, L. Optical properties and antimicrobial effects of silver nanoparticles synthesized by femtosecond laser photoreduction. Opt. Laser Technol. 2018, 103, 233-238.

[35] Brown, P.K.; Qureshi, A.T.; Moll, A.N.; Hayes, D.J.; Monroe, W.T. Silver nanoscale antisense drug delivery system for photoactivated gene silencing. ACS Nano 2013, 7, 2948-2959

[36] Bahram Bahrami-Teimoori, YaserNikparast, Mostafa Hojatianfar, Mahdi Akhlaghi, Reza Ghorbani& Hamid Reza Pourianfar (2017). Characterisation and antifungal activity of silver nanoparticles biologically synthesised by Amaranthusretroflexus leaf extract, Journal of Experimental Nanoscience, 12(1): 129-139.

[37] Sang Woo Kim ,JinHee Jung , Kabir Lamsal , Yun Seok Kim , Ji Seon Min and Youn Su Lee (2012). Antifungal Effects of Silver Nanoparticles (AgNPs) against Various Plant Pathogenic Fungi, Mycobiology, 40(1): 53-58.

[38] Wafa I Abdel-Fattah, Ghareib W Ali (2018). On the anti-cancer activities of silver nanoparticles, Journal of Applied Biotechnology & Bioengineering, 5(1):43-46.

[39] Zhou Y, Kong Y, Kundu S, Cirillo JD, Liang H (2012). Antibacterial activities of gold and silver nanoparticles against *Escherichiacoli* and *Bacilluscalmette-guerin*. Journal of Nanobiotechnology, 10: 19.

[40] Ardhendu K M (2017). Silver Nanoparticles as Drug Delivery Vehicle against Infections. Global Journal of Nanomedicine, 3(2): 555607.

[41] SlentersT V, Hauser-Gerspach I, Daniels A U,Fromm K M (2008., Silver coordination compounds as light-stable, nano-structured and anti-bacterial coatings for dental implant and restorative materials, Journal of Materials Chemistry, 18(44): 5359-5362.

[42] Zhao L, Chu P K, Zhang Y and Wu Z (2009),Antibacterial coatings on titanium implants, Journal of Biomedical Materials Research Part B: Applied Biomaterials, 91(1):470-480.

[43] Li F, Weir M D, Chen J and Xu H H K (2013), Comparison of quaternary ammonium-containing with nano-silver-containing adhesive in antibacterial properties and cytotoxicity, Dental Materials, 29(4): 450-461.

[44] Lindholm C and Searle R (2016). Wound management for the 21st century: Combining effectiveness and efficiency. International wound Journal, 13:5-15.

[45] Möhler J S, Sim W, Blaskovich M A T, Cooper M A and Ziora Z M (2018). Silver bullets: A new lustre on an old antimicrobial agent. Biotechnology Advances, 8(36):1391-1411.

[46] Gallo A L, Pollini M and Paladini F (2018). A combined approach for the development of novel sutures with antibacterial and regenerative properties: The role of silver and silk sericin functionalization. Journal of Materials Science. Materials in Medicine, 9:133.

[47] Bharti A. Singh S, Meena V K, Goyal N (2016). Structural characterization of silver-hydroxyapatite nanocomposite: A bone repair biomaterial. Materials Today Proceedings, 3:2113-2120.

[48] Kora A J and Sashidhar RB (2018), Biogenic silver nanoparticles synthesized with rhamnogalacturonan gum: Antibacterial activity, cytotoxicity and its mode of action. Arabian Journal of Chemistry, 11:13-323.

[49] Chen X and Schluesener H J (2008). Nanosilver: a nanoproduct in medical application, Toxicology Letters, 176(1):1-12.

[50] John C. Reed (2000) , Mechanisms of Apoptosis, Am J Pathol. 2000 Nov; 157(5): 1415-1430

Chapter 6

Nanosilver in Biomedicine: Advantages and Restrictions

Olga V. Morozova and Dmitry V. Klinov

Abstract

Nanosilver (in a range 1–100 nm) binds with thyol-, amino- and carboxy-groups of aminoacid residues of proteins and nucleic acids, thus providing inactivation of pathogenic multidrug-resistant microorganisms. Besides antibacterial, antiviral, antifungal and anti-cancer properties Ag-based nanomaterials possess anti-inflammatory, anti-angiogenesis and antiplatelet features. Drug efficacy depends on their stability, toxicity and host immune response. Citrate coated Ag nanoparticles (NPs) remain stable colloid solutions in deionized water but not in the presence of ions due to replacement of Ag^+ by electrolyte ions, potential formation of insoluble AgCl, subsequent catalyzed oxidative corrosion of Ag and further dissolution of surface layer of Ag_2O. Protein shells protect core of AgNPs from oxidation, dissolution, aggregation and provide specific interactions with ligands. These nanoconjugates can be used for immunoassays and diagnostics but the sensitivity threshold does not exceed 10 pg Cytotoxicity of AgNPs conjugated with proteins is associated with the rate of intracellular Ag^+ release, a 'Trojan horse' effect, and exceeds one of Ag^+ because of endocytosis uptake of NPs but not ions. Relatively toxic nanosilver causes immunosuppression of the majority of cytokines with a few exceptions (IL-1β, G-CSF, MCP-1) whereas $AgNO_3$ additionally activate TNFα and IL8 gene expression.

Keywords: silver ions, nanoclusters, nanoparticles, nanoconjugates with silver core and protein shells, stability, cytotoxicity, immunosupression

1. Introduction

Nanosilver is a generic term that refers to nanoscale Ag materials that have at least one dimension less than 100 nm, and which are commonly in the form of particles called silver nanoparticles (AgNPs). They remain the most used nanostructures in commercialized products. Approximately 320 tons of AgNPs are manufactured each year [1]. There are nearly 500 consumer products that claim to contain nanosilver. At present they are included in nanomedical devices, as tools for medical imaging and biosensing [2] which are used for diagnostics. AgNPs are also employed as antifungal, antibacterial and antiviral drugs [3], for wound dressings and long-term burn care products, anti-bacterial cosmetic lotions for both treatment and supplementary drug and/or nutrient delivery [2]. Besides broad implementation of the nanosilver in health care systems for diagnostic and therapy purposes, medical device coating, medical textiles, contraceptive devices, Ag-containing nanostructures are currently used in cosmetics, clothing, household and food products.

The antimicrobial mechanisms of AgNPs include adhesion to cell surface altering the membrane properties, the formation of free radicals damaging the bacterial membranes and viral envelopes, interactions with DNA, and enzyme deterioration [4]. Besides that oxidative stress induction, heavy metal ion release that occurs in aqueous solutions, producing biologically active Ag^+ [5] and non-oxidative mechanisms were suggested for silver nanostructures [6]. The generation of reactive oxygen species (ROS) inhibits the antioxidant defense system and causes mechanical disruptions of the viral envelopes and cellular membranes. Metal ions are slowly released from metal oxide and are absorbed through the cell membranes or viral envelopes, followed by direct interaction with the functional groups of proteins and nucleic acids, such as mercapto (–SH), amino ($–NH_2$), and carboxyl (–COOH) groups, damaging enzyme activity, changing their structure, affecting the normal physiological processes, and ultimately inhibiting the pathogens of different origin. Currently additional mechanisms of Ag^+ antimicrobial action are becoming evident. Ag^+ ions may react with phosphorus and sulfur groups of surface proteins of the cellular membranes, bacterial cell wall as well as virions after posttranslation modification. Ag^+ binds to negative parts of the membranes including viral envelopes, making a hole. Ag^+ ions damage cytochrome of electron transport chain, impass and destroy RNA and DNA. Ag^+ hinders DNA replication. Ag^+ prevents translation of protein due to damage of ribosomal 30S subunits. Ag^+ ions are sources for the formation ROS that have harmful effect to both eukaryotic and bacterial cells. However, the impact of metal ions on the pH inside membrane coated vesicles is small and has weak antimicrobial activity. Therefore, dissolved metal ions are not determined the main antimicrobial mechanism of AgNPs. Moreover, heavy metal ions can indirectly act as carriers of antimicrobial substances [6]. Thus, disruption of the cellular membranes and viral envelopes, interactions with proteins and nucleic asids [6] are the majors known processes of silver-induced disinfecting activity. These three independent mechanisms take place simultaneously with reversible equilibrium between AgNPs with permanent liberation of Ag^+ ions and reverse deposition of AgNPs from recovered ions and nanoclusters in cells. The numerous mechanisms of action against infectious agents would require multiple simultaneous gene mutations for resistance to develop; therefore, a resistance to silver-containing compounds and nanostructures is hardly possible [6].

Despite the evergrowing presence of Ag-containing products in the market and extensive reports on the antimicrobial activity of AgNPs, insufficient data are currently available about the principal restrictions for the nanosilver to use as diagnostic and therapeutic agents. Inevitably, from the rapid growth in its manufacture and utilization follows an increased environmental and human exposure, whereas the potential acute and chronic toxicity has yet to be fully addressed.

Current research is to analyze stability, cytotoxicity and immunomodulation potential of Ag^+ ions and NPs.

2. Stability of Ag^+ ions, citrate coated AgNPs and their nanoconjugates with proteins

AgNPs in the presence of ions and especially after addition of EDTA are not stable due to oxidation, dissolution and aggregation during a few hours. UV–visible spectroscopy, dynamic light scattering (DLS) and scanning electron microscopy (SEM) revealed that the citrate coated AgNPs remained stable colloid solutions in deionized water at room temperature for decades but not in the presence of ions. Citrate coated AgNPs with the surface Ag_2O layer are not stable in the presence of phosphate buffer solution (PBS) (0.01 M Na_2HPO_4/KH_2PO_4,

0.15 M NaCl/KCl) during 1 hour at room temperature (**Figure 1**) due to replacement of Ag^+ by electrolyte ions, potential formation of insoluble AgCl, subsequent catalyzed oxidative corrosion of Ag and further dissolution of surface layer of Ag_2O [7, 8]. To prevent AgNPs dissolution and aggregation various surfactants and polymers are introduced during or after synthesis [7]. Coating layers may enhance electrostatic and steric repulsion. Adsorption of polymers or nonionic surfactants provides steric hindrances depending upon the thickness of the adsorbed layer [7]. Nanosilver like other NPs immediately after administration into organism becomes wrapped by serum and cellular proteins constituting the protein corona. This protein shells decrease the efficiency of targeting by directing the NPs to the reticuloendothelial system, by masking the active targeting moieties and decreasing their ability to bind to their target receptor, but may re-direct NPs towards endogenous targets. The NPs stability depends on the affinity of coating molecules to the particle surface, repulsion from neighboring molecules, loss of chain entropy upon adsorption, and also nonspecific dipole interactions between the macromolecule, the solvent, and the surface. Protein corona protect AgNPs from dissolution and aggregation (**Figure 1**). The nanoconjugates of the noble metal NPs with proteins remain stable at +4 °C for several months [8]. Surface of nanosilver dynamically adsorbs proteins forming a robust rapidly exchanging "biocorona". A hard corona with long-term stability can be formed with immunoglobulins IgG/IgM and fibrinogens and may alter NPs size, shape, surface charge and agglomeration state, as well as cellular toxicity and internalization, trafficking, opsonization and eventually pattern of biodistribution [8]. Colloids of Ag possess high affinity for binding with serum albumins, their ability to bind with *Staphylococcus aureus* protein A is less efficient, whereas a number of proteins (for example, human immunodeficiency virus (HIV-1) envelope antigen) cannot attach to AgNPs at all. Despite known chemical affinity of sulfur atoms to precious metals direct correlation between cystine disulfide bridge content and binding with AgNPs was not observed perhaps because of strong bonds between two cysteines that stabilize protein conformation.

$AgNO_3$ and its water solutions should be stored in the dark because of possible recovery of silver atoms with formation of nanostructures. Thus, fluorescent metal nanoclusters with sizes less than 2 nm consisting of a few silver atoms can be recovered from Ag^+ in the presence of proteins (albumins, immunoglobulins of different classes and origin and $NaBH_4$ (unpublished data).

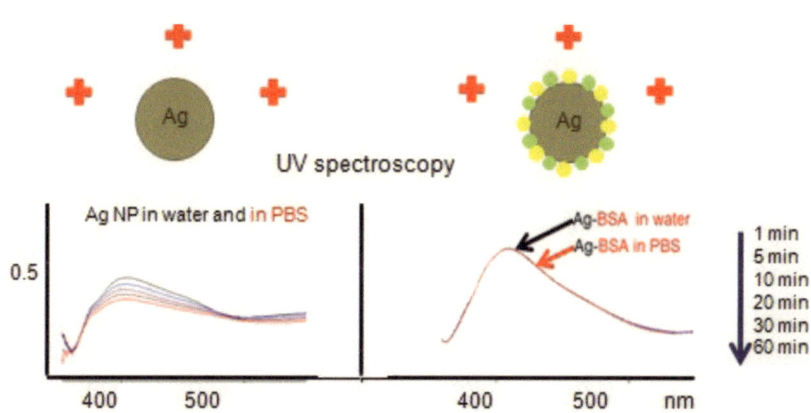

Figure 1.
Stability of citrate-coated and protein-coated AgNPs in deionized water and phosphate buffer solution (PBS) (0.01 M Na_2HPO_3/K_2HPO_4, 0.15 M NaCl/KCl).

3. Nanosilver in immunodiagnostics

Physicochemical features of the nanosilver determine possible implementation in diagnostics. Typical size range of AgNPs 30–80 nm provides high surface to volume ratio. Binding of AgNPs with NH_2- and SH-groups of proteins is weaker compared to AuNPs but protein corona can be formed with the majority of proteins including the main blood proteins. However, leaking Ag^+ cations may damage proteins of envelopes. Extinction, light scattering, surface plasmon resonance (SPR) and SERS of AgNPs exceed those of AuNPs in 10 and 100 times, respectively. Relatively low price is also an advantage of the nanosilver.

The stable nanoconjugates of AgNPs with immunoglobulins of different origin, classes and specificity including both polyclonal and monoclonal antibodies were constructed by: (1) direct binding of AgNPs with purified IgG or IgM [8]; (2) nanoprecipitation of proteins from their solutions in fluoroalcohols [9]; (3) physisorption of proteins on the AgNPs surface treated with poly(allylamine)s; (4) encapsulation of AgNPs into SiO_2 envelope and functionalization with organosilanes. Adsorption of proteins on surfaces of AgNPs is reversible and up to 70% of the attached proteins can be eluted. AgNPs possess high affinity for binding with immunoglobulins but not with any protein. SiO_2 layer on surfaces of metal NPs is suitable for silanization and covalent attachment of any protein. Protein corona prevents AgNPs from oxidation, dissolution and aggregation. The developed methods of fabrication of AgNPs with protein shells permit to attach any protein at different distances from metal core to avoid possible inactivation of proteins, to reduce fluorescence fading and to stabilize the nanoconjugates [8].

To detect binding of immobilized antigens in chip with nanosilver conjugated with IgG the analyzer based on light scattering of dark field laser of total internal reflection with the wave length 532 nm and corresponding software were used. The sensitivity limit of the nanosilver-based immunodiagnostic systems was nearly 10 pg/dot for direct binding of AgNPs with immobilized IgG and 100 pg for 3-layer sandwich immunoassay. For comparison, thresholds of commercially available conventional ELISA and xMAP multiplex immunofluorescent analysis with fluorescent magnetic microspheres were 1 pg/ml. Specificity of Ag nanoconjugates is limited due to their binding with the major blood serum proteins: IgM, IgG, fibrinogen and albumins with increased background level. Protein dots on NH_2- and COOH-modified surfaces of chips are not homogenous causing problems of dot-to-dot reproducibility.

Taken together, immunodiagnostics based on AgNPs covered with IgG shells yields to specificity and sensitivity of the widely used ELISA and xMAP in 10–100 times. Specificity of immunodetection and ratio of signal to background are limited because of binding between AgNPs and blood proteins. Besides the nanoconjugates of AgNPs with protein shells, fluorescent silver nanoclusters containing a few recovered Ag atoms with sizes less than 2 nm can be used in immunofluorescent diagnostics.

4. Cytotoxicity of Ag ions and nanoconjugates of AgNPs with major blood proteins

The common mass-only dose metric model employed in toxicology for traditional substances is not convenient for engineered nanomaterials. Alternative dose metrics include particle number, ion release (kinetics, equilibrium), and the total particle surface area. Nevertheless, polydisperse particle suspensions, the ambiguity in the surface area and concentrations will obscure the analysis. Therefore, Organisation for Economic Cooperation and Development recommended that particle number, surface area, and mass should all be measured when possible to enable calculation of alternative dose metrics. For AgNPs, both surface area and ion release have been reported as effective alternative dose metrics for nanotoxicological assessment.

Silver in ionic, nanoparticulate, and bulk forms behave very differently. Due to large surface area AgNPs are able for rapid oxidation, dissolution, reactive capacity and binding with biomolecules [10]. When the size of metallic silver is shrunk to nanometre scale, it can enter the cells and cause adverse health effects [10]. AgNPs enter into eukaryotic cells either by endosomal uptake or by diffusion. They can penetrate in living organisms via several routes including inhalation, oral ingestion, intravenous injection, and dermal contact. The American Conference of Governmental Industrial Hygienists has established threshold limit values for metallic silver (0.1 mg/m^3) and soluble compounds of silver (0.01 mg/m^3). Long exposure of humans to the nanosilver from cations to NPs through oral and inhalation routes can lead to argyria, or skin discoloration, and argyrosis, or discoloration of the eyes, as soluble silver incorporates into the tissues with permanent damage of skin microvessels and eyes [11]. Studies *in vivo* with experimental animals have revealed AgNPs accumulation in their liver, spleen, and lung. Similarly, AgNPs-mediated cytotoxicity in mammalian cells depends greatly on the nanoparticle size, shape, surface charge, dosage, oxidation state, and agglomeration condition as well as the cell type. Smaller AgNPs cause more toxicity than larger ones owing to their larger surface area and reactivity [11]. However, currently available data about toxicity of silver nanowires (AgNW) (micron-range long with diameters <100 nm) remain contradictory [11]. For both short (1.5 mm) and long (10 mm) AgNW after inhalation lung inflammation at day 1, disappearing by day 21 has been described, and in bronchoalveolar lavage fluid, long AgNW cause neutrophilic and macrophagic inflammation [12].

Exposure to different forms of the silver leads to distinct outcomes. Whereas elemental silver exposure is not associated with health effects, soluble silver is associated with lowered blood pressure, diarrhoea, respiratory irritation, and fatty degeneration in the liver and kidneys. Furthermore, after different routes of administration including intravenous, intraperitoneal, and intratracheal ways the AgNPs can cross the brain blood barrier *in vivo* and tend to accumulate in liver, spleen, kidney and brain [9].

Respiratory tract, gastrointestinal tract, skin, and female genital tract are the main entry portals of nanosilver into the human body through direct substance exchange with the environment. Additionally, systemic administration is also a potential route of entry, since colloidal silver nanoparticles have been exploited for diagnostic imaging or therapeutic purposes. Inhalation and instillation experiments in rats showed that low concentration, but detectable, ultrafine silver (14.6 ± 1.0 nm) appeared in the lung and was subsequently distributed to the blood and other organs, such as heart, liver, kidney, and even brain. Nanosilver accumulates in blood, liver, lungs, kidneys, stomach, testes, and brain. AgNPs less than 12 nm affect early development of fish embryos, cause chromosomal aberrations and DNA damage.

Animal and human studies indicate that it is difficult to remove silver completely once it has been deposited in the body; however, nanosilver can be excreted through the hair, urine, and feces.

Human liver cells may develop a metabolic-based protection mechanism against AgNPs and Ag^+. The nanosilver penetration through the blood–brain barrier is still debatable. However, even in the absence of Ag in cerebrospinal fluid, Ag-mediated neurotoxic complications such as hypoactivity or reverse increased vivacity, changes in noradrenaline, dopamine and 5-HT concentrations in the brain were observed. Upon oral exposure to $AgNO_3$, the main target organs include liver and spleen, followed by testes, kidney, brain and lungs, and AgNPs are formed *in vivo* from Ag^+ ions and they are probably composed of silver salts. The elimination of silver from brain and testes is extremely slow [12]. AgNPs may translocate into the central nervous system through damaged blood–brain barrier, nerve afferent signaling and eye-to-brain ways, and even through olfactory receptors of the brain neurons. NPs could stimulate the activation of glial cells to release proinflammatory cytokines and generate reactive oxygen species and nitric oxide production, resulting in the neuroinflammation, including several immune response relevant signaling pathways [13]. While Ag_2S deposits have been seen in the region of cutaneous nerves, there is no indication of toxic risk of silver to the peripheral nervous system [12]. *In vitro* data indicate that silver ions alter mitochondrial function, resulting in the release of apoptogenic signals and subsequent cell death. Moreover, other studies show dose-dependent effects of silver ion on cell replication and other developmental endpoints in mammalian cells [12].

Evergrowing production of AgNPs increases their release into aquatic environments. Once AgNPs get to freshwater, sea or underground water, they oxidize into Ag^+ ions that are toxic to aquatic organisms. Later on, silver cations can bind with abundant anions available in environment with formation of sparingly soluble salts AgCl or Ag_2S. Marine inhabitants (shrimps, prawns, crabs, lobsters and crayfish), are known to be much more vulnerable to the impacts of silver than bacteria [12]. By accumulating in aquatic organisms, AgNPs can enter the human body.

AgNPs have been reported to be toxic to human cell lines [11]. Cellular uptake of AgNPs takes place either via diffusion (translocation), endocytosis or phagocytosis. Upon entering the cytoplasm, AgNPs themselves or Ag^+ ions can generate ROS, leading to DNA damage, protein denaturation, and apoptosis. AgNPs of different sizes and shapes tend to accumulate in the mitochondria, thereby inducing mitochondrial dysfunction, i.e., a reduction in mitochondrial membrane potential (MMP), and promoting ROS creation [11]. AgNPs cytotoxicity in mammalian cells depends on the NPs sizes, shape, surface charge, dosage, oxidation state, agglomeration condition and cell type. They induce a dose-, size- and time-dependent cytotoxicity, particularly for NPs with sizes less 10 nm.

Surface charge of AgNPs stabilized with citrate anions or protein envelopes is a parameter responsible for cellular uptake. In particular, high-level toxicity of positively-charged nanoconjugates versus negatively-charged coatings has been reported. It can be caused by the adhesion of AgNPs onto the negatively charged cell membranes, their consequent entry to the cell, potential release of Ag^+ inside the cell, damage of cellular proteins and nucleic acids and other cytotoxic effects. For instance, the following coatings posses surface charges: (1) positive: polyethylenimine, chitosan, poly-L-lysine and cetyltrimethylammonium bromide; (2) negative: bovine serum albumin (BSA), citrate, odium bis(2-ethylhexyl)-sulfosuccinate; (3) neutral: polyvinylpyrrolidone.

Ag^+ ions alter mitochondrial function, resulting in the release of apoptogenic signals and subsequent cell death. They may destroy DNA-dependent DNA replication, RNA transcription and translation.

Various cellular defense mechanisms, innate immunity of vertebrates and accumulation in certain organs for metabolic-based degradation and subsequent elimination of the nanosilver provide relative protection.

Endocytosis and exocytosis of AgNPs occur simultaneously and depend on physicochemical properties of NPs and protein corona [14]. All nanostructures preferentially accumulate in tumor cells due to the enhanced permeability and retention (EPR) effect. The tumors possess a leaky vasculature and lack lymphatic drainage allowing the drug-loaded NPs to reside at the tumor site for a longer duration compared to the free drug molecules [14].

The formation of protein corona on NPs' surfaces upon the *in vivo* administration is inevitable. Protein shells of all nanostructures after their systemic administration may explain the lack of the *in vitro-in vivo* correlation and the preclinical to clinical extrapolation. Protein shells provide stability of the nanocongugates, decrease their cytotoxicity and determine the interaction of NPs with the target and non-target cells. The chemical composition of protein corona may serve as a fingerprint for NPs of certain type since different NPs tend to recruit cellular and serum proteins to variable extents. Vitronectin mediates accumulation in integrin receptor-expressing melanoma cells both *in vitro* and *in vivo*, while complement 3 protein (C3) and, as an opsonin and dysopsonin, regulate the balance between the the reticuloendothelial system uptake and blood circulation [14].

The cytotoxic effect and oxidative stress of silver ions (Ag^+) on mouse lung macrophages cells resulted in necrotic rather than apoptotic cell death by reducing functional sulfhydryl groups in the cells [12].

Other defense mechanism performed by the most abundant leukocytes - polymorphonuclear neutrophils (PMN) is based on neutrophil extracellular traps (NETs) [15]. NETs are extracellular fibers consisting of DNA with histones (H3), myeloperoxidase and neutrophil elastase. NETs form a barrier that hinders the transmission of pathogens and due to high local concentrations of antimicrobial proteins degrade virulence factors and kill bacteria. However, high concentrations of active proteins may cause host immune damage, contribute to platelet aggregation and cause thrombosis [15, 16]. The nanosilver activates polymorphonuclear neutrophils to release NETs did not alter the extracellular lactate dehydrogenase level [16].

Noteworthy, that the nanosilver toxicity exceeds one of Ag^+. For instance, cytotoxic concentration of $AgNO_3$ for 50% of human larynx carcinoma HEp-2 cells (CC_{50}) was 50 μg/ml, whereas for the nanoconjugates of AgNPs with surface proteins CC_{50} = 0.1 o.u. = 1.4 μg/ml (calculations are based on the calibration curve of atomic absorption spectroscopy (AAS) data). One of possible explanations is AgNPs limited cellular uptake. At present there is not evidence of the efficient uptake or intracellular localization of the citrate coated AgNPs conjugated with fluorescent proteins. Intracellular Ag^+ release appears to be responsible for the toxicity since the cultural media after treatment of cells with the nanosilver do not cause any cytotoxicity. Thus, AgNPs were toxic for mammalian cells and the cytotoxicity is associated with the rate of intracellular Ag release, a 'Trojan horse' effect [1, 9].

5. Inhibition of innate immunity with the nanosilver

Cytotoxicity of the nanosilver evidently determines host defence such as apoptosis, necrosis [2] and NETs formation [16]. Moreover, AgNPs, but not Ag^+ ions, decrease the viability and the cytotoxic potential of natural killer (NK) cells secreting cytokines and killing damaged cells [12].

Nanosilver possess anti-inflammatory properties in both animal models and in clinic. Thus, AgNPs inhibit the expression of proinflammatory cytokines transforming growth factor (TGF) -β and tumor necrosis factor(TNF) -α. Nanosilver administration attenuates nasal symptoms of allergic rhinitis in mice and inhibites immunoglobulin IgE, IL-4, IL-10. In clinical study, wound dressing containing AgNPs promoted the healing of chronic leg ulcers due to antibacterial effect in the wound and by decreasing inflammatory response. Ability of nanosilver to reduce cytokine release and production of matrix metalloproteinases, decrease lymphocyte and mast cell infiltration, and induce apoptosis in inflammatory cells may explain their anti-inflammatory mechanisms [17].

Innate immunity induction with $AgNO_3$ and AgNPs conjugated with the major blood proteins – albumin, fibrinogen, immunoglobulines was assayed by xMAP with fluorescent magnetic microspheres.

Inflammation biomarkers	Th1							Th2					Th17		Others		
	IFNγ	TNFα	IL-1β	IL-12(p70)	IL-2	IL-4	IL-5	IL-6	IL-7	IL-8	IL-10	IL-13	IL-17A	G-CSF	GM-CSF	MCP-1 (MCAP)	MIP-1β
$AgNO_3$	1.03	1.15 ↑	1.38 ↑	0.70	0.74	0.84	0.88	0.85	0.70	3.71 ↑	0.78	0.81	0.97	1.12 ↑	0.92	1.19 ↑	1.17 ↑
AgNP-BSA	0.76	0.72	0.70	0.70	0.82	0.74	0.70	0.75	0.50	0.64	0.78	0.69	0.76	0.78	0.79	0.85	0.76
AgNP-Fb	0.72	0.65	1.00	0.72	0.78	0.75	0.82	0.75	0.55	1.00	0.77	0.83	0.84	1.00	0.83	1.15 ↑	0.97
AgNP-hIgG	0.68	0.51	2.11 ↑	0.54	0.48	0.43	0.60	0.41	0.44	0.34	0.54	0.50	0.97	1.20 ↑	0.83	0.76	0.66

Table 1.
Multiplex immunofluorescent xMAP data for 17 inflammation biomarkers in HEp-2 cells in 2 days posttreatment with 5 μg/ml $AgNO_3$ and 0.02 o.u. of nanoconjugates of AgNP-with BSA, fibrinogen (Fb) or hIgG. Normalization was carried out as a ratio of mean fluorescence intensity (MFI) of fluorescent magnetic beads after treatment with Ag ions or nanoconjugates to MFI of control intact HEp-2 cells.

The maximal production of all 17 studied biomarkers including T-helper (Th)1 cytokines: interferon (IFN) γ, tumour necrosis factor (TNF) α, interleukin (IL-1β), IL-12 (p70); Th2 cytokines: IL-2, 4, 5, 6, 7, 8, 10, 13; Th17 – IL-17A and other inflammation biomarkers: granulocyte colony-stimulating factor (G-CSF), granulocyte macrophage colony stimulating factor (GM-CSF), Monocyte Chemotactic Protein 1 (Monocyte Chemotactic And Activating Factor) (MCP-1 (MCAF)), Macrophage Inflammatory Protein 1β (MIP-1β) was registered during first two days posttreatment. In HEp-2 human cell line in 2 days posttreatment with 5 μg/ml AgNO3 (for comparison, cytotoxic concentration for 50% cells (CC_{50}) of $AgNO_3$ for HEp-2 cells is 50 μg/ml) the significant up regulation was registrated for IL-1β (slight increase in 1.38 times up to 0.08 pg/ml), IL-8 (significant growth in 3.7-times up to 0.94 pg/ml) whereas steady production of TNF α (growth in 1.15 times), G-CSF (up in 1.12) MCP-1 (MCAP) (increase in 1.17 times up to 0.39 pg/ml) was observed; down regulation – for regulatory IL-10 (below the IL-10 production in control intact cells with less than 0.05 pg/ml and therefore undetectable in ELISA with detection limit 1 pg/ml) (**Table 1**). IL production is known to depend on the origin of human cells. Thus, IL-1β is mainly produced by macrophages and monocytes but HEp-2 cell line is derived from larynx carcinoma cells and HT-29 – from human colorectal adenocarcinoma cells. IL-1β is responsible for initiation and regulation of inflammation, stimulation of acute phase cytokines such as IL-2, −3, −6, TNF-α as well as for temperature growth and fever. Therefore, inflammation is hardly possible with low level of IL-1β and stimulated cytokines. The only exception was IL-1β-induced up-regulation of IL-8 (**Table 1**) which is associated with acute and chronic inflammation. Silver ions at concentration lower CC_{50} were not toxic for the human cell line and could not penetrate through membranes in cells. Therefore, the observed immunomodulation with $AgNO_3$ was so modest (if any).

Immunomodulation with nanoconjugates of AgNPs covered with the major blood proteins: albumins, fibrinogen, or immunoglobulins differed (**Table 1**). Significant increase in 2.11-times of IL-1β production (up to 0.12 pg/ml) and G-CSF (in 1.20-times) was detected in 2 days after treatment of HEp-2 cells with AgNPs-hIgG simultaneously and similar to its growth after $AgNO_3$ addition. However, the growth did not stimulate production of other cytokines (**Table 1**). Growth of MCP-1 (MCAP) secretion (up to 0.37 pg/ml) was caused by AgNPs-fibrinogen. Inhibition of IL-8 and regulatory IL-10 production in the presence of all nanoconjugates with Ag core and protein shells resulted in slight changes (if any) of 17 biomarkers. Noteworthy that AgNPs-BSA caused decreased production of all inflammation biomarkers studied (**Table 1**). However, the lowest concentrations below 50% of the corresponding values in control cells were found after treatment with AgNPs-hIgG perhaps due to reverse regulation of innate immine response with antibodies.

6. Conclusion

Limited stability of nanosilver without stabilizing envelopes in biologically relevant media, cytotoxicity for both eukaryotic and bacterial cells, negligible cellular uptake restrict their further implementation for combined antiviral and antibacterial therapy. However, spontaneous binding with the major blood proteins and anti-inflammatory properties with inhibition of cytokine production at the early stages after treatment may be helpful in prevention of cytokine storm caused by RNA-containing viruses.

Conflict of interest

The authors declare no conflict of interest.

Author details

Olga V. Morozova[1,2*] and Dmitry V. Klinov[1,3]

1 Federal Research and Clinical Center of Physical-Chemical Medicine of Federal Medical Biological Agency, Russia

2 Ivanovsky Institute of Virology of the National Research Center of Epidemiology and Microbiology of N.F. Gamaleya of the Russian Ministry of Health, Russia

3 Moscow Institute of Physics and Technology, Russia

*Address all correspondence to: omorozova2010@gmail.com

References

[1] Gliga AR, Skoglund S, Wallinder IO, Fadeel B and Karlsson HL. Size-dependent cytotoxicity of silver nanoparticles in human lung cells: the role of cellular uptake, agglomeration and Ag release. Particle and Fibre Toxicology. 2014;11:11 http://www.particleandfibretoxicology.com/content/11/1/11.

[2] Farouk MM, El-Molla A, Salib FA, Soliman YA, Shaalan M. The Role of Silver Nanoparticles in a Treatment Approach for Multidrug-Resistant Salmonella Species Isolates International Journal of Nanomedicine 2020:15 6993-7011.

[3] Munir MU, Ahmed A, Usman M, Salman S. Recent Advances in Nanotechnology-Aided Materials in Combating Microbial Resistance and Functioning as Antibiotics Substitutes International Journal of Nanomedicine 2020:15 7329-7358.

[4] Akbarzadeh A, Kafshdooz L, Razban Z, Dastranj Tbrizi A, Rasoulpour S, Khalilov R, Kavetskyy T, Saghfi S, Nasibova AN, Kaamyabi S, Kafshdooz T. An overview application of silver nanoparticles in inhibition of herpes simplex virus. Artif Cells Nanomed Biotechnol. 2018;46(2): 263-267. DOI: 10.1080/21691401.2017.1307208.

[5] Brown AN, Smith K, Samuels TA, Lu J, Obare SO, Scott ME. Nanoparticles functionalized with ampicillin destroy multiple- antibiotic-resistant isolates of *Pseudomonas aeruginosa* and *Enterobacter aerogenes* and methicillin-resistant *Staphylococcus aureus*. Appl Environ Microbiol. 2012;78(8):2768-2774. DOI:10.1128/AEM.06513-11.

[6] Wang L, Hu C, Shao L. The antimicrobial activity of nanoparticles: present situation and prospects for the future. International Journal of Nanomedicine 2017:12 1227-1249. https://www.dovepress.com/ by 195.178.203.153

[7] Li X, Lenhart JJ, Walker HW Aggregation kinetics and dissolution of coated silver nanoparticles. Langmuir. 2012;28(2):1095-1104. DOI: 10.1021/la202328n.

[8] Morozova OV, Volosneva ON, Levchenko OA, Barinov NA and Klinov DV. Protein corona on gold and silver nanoparticles. Materials Science Forum. 2018; 936: 42-46. DOI: https://doi.org/10.4028/www.scientific.net/MSF.936.42.

[9] Morozova O.V., Pavlova E.R., Bagrov D. V., Barinov N.A., Prusakov K.A., Isaeva E.I., Podgorsky V.V., Basmanov D. V., Klinov D.V. (2018) Protein nanoparticles with ligand-binding and enzymatic activities. International Journal of Nanomedicine 2018:13; 6637-6646. ISSN: 1178-2013, DOI:10.2147/IJN.S177627.

[10] Hansen SF, Baun A.When enough is enough. Nat. Nanotechnol. 2012;7 :409.

[11] Liao C, Li Y, Tjong SC. Bactericidal and Cytotoxic Properties of Silver Nanoparticles. International Journal o f Molecular Sciences Int. J. Mol. Sci. 2019;20:449; DOI:10.3390/ijms20020449.

[12] Rezvani E, Rafferty A, McGuinness C, Kennedy J. Adverse effects of nanosilver on human health and the environment. Acta Biomaterialia. 2019;94:145-159. DOI:10.1016/j.actbio.2019.05.042.

[13] Wu T, Tang M. The inflammatory response to silver and titanium dioxide nanoparticles in the central nervous system. Nanomedicine. 2017;13(2) DOI:10.2217/nnm-2017-0270.

[14] Sebak AA, Gomaa IEO, ElMeshad AN, Farag MH, Breitinger U, Breitinger H, AbdelKader MH. Distinct Proteins in Protein Corona of Nanoparticles Represent a Promising Venue for Endogenous Targeting – Part II: In vitro and in vivo Kinetics Study. International Journal of Nanomedicine. 2020;15: 9539—9556 DOI:10.2147/IJN. S273721.

[15] Brinkmann V, Reichard U, Goosmann C, Fauler B, Uhlemann Y, Weiss DS,Weinrauch Y, Zychlinsky A. Neutrophil Extracellular Traps Kill Bacteria. Neutrophil extracellular traps kill bacteria. Science. 2004;303: 1532-1535. DOI: 10.1126/science.1092385.

[16] Wang C, Liu X, Han Z, Zhang X, Wang J, Wang K, Yang Z, Wei Z. Nanosilver induces the formation of neutrophil extracellular traps in mouse neutrophil granulocytes. Ecotoxicology and Environmental Safety. 2019;183:109508. DOI:10.1016/j. ecoenv.2019.109508

[17] Ge L, Li Q, Wang M, Ouyang J, Li X, Xing MMQ. Nanosilver particles in medical applications: synthesis, performance, and toxicity. International Journal of Nanomedicine 2014;9:2399-2407. DOI:10.2147/IJN.S55015.

Chapter 7

Silver Nanoparticles - Preparation Methods and Anti-Bacterial/Viral Remedy Impacts against COVID 19

Lyubomir Lazov, Risham Singh Ghalot and Edmunds Teirumnieks

Abstract

Silver has been an influential segment of pharmaceutical utilization for remedies & hygiene in the latest era. The first topic reviews the study on air sanitization ventilation & air sanitizer systems using laser ablated silver nanoparticles (inspired by 2020 Pandemic) directing to contamination of deadly biological particles. Intention of this investigation is to validate possible antiviral silver nanoparticles construction to be distributed by retention, to abate the aggravation of breathing organs flu. The underlying description of investigation consists of bibliometric reasoning of the review of the outcome of silver nanoparticles on the sterilization of viral ailments. The investigation will deliberate the approach of use of laser ablated silver nanoparticles for anti-actions. The chapter outcomes in the fascinating utilization of silver nanoparticles for pharmaceutical purposes for contagious diseases, viruses or bacteria and devotes to the upgradation of therapeutic education to safeguard health care workers from threatening viruses at therapeutic organizations. Morally, the investigation will obtain a hygienic scheme, which might be installed at every communal or individual places cost-effectively including silver nanoparticles (because of their therapeutic properties). The second section of investigation considers distinct techniques for manufacturing silver nanoparticles. The various schemes have been compared based on their pros & cons. The method of laser ablation for generating nanoparticles underwater is briefed. The intention of this part is to disclose the current & anticipation probabilities of the process - laser ablation, as a profitable and eco-favorable innovation for manufacturing silver nanoparticle in liquid solutions. The chapter is motivated by two of our reviewed papers i.e., "*Antibacterial and anti-viral effects of silver nanoparticles in medicine against covid 19*" and "*Methods for obtaining silver nanoparticles*".

Keywords: Air disinfectant, Air filters, Antibacterial, Antimicrobial, Antiviral, COVID-19, Laser Ablation, Pulsed Laser Ablation in Liquid, Silver nanoparticles

1. Introduction

The 2020 circumstances of COVID -19 pandemic, health & sanitization issues in communal buildings have increased recently. Infectious diseases resulted with the Corona Virus expansion were dominating challenge to humanity. That is why innovation into advance strain of practical sterilization strategies is the priority.

Disinfectants are reagents used to coat surfaces to reach out for microbes that either contaminate or eradicate them. Reagents are mostly used for the sanitization of diverse facades in communal premises and medical equipment. Disinfectants used in daily life can be handy to keep interacting surfaces clean by destroying biological microorganism on them and still pose no threat to human lives. Also, sterilizers must have a plentiful quantity for repeated utilization in short periods [1].

Presently, synthetic segments & oxidants like aldehydes, alcohols & sodium hypochlorite, hydrogen peroxides, iodine, and others, respectively, are favorably used for sanitization processes. In some events, these elements present numerous cons like metal corrosion, resistance to various bacteria, as well as other unfavorable environmental effects.

With the invention & promising features of nanomaterials, a wide range of options have been disclosed to overwhelming flaws. As interpreted by the International Organization for Standardization (ISO), "nanomaterials" are somewhat material particles of diameter varying between 1 nm & 100 nm. In present-day, distinct material nanoparticles are favorably utilized as efficient sanitizers. The preeminent attempts of scientists from nanotechnology, at the present day, are concentrated on enhancing their physiochemical features & increasing the results of their sanitizing ramification on emerging or modified viruses. Distinct nanoparticles of metals like copper (Cu), silver (Ag), and gold (Au), are now strongly preferred as antiseptics.

Products based on Ag ions and its other antimicrobial works are known since 1000 BC. They were commonly adopted by the archaic Indians in food & medicinal products as a remedy [2].

Antimicrobial effects of silver nitrate ($AgNO_3$) are generally well-known, but with Ag NPs onset, its beneficial properties have raised undoubtedly. This is briefed by the advanced physicochemical features that Ag NPs consists of. With a higher

Component	Physicochemical features of Ag NPs	Disinfectant results
Ag/TiO_2	Contraction in bandgap points to the noticeable active material	Noticeable active antibacterial results [3, 4]
Ag ZnO	O_2-, OH	Photo-oxidative elimination of bacteria [5, 6]
Ag NPs	Surface functionalization Anti-viral	Photocatalytic, self-sanitization bacterial deactivation [7–10] Eradication of aerosolized bacteriophage MS2 virus particles [11]
Ag polyamide	40–60 nm of size	Viable discharge of Ag^+ ion for antibacterial ramification [12].
LA/ZnO:Cu/Ag Bio-nanocomposites	Mechanical/structural, antibacterial & railing characteristics to UV light	Strengthening the lifespan of food [11]
Ag Co-NPs	Magnetic & antibacterial	Water Cleaning [13]
AgNP-SiO2	Spherical morphology	Propelling & synergic antibacterial action of air sanitization [14]
Ag NPs/Chitosan	Porosity, moisture retention efficiency, blood-clotting proficiency	Injury healing strength [15]
Ag/BC	Porosity	Injury healing strength [8]

Table 1.
Prominent physicochemical features of silver nanoparticles & products as a preservative.

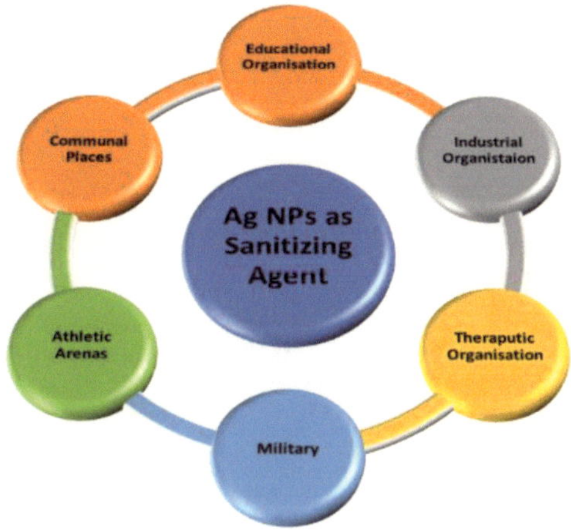

Figure 1.
Utilization of silver nanoparticles as a sanitizing agent at different communal places.

surface/volume, ratio they have a great synergy with microbe surfaces and leads to great antimicrobial safety. The size, shape of Ag NPs portrays a vital aspect in this mechanism. These physiochemical features of Ag NPs are necessary and compatible to the seek for new technics to tackle any future pandemics like COVID-19.

These substantial physiochemical features of Ag NPs are also universally accepted in fields like ecology, industry, and many more. The table beneath briefs the investigation of distinct research teams, demonstrating their fields of utilization.

According to **Table 1**, Ag NPs play a crucial function in air and water sanitization, as well as a medical assistant in a few therapeutic actions. They radiate highly significant behavior in interaction with various bacteria & viruses like *Staphylococcus aureus*, *Pseudomonas aeruginosa* coli and many others [16]. The extraordinary activities of Ag NPs are due to binding capability to sulfur (S) and phosphorus (P), as well as functionalized biomolecules, directing to a rise in their physicochemical features and the competence to destroy various viruses [17].

Figure 1 represents simplified promising operations of Ag NPs. Because of the multi-sphere utilization of Ag NPs, the vast sphere of the biomedical line, as an ingenious organization connected to the 21st century COVID-19 pandemic, that occurred globally and bringing the world to a halt. This review chapter is notably directed to the performance of Ag NPs as a purifying agent for isolating the escalation of Coronavirus, to quarantine Internal airborne ailments in which humans are likely to be contaminated with the viral atoms at communal places.

2. Silver nanoparticles

Silver is the exceptionally integral anti-microbial/bacterial (**Figure 2**) agent accessible way before the initiation of antibiotics [19]. Ag NPs are nanoparticles (NPs) of Ag constructed via nano-techniques of distinct size & shape in any proportion [20]. Approximately, three hundred and twenty tons of tiny silver is produced in a year and used worldwide at present [21]. For the past few decades, Ag NPs is the prime investigation theme because of their size, shape, antimicrobial effects, chemical stability, catalytic activity high conductivity, and other extraordinary features. Processes & methods like laser ablation, microwave processing, gamma irradiation, electron irradiation, chemical reduction, photochemical

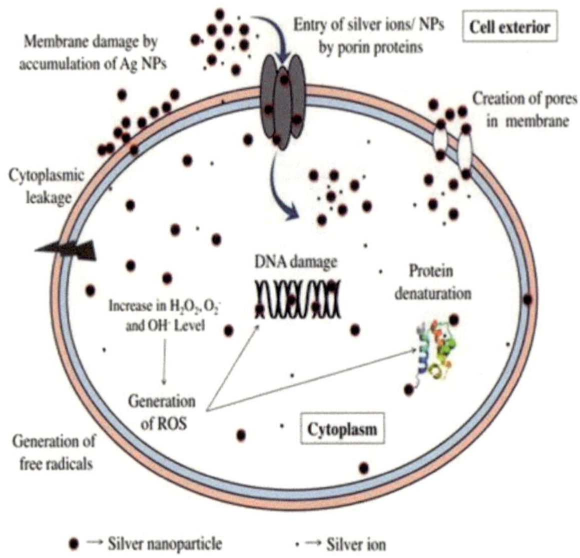

Figure 2.
Biological process of antibacterial impacts of silver nanoparticles & silver ions [18].

methods, and synthetic biological procedures are some prior researched techniques for constructed Ag NPs [22].

Nanotechnology is a fundamental research attention of present-day, managing design, synthesis, and molecule structures varying not more than 100 nm. NPs have several functions in medication, cosmetics, Eatables, optics, biomedical sciences, space industries, electronics, and in many other broad areas. Synthesized Ag NPs are valued for numerous medical functions such as [22, 23].

- Remedy to ulcerative colitis, acne & dermatitis
- Inhibition of HIV-1 replication
- Molecular imaging of cancer units
- Exposure of viral arrangements (SERS & Ag nanorods)
- Antimicrobial impacts for infectious organisms
- Lacquering of surgical mesh for pelvic repair & breathing mask patent, used in COVID-19.

Relying on the reducing agents and procedure there are 3 heyday mechanisms to develop Ag NPs.

- Physical techniques generally strive with laser impulse energy to dwindle Ag from volume to atoms and ions.
- Synthetic techniques utilize reducing chemical agents to reduce Ag ions from the bulk of metallic Ag.
- Biological techniques utilize reducing organic agents to reduce Ag ions from the bulk of metallic Ag [24].

3. Air decontamination

Currently, several human acts while the discrete modification practices direct a negative impact on air quality. Nowadays, the preeminent threat is to obtain a healthy air quality exempted from harmful particles like airborne bacteria/viruses, fine dust fragments, capricious biological elements & noxious gases. The eviction of threatening airborne microorganisms fascinate scientists because they are liable for chronic contagious disease and are genuine threats of airborne toxicity. Such microbes generate a need for air cleaning systems, which will lower the stability of several microorganisms in the air and will provide healthy air for inhalation. Mentioned microbes are the reason for several diseases like severe acute respiratory syndrome (SARS), anthrax, asthma etc.

Studies confirm that bioaerosols gather on air conditioner filters in massive amount & breed with high moisture [25]. Presently, several engineering solutions are available for bioaerosol removals like photocatalytic oxidation, UV irradiation with germicides, and aerial ozonolytic methods.

It is believed that a fruitful remedy for pandemics like COVID-19 also associates with the utilization of Ag NPs & Cu NPs (Copper nanoparticles) as a sanitizer agent. Enhanced air quality is obtained with Ag NPs sprayed on air filters [26]. Therefore, a belief is born, that nanomaterials & air filters coated with Ag NPs can be an engineering solution to pandemics.

Ag NPs coated with silica, in symmetric distribution, perform symbiotic antibacterial actions with gram-positive/negative, microbe in the air refining arrangements. Moreover, Ag NPs & SiO_2 infused together certifies a six-month steady solution & performs 99.99% antibacterial actions. Generating Ag aerosol NPs in a puffer as an antiviral assistant to Hay Bacillus & COVID-19 aerosols under expert medical guidance are favored applicable. Overall, we can claim that combined elements are beneficial for installing with air purifiers (**Figure 3**) in air refining devices [11, 14].

Figure 3 represents a schematic of the anti-viral impacts of Ag NPs coated on an air refiner. In the research paper, Herzong et al. manifested the probability of manufacturing an airway barrier with aerosolized Ag NPs featuring basic cytotoxicity [27]. Our research in this line confronts the Ag NPs originated products have a greater impact in safety and avoidance of viral ailments for upper respiratory section. It happens because, transfer of viral flu is observed primarily through inhalation, ingestion, and dermal routes.

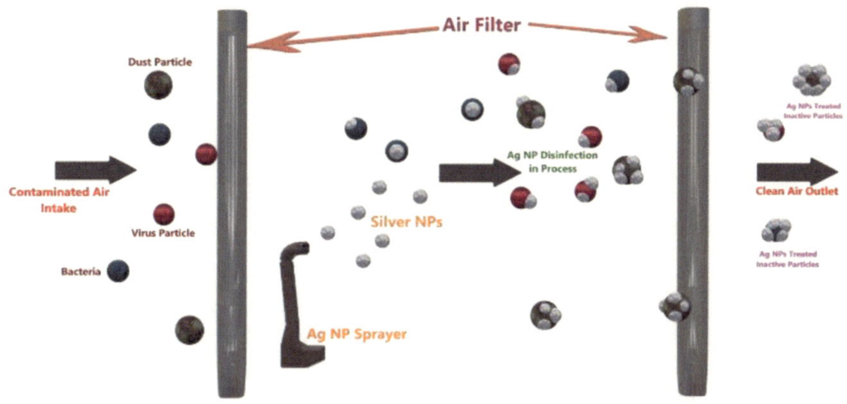

Figure 3.
Anti-viral impact of silver nanoparticles on bioaerosols impregnated in the air filter. (Regenerated from reference [27]).

So, sprayers originated from nanotechnology discharging airborne particles & Ag compounds have proven to boost good human health. Such ways substantiate a better air quality in communal places & provides enhanced protection against COVID-19 at its primary stage. Thus, mobile oxygen generating system & air coolers can easily be equipped with Ag NPs coated filters with improved efficiency & without being overpriced [28, 29].

4. Approaches for obtaining NPs

Many convincing procedures have been awarded for processing Ag NPs in the past years [30]. NP production is briefed under two elemental categories (**Figure 4**):

Top-down Approach: - Solid bulk material is applied with external force, forcing its break down into nanostructures.

Bottom-up Approach: - Its initiates from atomic scale & generates NPs, until appropriate size and shape is gathered [32].

Physical, Chemical, & Biological Methods to process the Ag NPs are explained below:

4.1 Physical methods

These could be defined as the green mechanism because of non-synthetic liquids in the tidy, thin films. Moreover, high consistency of NP distribution, is one of the pros of physically processed NPs as identify with any other methods [22]. This method consists of mechanisms like [30]:

- *Arc Discharge* - Condensation, nucleation, & construction of NPs pursues when marked component is volatilized via an arc discharge between electrodes [33].

- *High-Pressure Magneton Sputtering* - It falls under bottom-up process & inhabits in abridgment of super-saturated vapor via cold inert gas flow. This vapor atmosphere is acquired by discrete processes like thermal evaporation, laser ablation, magnetron plasma sputtering, and many other [34].

- *Laser Ablation* – A laser beam radiates to separate top layers of the bulk material with an absolute accuracy [35].

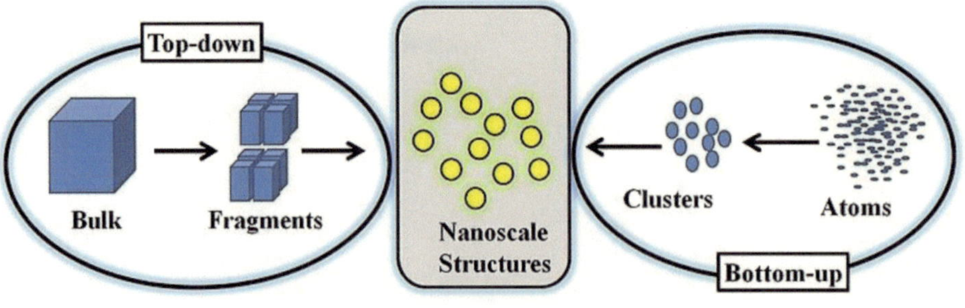

Figure 4.
Representation of two elemental categories, 'top-down' & 'bottom-up' approaches for the nanoscale material processing [31].

4.2 Chemical methods

A chemistry professor from Princeton - John Turkevich's work in the 1950s, gave a highlight to chemical methods. The chemical processing by natural and artificial reducing surfactants is the preferred method for constructing Ag NPs. Reducing agents like sodium citrate, ascorbate, elemental hydrogen, are generally employed for Ag^+ ions reduction in aqueous or non-aqueous solution. These agents shrink Ag^+ ions forming Ag^0 [36]. Protective agents are obligatory to stabilize breaking-up of NPs during their preparation and to restrict them to stay onto material surfaces [37]. To stabilize particle hike and to sustain them from sedimentation, accumulation, or losing surface properties, comprising functionalities to unite with molecule skin, the presence of surface-active agents like acids, amines, or alcohol are must [22]. This method comprises of [30, 38]:

- *Photochemical* – In an electron-contributing reagent, it requires UV irradiation of the solid forerunner solution. This process produces NPs subjected solid-state status & at low temperatures [39].

- *Microemulsion* – It is thermodynamically balanced colloidal diffusion of oil in water or vice-versa with a stabilizer. Ultralow interfacial tension, large interfacial area & monodispersed NPs, are pros of this technique [40].

- *Hydrothermal* - It is a solution reaction-originated approach which produce NPs at room-temperature to extremely high-temperature [41].

- *Chemical Reduction* - A usual process utilizing natural & artificial reducing agents like sodium citrate, sodium borohydride (NaBH4), elemental hydrogen, N-dimethylformamide (DMF), & poly (ethylene glycol)-block copolymers for Ag^+ ion break-down [22].

- *Electrochemical Mechanism* - A common process, containing a two-electrode mechanism for the electrochemical fusion. It has utmost preference over other bottom-up approaches because of pure quality of NPs received [42].

4.3 Biological methods

These are categorized under green synthesis and comprises polysaccharides, organic & irradiation techniques, preferred over other traditional practices requiring synthetic agents with hefty harms [43]. Because of benefits like organic surfactants (bacteria, fungi, yeast, and plant), cheap and eco-friendly techniques, a vast interest is gained by these methods. Even with natural surfactants, it has a usual reduction process [30]. Green synthesis pursues with:

- *Bacteria* – It is capable to weaken heavy metal ions & has its preferences for generating NPs. E.g.: *Pseudomonas stutzeri* and *Pseudomonas aeruginosa* [44].

- *Plants* - Metabolites in plant crude extract are capable of metal ions reduction into metallic NPs. E.g.: *A. indica*, *A. sessilis* [45].

- *Fungi* - They offer high tolerance NPs and are easy to handle. Its extracellular protein extract can stabilize NPs. E.g.: Fusarium oxysporum [46].

- *Algae* - These are single/multi-cellular organisms endured in surroundings. Processing at low temperature with high energy efficiency, less noxious to the environment are its pros. E.g.: *Chlorella Vulgaris*, Spirulina platensis [47].

5. Topnotch favored approaches

Pulsed Laser Ablation in Liquid (PLAL) subordinated to physical methods is the top-notch method for producing Ag NPs since it is a green method which utilize the least time & finances [48, 49]. Because of nucleation, advancement & cluster formation, it is a subordinate to the bottom-up process. Interacting with ultrashort laser beam pulses i.e., $10^{-13}\ to\ 10^{-8}\ \text{sec}$, generates nanostructures from the metal bulk substrate in any medium, is called a laser ablation process [32, 50, 51]. It depends on aspects like laser wavelength (λ), repetition rate, frequency, pulse duration, the light absorption efficiency of material, transmission, and chemical composition of the liquid. The monochromaticity and directionality of the laser beam are vital characteristics for this process.

Moreover, produced Ag NPs are pure and are free of metal colloids contamination and have great quality because of non-utilization of any artificial or natural surfactant. Efficiency and features of produced Ag NPs are based on specifications, consisting of "λ", pulse duration, ablation time & the laser beam interaction with each NP inside the liquid [52, 53]. Spatial profiling of laser beam intensity or fine balancing of "λ" controls size distribution [54]. Essential specifications are briefed be/neath:

- *Pulse Duration*: In PLAL it can vary from *Femto-Nano* seconds. NPs originated via sputtering of the molten surface layer by the liquid's recoiling pressure around the processing metal. The temperature "*T*" of the laser beam spot is denoted with simple heat balancing equation:

$$T \approx \frac{Aj}{cph} \tag{1}$$

Here, A - absorptivity of marked material (A = 1- R) [R - reflectivity coefficient].

c - heat capacity of marked material,

ρ - density of marked material,

h - heat diffusion length inside marked material,

The heat diffusion length "*h*" is dependent on the heat diffusivity of the processed sample, expressed as:

$$h \propto \sqrt{at_p} \tag{2}$$

wherein, $a = \frac{k}{c}$, [*k* - heat conduction coefficient of marked material & "t_p" - laser pulse duration].

Larger the t_p, bulky is the layer of material which is heated by laser energy absorption. Moreover, for heating & evaporation of liquid adjacent to the laser beam spot, an unwanted bulk of energy is drained approximate to the one absorbed due to reduced thermal conductivity of liquids [54].

- *Laser Wavelength*: As most of constructed NPs can consume laser radiation (which can melt material), so, UV ranging "λ" are least preferred in PLAL [54].

- *Repetition Rate*: With every laser pulse, NPs are discarded from the processing sample surface. Therefore, higher the repetition rate of laser pulses, higher is the NP production [54].

In PLAL, laser beam is directed to a processing sample top-layer plunged into liquid. Supposedly, because of transparency and non-absorption of liquid, the laser beam radiation can focus directly on the target. The ideal facade of the liquid avoids surplus reflection at the interface "covering glass/air". However, the use of unstable liquids like acetone, ethanol, etc., demands covering the vessel with a transparent lid [54].

The laser ablation is successful only when tiny structures of processed metal are constructed by PLAL, see **Figure 5** [55]. In **Figure 5**, an *Ag* sample is immersed in distilled water and is synthesized under pulsed laser. Different types of lasers like Solid, Gas or chemical laser, can be used with distinct "λ" ranging from pico-femtosecond. Generated NPs have extraordinary features and are irreproducible by any other traditional methods [56].

Figure 5 demonstrates the Ag NPs generation out of a bulk silver plate via high-power pulsed lasers in PLAL. Material properties affect the ablation at its best. Various laser parameters are used to composite NP characters like size, shape, surface properties, aggregation state, solubility, structure, and chemical composition, see **Table 2** [30].

When compared,

- The generation efficiency & the size of colloidal components of "*ns*" laser was larger than "*fs*" laser.
- The manufactured colloids by "*ns*" pulses were extra scattered than "*fs*" pulses.

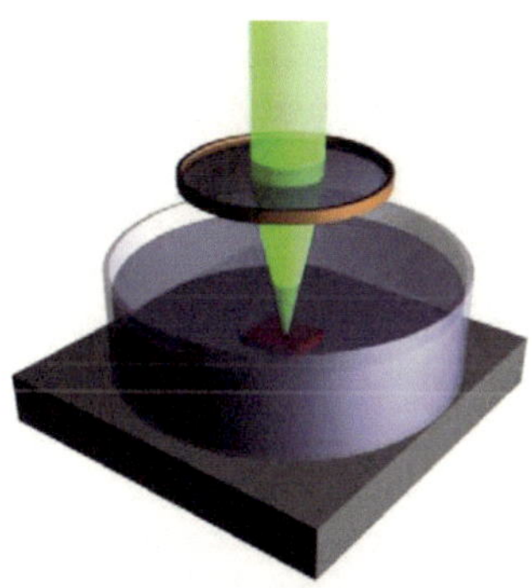

Figure 5.
Diagrammatic representation of PLAL process with metal immersed in distilled water [54].

Wavelength	532 nm	800 nm	1064 nm
Laser Type	Nd: YAG (ns*)	Ti: Sapphire (fs**)	Nd: YAG (ns*)
Mean Diameter of NPs	2–3 nm***	400 nm***	2–5 nm***
Absorbance	0.28	> 0.05	< 0.1
Quantity & Type of Liquid	170 ml**** (Pure water)	5 ml**** (Deionized water)	25 ml**** (Pure water)

**Nanosecond*
***Femtosecond*
****Nanometer*
*****Milliliter.*

Table 2.
Different laser parameters for obtaining Ag NPs via PLAL, based on the theoretical research [57–59].

- The ablation demonstration for *"fs"* ablation in air was lessened than distilled water but *"ns"* pulses did not have any change of ablation energy in either of the mediums.

- The self-absorption capability of Ag is higher at lower wavelengths [57, 60].

6. Nosocomial epidemic (hospital-acquired infections)

The predominating threats in the biomedical line are the penicillin-immunity of the living beings, product advancement behavior and its services in terms of hazardous, healing timespan & aftereffects on the mortal cells. Also, the revelation of infection-inducing non-bacterial microorganisms, supervising infection administration & avoiding nosocomial infections (**Figure 6**) are main confronting puzzles for the science [61].

A survey by WHO concluded that moderately 8.7% of world population has suffered from nosocomial disease, and approximately 1.4 million people suffered Hospital-Acquired Infections (HAI) [62]. Eastern Mediterranean & south-east Asian regions are HAI sensitive rather than the other regions. Nosocomial Infections covers categories likes surgical wounds, urinary & lower respiratory tract infections. The latter was particularly pronounced for the 2020 COVID-19 infections.

Avoidance of HAI demands a coordinated & overseen plan in which peculiar features must be calculated. The transmission of bacterial/viral microorganism from person to person is diminished by individual cleaning like handwashing, hand gloves, masks, working clothes, shoes, & sterilization of medical apparatus [63].

Also, higher the bacterial/viral protection, higher is the impact on wellbeing. Many theories have been executed worldwide for restricting HAI's. It is noted that Ag NPs utilization is highly emerging solution to quarantine the Nosocomial Infections, also, an efficient nano-weapon to multidrug resistance microorganisms [64].

The unmanaged & enormous use of penicillin and their bacterial hostility is a present-day menace for medicinal industry, which also consists of Ag NPs induced antibiotics [65]. Ag NPs obtained from bio-methods using *B. marisflavi* resulted high antibacterial actions against bacteria leading to nosocomial infections [66].

In therapeutic institutions, medical apparatus is the usual source of spreading infection. The quick fix to this problem could be the coating apparatus with Ag NPs to bypass bacterial contagion. Ag NPs are productively utilized in catheters

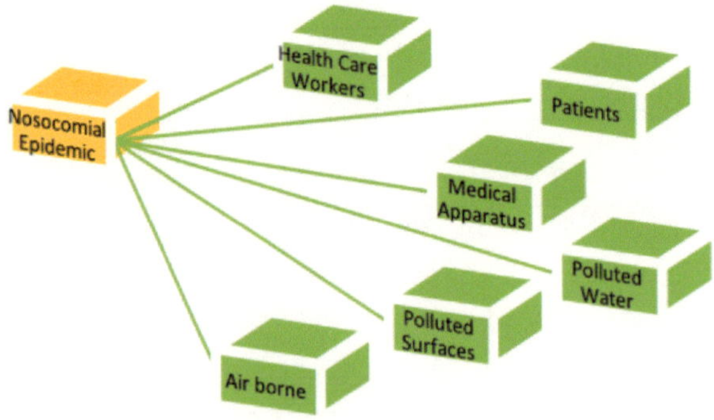

Figure 6.
Elements subjected to nosocomial infections.

for enhanced antimicrobial actions & zero thrombogenicity [67]. The writers also questioned the impact of Ag NPs on blood clotting, and the outcomes are fascinating. It should also be understood that the antimicrobial impact of Ag NPs coated titanium instruments is modified by UV light exposure [68].

Exploration in this branch had concluded an overall bacterial termination of Methicillin-resistant *staphylococcus aureus* (MRSA) as a result of Ag NPs synergic effect as a photocatalyst & visible light was accomplished [5, 69]. An identical energetic bactericidal effect was detected with stents & catheters used in cardiovascular operations [70]. So, it can be believed that the Ag NPs are advantageous ingredient to isolate differential ailments occurring in therapeutic organizations. Also, precautions should be taken to avoid their noxious impact on humans.

7. Anti-action operations of AgNPs

Many researchers have examined & published the antibacterial performances of Ag NPs. The bacterial cell membranes consist of sulfur (S) comprising proteins & involves amino acids. On inner & outer membrane Ag can blend in with them preventing bacterial expansion. Ag NPs also produce Ag ions which blends in with Deoxyribonucleic acid's (DNA's) phosphorus (P) and with "S" consisting of proteins, inhibiting the enzymes movement. Characteristics like "size" & "shape" of NPs guides to evaluate Ag NP's antimicrobial behavior. NPs of size <20 nm will execute a better attachment of "S" comprehending protein of membrane ending in cell death of bacteria as a result of maximum permeability via membrane [71].

Figure 7 represents the interaction mechanism of Ag NPs surface-to-volume ratio on bacteria/virus. Nanomaterials has a nature, that smaller the NPs, greater surface interaction is exposed and contributes to better microbicidal effect [72, 73].

Shape of NPs also plays avital role to interact with the membrane walls. Analysis in [74] directs to large anti-activities towards *E. coli* bacteria with abbreviated triangles of Ag NPs better than those of spherical & rod-shaped NPs. It is also mentioned that Ag NPs <10 nm creates holes in the cell wall as a result of these holes the cytoplasmic is dispensed into the medium, responsible for controlling cell death without interaction between intracellular and extracellular proteins and bacterial nucleic acids. The writers of [75] declares that the interaction of Ag NPs with virus cells directs to an escalation in actions of their programmed cell death i.e., apoptosis.

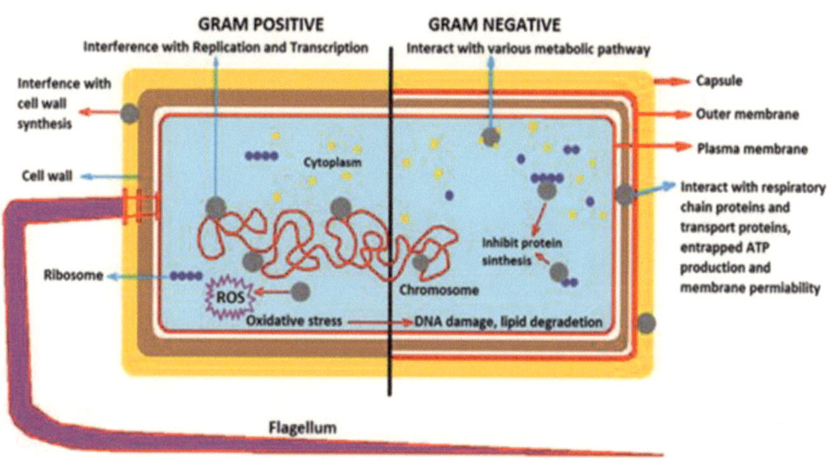

Figure 7.
Demonstrating antibacterial features of silver nanoparticles.

In simple words, Ag NPs are fascinating disinfectants better than merchandise like Acticoat™ & Silverline® for wound dressing & polyurethane ventricular catheter, respectively [70]. Moreover, Ag NPs are utilized in several merchandise-like surgical masks, toothpaste, hand wash, shampoo, detergent as well as humidifiers but their toxicity to human lives is still a top concern.

8. Early-stage treatment via respiration approaches with Ag NPs for COVID-19

Numerous drug manufacturing experiences the well-known antimicrobial features of Ag NPs. The anti-bacterial/viral features of Ag are very well researched in the scientific world [76, 77].

To investigate and validate the anti-action features Ag NPs as a purification segment for medicinal benefits, and their functions for treatment & anticipation of viral disease for inhalers, was the prime intuition of this investigation. Viral & Bacterial ailments threats to respiratory supported patients in intensive care units (ICU) due to ventilator-associated pneumonia (VAP), are basic scenarios that are present day discussion topic. The nasopharynx and/or bronchial tree of the breathing organs (respiratory system) are vulnerable to be infected by any infections even by a regular bacteria/virus (**Figure 8**) [78, 79].

The sinking of the microorganism to the lower vent of the breathing organs aggravates the conditions, intensifying the immune response and consequence huge damage (**Figure 9**) [81, 82]. Therefore, a superior proficient approach to diminish the bacterial/viral growth at the very starting i.e., in the upper breathing organs. For example, patients with slight symptoms or patient's arrival at ICU in the therapeutic institution before any VAP infection.

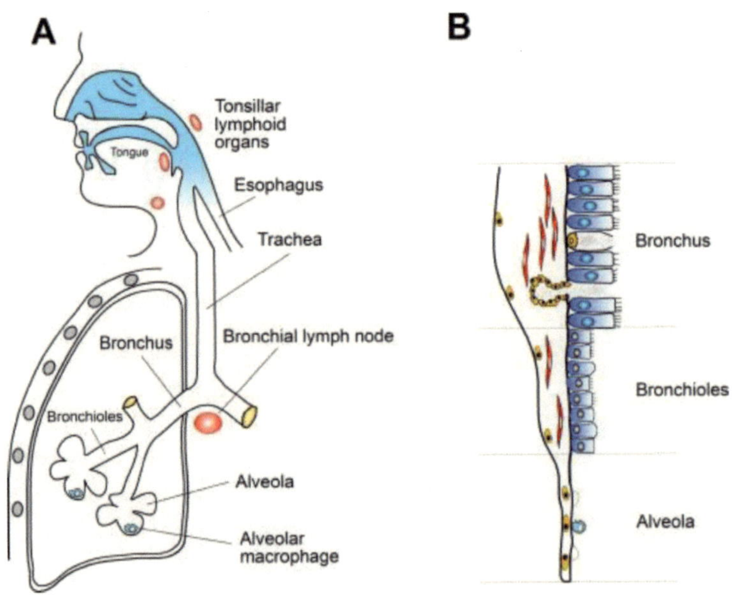

Figure 8.
Entry points of influenza into respiratory system. (A) The description of the human airways. Microbes first infects the upper vents and the rough cells in the bronchus and bronchioles. The adaptive resistance is initiated in lymph nodes along the airways. (B) The respiratory epithelia are especially equipped to defend from incoming pathogens by a layer of mucus (bronchus), ciliated cells (bronchus and bronchioles), and alveolar macrophages (alveoli).

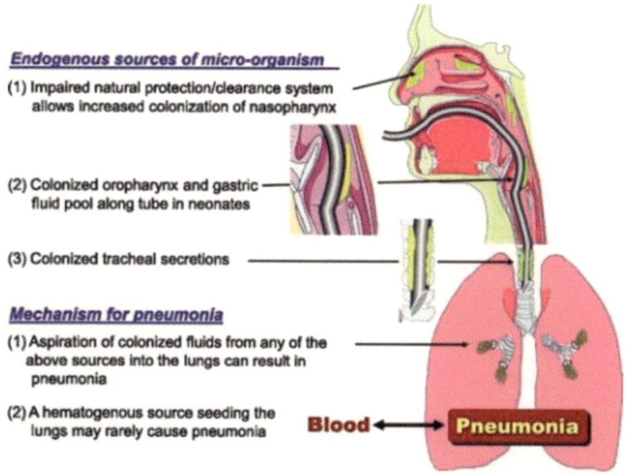

Figure 9.
Microbes of HAI - VAP [80].

An adequate solution for upper & lower respiratory system remedy is briefed in the paper. It is mandatory that for intake, the difference between ionic Ag solution & colloidal Ag and Ag particles suspensions/solution in water has the relevant structure.

- In ionic Ag, the atomic Ag ions are soluble in water.
- Colloidal Ag are nanosized Ag pieces ranging from 1 to 100 nm diameter [80].

There is a debate on colloidal particles being more persuasive than ionic Ag for anti-viral features [83], as described in **Table 3** for Human Immunodeficiency Viruses (HIV). Evaluation the anti-action of colloidal Ag will help understand the quantity of Ag NPs for healthy inhalation.

Minimum Inhibitory Concentration (MIC) should be developed to calculate the authorize dosage of any effective agent. The delicacy of MIC (μg/ml) to NP size, is problematic to settle conclusions from disclosed studies.

NPs >10 nm are ultimately resulting than those >25 nm [83, 84]. Higher density was noticed with the small sizes of NPs which leads extraordinary synergy with microorganisms. That is why, the MIC of NPs > 10 nm must be tiny than that of large sized NPs and is experimentally absolute (see **Table 4**) [80]. With every publication the value to MIC has varied with the NP size. It has been stated in the research of the effect of Ag NPs size on anti-viral efficacy in HIV treatment. These results are concluded by assuming that because the virus size is ≥ 100 nm for e.g., HIV has a size of 120 nm. And to interact and act on the virus the AG NPs must be tiny for the virus particle (approx. of 10 nm). These assumptions were made by direct imaging of NPs [85].

Element	IC50*
Silver Nanoparticles	0.44 mg/mL (± 0.3)
Silver sulfadiazine (ions)	39.33 μg/mL (± 14.60)

**The half maximal inhibitory concentration.*

Table 3.
Anti-action impacts of Ag ion & Ag NPs against HIV-1 [83].

Specimen	MIC (μg/mL)	
	Bacteria	
	E. coli	*S. aureus*
7 nm Ag NPs	6.25	7.5
29 nm Ag NPs	13.02	16.67
89 nm Ag NPs	11.79	33.71

Table 4.
MIC of colloidal Ag for antibacterial effects [80].

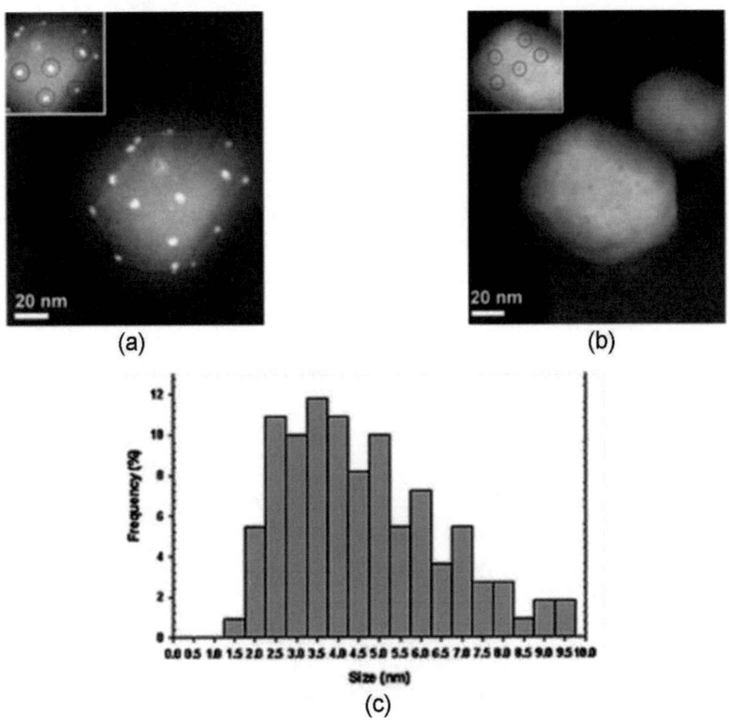

Figure 10.
a) HIV-1 virus commuting silver; b) HIV-1 viruses without commuting silver; c) complex size distribution of Ag NPs commuted with HIV-1 virus, resulted via tested preparations [85].

Fascinating fact is that the culminating virus attachment for NP impact ranged between 3 and 7 nm (see **Figure 10**) [85]. Hence, the size-dependency for NP interaction to virus is stated.

In this chapter we have discussed miniscule measure of Ag NPs utilization in the biomedical work, whereas worldwide testing rooms are actively researching on it. In overall it could be said that, presently, the anti-bacterial/viral features of colloidal Ag are definite but in case of respiratory disease, a lot of must be researched deeply. Also, the MIC size of colloidal Ag ranging between 10 μg / ml and 25 μg/ml (indirect conclusions, requires deep study to specify) & 3–7 nm, respectively, have impressive suppression on viral infections.

Expecting, these formulations can be persuasively precautionary & remedy in beginning of respiratory viral ailments, including COVID-19/SARS-CoV-21. Such remedy could be helpful for bacterial ailments like prevention of clinical VAP & in ICU's and only half dosages compared to other anti-bacterial.

Until now, high assumptions are made from Ag NPs applications, but their over concentration can be threatening to human lives [86]. For example, excessive concentration can cause high blood pressure [87]. However, the hazardous of AG NPs is not that higher because of their nano size and possibly end in the urine [88].

The adequate & conservative utilization of Ag NPs should be concrete, as their catastrophic consequences are known only through animal & In vitro tests.

9. Ending

Because Ag NPs have chemical stability and retains uncommon features, they are used in fields like therapeutic, environment, food industry, biotechnology, microelectronics, and many other. Moreover, there antimicrobial assistance has compelling vigilant ramifications against microbes. Ag NPs are an elemental ingredient to configuration of products with antimicrobial actions due to their non-toxicity to mortal bodies but only in low mixtures. These uncommon features of Ag NPs have interested many researchers for innovating modern techniques to obtain their different size & shapes. Many leading research teams are enhancing the Ag NPs manufacturing methods using chemical reduction, laser ablation, electron beam exposure, concentrated microwave exposure, and other.

The PLAL (branch of Physical methods) tends to occur any exceptional environmental side-effects by producing pure & clean metal NPs with uncommon characteristics, without any organic/inorganic chemical agents. Laser wavelength (λ) & intensity are vital parameters required for NP construction. For example, green lasers can construct NPs with better efficiency than infrared (IR) lasers and "*ns*" pulsed laser generates strong heat.

The hasty incorporation & flexibility of chemical methods into distinctive surroundings is the motivation for scientists to keep up with the mechanistic conditions of the anti-microbial/viral/inflammatory impacts of Ag NPs. This takes place with the alteration in the form, NP size by altering the reaction requirements, reagents and stabilizing surfactants used.

Whereas biological methods for NP synthesis possess vast alternatives for using natural & artificial stabilizing surfactants. Use of bio-organisms for NP extraction might commit to the surrounding purification. Surface chemistry & morphology, size, shape, coating agent, NP agglomeration & dissolution rate, particle reactivity in solution, and ion release efficiency, are the dependencies for biological actions of Ag NP.

The fascinating use of Ag NPs in medicinal purposes especially against contagious ailments as a result of their extraordinary microbicidal spectrum. In recent years, studies have confirmed Ag NPs impact on viruses, which was expected to be opposing. Contact infection between Health Care Workers and patients could be prevented effectively with Ag NPs, must be advised. Optimistic results have been obtained for Ag NPs contribution in enhancing the microbicidal effects of biocompatible medical devices. Expectations have aroused for contribution of effectual utilization of Ag NPs in modification of medical science for treatment of contagious ailments.

With the time the research in this line has reached to a far point but there is a lot more to be researched like to improve & interpret the last-longing benefits of NPs on mortal souls. Also, the optimal accumulation on the distinct points of the ailments in sufferer without noxious impact on living beings. It requires all hands joined from around the globe to put their efforts for research in enhancing technologies for generating NPs with various features, as well as carrying out therapeutic observations & particular objective investigation.

Author details

Lyubomir Lazov*, Risham Singh Ghalot and Edmunds Teirumnieks
Rezekne Academy of Technology, Latvia

*Address all correspondence to: lyubomir.lazov@rta.lv

References

[1] W. D. Rutala W.A., *Guideline for Disinfection and Sterilisation in Healthcare Facilities,* 2008.

[2] S. S. Gupta A., *Molecular genetics: silver as a biocide: will resistance become a problem?,* 16 ed., vol. 888, Nat. Biotechnol, 1998.

[3] S. M. V. K. e. a. S.P. Deshmukh, *Ag Nanoparticles Connected to the Surface of TiO2 Electrostatically for Antibacterial Photoinactivation Studies,* vol. 94, Photochem. Photobiol., 2018, p. 1249-1262.

[4] P. A. C. M. L. N. C. C. L. a. L. B. T. A. Cozmuta Mihaly A., *Active Packaging System Based on Ag/TiO2 Nanocomposite Used for Extending the Shelf Life of Bread. Chemical and Microbiological Investigations,* vol. 28, 2015, p. 271-284.

[5] P. K. A. Z. a. Dunnill C.W., *Nanoparticulate silver coated-titania thin films—photo- oxidative destruction of stearic acid under different light sources and anti- microbial effects under hospital lighting conditions,* vol. 220, J. Photochem. Photobiol., 2011, p. 113-123.

[6] G. S. S. N. S. N. R. R. Manna J., *Biomimetic method to assemble nanostructured Ag ZnO on cotton fabrics: application as self-cleaning flexible materials with visible-light photocatalysis and antibacterial activities,* vol. 7, ACS Appl. Mater. Interfaces, 2015, pp. 8076-8082.

[7] C. M. H. S. D. S. Atiyeh B.S., *Effect of silver on burn wound infection control and healing: review of the literature,* vol. 33, Burns, 2007, pp. 139-148.

[8] N. R. S. M. L. A. Pal S., *Silver-functionalised bacterial cellulose as antibacterial membrane for wound-healing applications,* vol. 2, ACS Omega, 2017, p. 3632-3639.

[9] D. N. J. H. e. a. Wagener S., *Textile functionalisation and its effects on the release of silver nanoparticles into artificial sweat,* vol. 50, Environ. Sci. Technol. , 2016, p. 5927-5934..

[10] F. M. F. A. e. a. Zille A., *Size and aging effects on antimicrobial efficiency of silver nanoparticles coated on polyamide fabrics activated by atmospheric DBD plasma,* vol. 7, ACS Appl. Mater. Interfaces, 2015, p. 13731-13744.

[11] P. D. H. J. Joe Y.H., *Evaluation of Ag nanoparticle coated air filter against aerosolised virus: anti-viral efficiency with dust loading,* vol. 301, J. Hazard. Mater., 2016, p. 547-553.

[12] R. M. Ş. M. e. a. Vasile C., *New PLA/ ZnO: Cu/Ag bio nanocomposites for food packaging,* vol. 11, Express Polym Lett., 2017.

[13] M.-B. X. V. N. e. a. Alonso A, *Superparamagnetic Ag@co nanocomposites on granulated cation exchange polymeric matrices with enhanced antibacterial activity for the environmentally safe purification of water.,* vol. 23, Adv Funct Mater, 2013, p. 2450-2458.

[14] J. Y. S. M. L. K. H. J. W. K. Ko Y.-S., *Prompt and synergistic antibacterial activity of silver nanoparticle-decorated silica hybrid particles on air-filtration,* vol. 2, J. Mater. Chem., 2014, p. 6714-6722.

[15] L. Z. Y. H. G. J. C. R. Liang D., *Novel asymmetric wettable AgNPs/chitosan wound dressing: in vitro and in vivo evaluation,* vol. 8, ACS Appl. Mater. Interfaces, 2016, p. 3958-3968.

[16] P. I, *Application of silver nanoparticles in experimental physiology and clinical medicine: current status and future prospects,* vol. 37, Rev. Adv. Mater. Sci., 2014.

[17] S. R. S. K. e. a. Pandey J.K., *Silver nanoparticles synthesised by pulsed laser*

ablation: as a potent antibacterial agent for human enteropathogenic gram-positive and gram-negative bacterial strains, vol. 174, Appl. Biochem. Biotechnol., 2014, p. 1021-1031.

[18] G. R. P. J. Pareek V., "Do physico-chemical properties of silver nano-particles decide their interaction with biological media and bactericidal action? A review.," *Mater. Sci. Eng. C.,* vol. 90, p. 739-749, 2018.

[19] J. W. Alexander, "History of the Medical Use of Silver," *SURGICAL INFECTIONS,* vol. 10 (3), 2009.

[20] V. D. L. I. A. v. B. A. Graf Christina, "A General Method To Coat Colloidal Particles with Silica," *Langmuir,* vol. 19 (17), p. 6693-6700, 11 July 2003.

[21] S. F. L. f. M. S. a. T. (EMPA), "ScienceDaily," 31 January 2011. [Online]. Available: https://www.sciencedaily.com/releases/2011/01/110131133005.htm. [Accessed March 2021].

[22] K. H. M. S. Z. B. Iravani S, " Synthesis of silver nanoparticles: chemical, physical and biological methods," *Research in Pharmaceutical Sciences,* vol. 9, p. 385-406, 2014.

[23] S. S., *Biosynthesis and immobilization of nanoparticles and their applications,* University of pune, 2005, p. 1-57.

[24] Y. T. &. B. V. Shu Zhang, "A Review on Preparation and Applications of Silver-Containing Nanofibers," *Nanoscale Research Letters,* vol. 80, 09 February 2016.

[25] R. H. Schleibinger H., *Air filters from HVAC systems as possible source of volatile organic compounds (VOC)–laboratory and field assays,* vol. 33, Atmos. Environ., 1999, p. 4571-4577.

[26] B. J. P. J.-H. J. J. B. G. Yoon K.-Y., *Hwang J., Antimicrobial characteristics of silver aerosol nanoparticles against Bacillus subtilis bioaerosols,* vol. 25, Environ. Eng. Sci., 2008, p. 289-294.

[27] C. M. P. F. e. a. Herzog F., *Exposure of silver-nanoparticles and silver-ions to lung cells in vitroat the air-liquid interface,* vol. 10, Part. Fibre Toxicol., 2013, p. 11.

[28] H. T. M. C. e. a. Calderón L., *Release of airborne particles and Ag and Zn compounds from nanotechnology-enabled consumer sprays: implications for inhalation exposure,* vol. 155, Atmos. Environ., 2017, p. 85-96.

[29] D. S. P. D. A. G. M. K. V. D. S. D. Patil S. M, *Different strategies for modification of titanium dioxide as heterogeneous catalyst in chemical transformations,* vol. 21, Curr. Org. Chem, 2017, p. 821-833.

[30] M. I. A. V. M. C. R. A. P. A. A. P. M. L. G. P. a. N. C. Maria Chiara Sportelli, "The Pros and Cons of the Use of Laser Ablation Synthesis for the Production of Silver Nano-Antimicrobials," *Antibiotics (Basel),* vol. 7, 28 July 2018.

[31] N. Instruments, *Nanoparticle Synthesis.*

[32] B. B. B. C. R. e. Manish Kothakonda, "Core-Shell Nanoparticles for Energy Storage Applications," in *Pulsed Laser Ablation: Advances and Applications in Nanoparticles and Nanostructuring Thin Films*, Pan Stanford Publishing Pte. Ltd., 2018, pp. 277-316.

[33] A. M.A.Virji, "8.06 - A Review of Engineered Nanomaterial Manufacturing Processes and Associated Exposures," Comprehensive Materials Processing, vol. 8, pp. 103-125, 16 April 2014.

[34] L. G. C. E. M. e. Tomy Acsente, Tungsten Nanoparticles Produced by Magnetron Sputtering Gas Aggregation: Process Characterization and Particle Properties, London: IntechOpen, 2020.

[35] SPI LASERS LIMITED, [Online]. Available: https://www.spilasers.com/

application-ablation/what-is-laser-ablation/.

[36] R. W. G. L. B. H. M. a. D. M. Getahun Merga, "Redox Catalysis on "Naked" Silver Nanoparticles," *J. Phys. Chem. C,* vol. 111, p. 12220-12226, 2007.

[37] U. D. Z. D. Z. A. Oliveira M, " Influence of synthetic parameters on the size, structure, and stability of dodecanethiol-stabilized silver nanoparticles," *J Colloid Interface Sci.,* vol. 292, p. 429-435, 2005.

[38] J. Y. Song and B. S. Kim, "Rapid biological synthesis of silver nanoparticles using plant leaf extracts," *Bioprocess and Biosystems Engineering,* vol. 32, pp. 79-84, 26 04 2008.

[39] J. G. L. S.-Z. Qiao, "Chapter 21 - Synthetic Chemistry of Nanomaterials," in *Modern Inorganic Synthetic Chemistry (Second Edition)*, Elsevier B.V., 2017, pp. 613-640.

[40] A. A. S. C. P. Suk Fun Chin, "Size Controlled Synthesis of Starch Nano-particles by a Microemulsion Method," *Journal of Nanomaterials,* vol. 2014.

[41] A. H. J. Z. Y. X. C. M. L. Yong X. Gan, "Hydrothermal Synthesis of Nano-materials," *Journal of Nanomaterials,* vol. 2020.

[42] S. B. S. A. R. Singaravelan, "Electro-chemical synthesis, characterisation and phytogenic properties of silver nano-particles," *Applied Nanoscience,* vol. 5, no. 8, p. 983-991, 18 January 2015.

[43] A. M. Hussein, *Synthesis of silver nanoparticles,* Mansoura University, 2016.

[44] S. Iravani, "Bacteria in Nanoparticle Synthesis: Current Status and Future Prospects," vol. 2014.

[45] M. M. G. P. M. N. G. Palaniselvam Kuppusamy, "Biosynthesis of metallic nanoparticles using plant derivatives and their new avenues in pharmacological applications – An updated report," Saudi Pharmaceutical Journal, vol. 24, no. 4, pp. 473-484, July 2016.

[46] R. d. *L. Mariana* Guilger-Casagrande, "Synthesis of Silver Nanoparticles Mediated by Fungi: A Review," *Bioeng. Biotechnol,* vol. 7, 22 October 2019.

[47] S. V. M. A. e. Felix LewisOscar, Algal Nanoparticles: Synthesis and Biotechnological Potentials, London: IntechOpen, 2016.

[48] K. J. T. Y. K. T. S. H. Mafune F, "Structure and stability of silver nanoparticles in aqueous solution produced by laser ablation," *J Phys Chem B.,* vol. 104, p. 8333-8337, 2000.

[49] S. A. V. V. S. G. B.-V. F. Dolgaev SI, "Nanoparticles produced by laser ablation of solids in liquid environment," *Appl Surf Sci.,* vol. 186, p. 546-551, 2002.

[50] a.M. M. Amendola V., "Laser ablation synthesis in solution and size manipulation of noble metal nanoparticles," *Phys. Chem. Chem. Phys.,* vol. 11, p. 3805-3821, 2009.

[51] F. M. a. M. T. Sakamoto M., "Light as a construction tool of metal nanoparticles: synthesis and mechanism," j. Photochem. Photobiol. C, vol. 10, no. 1, p. 33-56, 2009.

[52] N. N. Kawasaki M, "1064-nm laser fragmentation of thin Au and Ag flakes in acetone for highly productive pathway to stable metal nanoparticles," *Appl Surf Sci.,* vol. 253, p. 2208-2216, 2006.

[53] B. C. N. B. E.-S. M. Link S, "Laser-Induced shape changes of colloidal gold nanorods using femtosecond and nanosecond laser pulses.," *J Phys Chem B.,* vol. 104, p. 6152-6163, 2000.

[54] G. A. Shafeev, "FORMATION OF NANOPARTICLES UNDER LASER ABLATION OF SOLIDS IN LIQUIDS,"

in *Laser Ablation: Effects and Applications*, New York, Nova Science Publishers, Inc., 2011, pp. 191-226.

[55] I. I. B. N. N. N. I. K. K. D. B. K. G. B. A. A. S. Nikolov, "Influence of the laser pulse repetition rate and scanning speed on the morphology of Ag nanostructures fabricated by pulsed laser ablation of solid target in water," *Applied Physics: A Material science and processing*, vol. 123, 27 October 2017.

[56] M. A. M. F. A. a. S. A. R. Amir Reza Sadrolhosseini, "Laser Ablation Technique for Synthesis of Metal Nanoparticle in Liquid," in *Laser Technology and its Applications*, 2018.

[57] K. I. N. W. M. T. Takeshi Tsujia, "Preparation of silver nanoparticles by laser ablation in solution: influence of laser wavelength on particle size," *Applied Surface Science,* Vols. 202 (1-2), p. 80-85, December 2002.

[58] B. S. H. A. K. Mohammad Zamakhsari Alhami, "Synthesis of silver nanoparticles using laser ablation Method utilizing Nd:YAG Laser," *AIP Conference Proceedings,* vol. 2202, no. 1, 27 December 2019.

[59] K. S. M. Y. e. Alexander Pyatenko, "Synthesis of silver nanoparticlesby laser ablation in pure water," Applied Physics A, vol. 79, no. 4, pp. 803-806, September 2004.

[60] K. T. T. M. Tsuji T, "Preparation of nano-size particle of silver with femtosecond laser ablation in water.," *Applied Surface Science,* vol. 206, p. 314-320, 2003.

[61] G. A. T. S. e. a. Mehta Y., *Guidelines for prevention of hospital acquired infections,* vol. 18, Indian J. Crit. Care Med., 2014, p. 149-163.

[62] F. J. N. L. Ducel G., *Prevention of Hospital-acquired Infections: A Practical Guide,* WH Organization, 2002.

[63] G. D. J. J. e. a. Shlaes D.M., *Society for Healthcare Epidemiology of America and Infectious Diseases Society of America Joint Committee on the prevention of antimicrobial resistance guidelines for the prevention of antimicrobial resistance in hospitals,* vol. 18, Infect. Control Hosp. Epidemiol. , 1997, p. 275-291.

[64] D. S. I. A. G. A. Rai M., *Silver nanoparticles: the powerful nano-weapon against multidrug-resistant bacteria,* vol. 112, J. Appl. Microbiol., 2012, p. 841-852.

[65] A.-H. S. R.-S. M. I. M. Rad-Sharifi J., *Antimicrobial synergic effect of Allicin and silver nanoparticles on skin infection caused by methicillin resistant Staphylococcus aureus,* vol. 4, Ann. Med. Health Sci. Res., 2014, p. 863-868.

[66] K. S. Nadaf N., *Antibacterial activity of silver nanoparticles singly and in combination with third generation antibiotics against bacteria causing hospital acquired infections biosynthesised by isolated Bacillus marisflavi YCIS MN 5,* vol. 10, Dig. J. Nanomater. Biostruct., 2015, p. 1189-1199.

[67] B.-C. O. B. v. d. E. e. a. Stevens K.N., *The relationship between the antimicrobial effect of catheter coatings containing silver nanoparticles and the coagulation of contacting blood,* vol. 30, Biomaterials, 2009, p. 3682-3690.

[68] G. A. M. A. N. K. S. A. G. V. B. L. M G Strakhovskaya, *Photoinactivation of coronaviruses: going along the optical spectrum,* vol. 17, Laser Physics Letters, 2020.

[69] D. A. V. S. D. S. D. S. Patil S., *Multi-applicative tetragonal TiO2/SnO2 nanocomposites for photocatalysis and gas sensing,* vol. 115, J. Phys.Chem. Solids., 2018, p. 127-136.

[70] M. Y. S. A. Chaloupka K., *Nanosilver as a new generation of nanoproduct in biomedical applications,* vol. 28, Trends Biotechnol., 2010, p. 580-588.

[71] E. J. C. A. e. a. Morones J.R., *The bactericidal effect of silver nanoparticles,* vol. 16, Nanotechnology, 2005, p. 2346.

[72] T. D. A. P. F. e. a. Wigginton N.S., *Binding of silver nanoparticles to bacterial proteins depends on surface modifications and inhibits enzymatic activity,* vol. 44, Environ. Sci. Technol., 2010, p. 2163-2168.

[73] S. N. K. D. D. S. C. S. A. P. Khanna P., *Water based simple synthesis of re-dispersible silver nanoparticles,* vol. 61, Mater. Lett. , 2007, p. 3366-3370.

[74] T. Y. S. J. Pal S., *Does the antibacterial activity of silver nanoparticles depend on the shape of the nanoparticle? A study of the gram-negative bacterium Escherichia coli,* vol. 73, Appl. Environ. Microbiol., 2007, p. 1712-1720.

[75] S.-S. B. Sondi I., *Silver nanoparticles as antimicrobial agent: a case study on E. coli as a model for Gram-negative bacteria,* vol. 275, J. Colloid Interface Sci., 2004, p. 177-182.

[76] A. BA., *Metal-Based Nanoparticles for the Treatment of Infectious Diseases,* 8 ed., vol. 22, Molecules. , 2017, p. 1370.

[77] E. A. R. M. e. a. Haggag EG, *Anti-viral potential of green synthesised silver nanoparticles of Lampranthuscoccineus and Malephora lutea,* vol. 14, Int J Nanomedicine., 2019, pp. 6217-6229.

[78] B. G. a. S. Matthias, *Pathogenesis and Immunology, in Influenza Report,* 2006.

[79] Garland S. Jeffery, *Ventilator-Associated Pneumonia in Neonates: An Update,* vol. 15, Neo Reviews, 2014, pp. e225-e235.

[80] M.-N. N. G.-M. F. e. a. Castañón-Martínez G.A., *Synthesis and antibacterial activity of silver nano-particles with different sizes.,* vol. 10, J Nanopart Res., 2008, p. 1343– 1348.

[81] R.-O. D. P. G.-G. L. H. D.-O. R. a. C.-G. C. Manjarrez-Zavala Ma. Eugenia, *Pathogenesis of Viral Respiratory Infection, Respiratory Disease and Infection – A New Insight,* 2013.

[82] V. S. B. Lucas D Dias, *An update on clinical photodynamic therapy for fighting respiratory tract infections: a promising tool against COVID-19 and its co-infections,* vol. 17, Laser Physics Letter, 2020.

[83] H. H. e. a. Lara, *Mode of anti-viral action of silver nanoparticles against HIV-1,* 1 ed., vol. 8, Journal of nanobiotech-nology, 2010.

[84] B. J. M. J. e. a. Elechiguerra J.L., *Interaction of silver nanoparticles with HIV-1,* vol. 3, J Nanobiotechnol, 2005.

[85] Galdiero, F. A., V. M., C. M., M. V. and G. S. M., *Silver Nanoparticles as Potential Anti-viral Agents,* vol. 16, Molecules, 2011, pp. 8894-8918.

[86] L. Ge, Q. Li, M. Wang, J. Ouyang, X. Li and M. Xing, *Nano-silver particles in medical applications: Synthesis, performance, and toxicity.,* vol. 9, Int. J. Nanomed., 2014, p. 2399-2407.

[87] L.-R. M.A., B.-A. P. and C.-M. e. al., *Evaluation of cardiovascular responses to silver nanoparticles (AgNPs) in spontaneously hypertensive rats.,* vol. 14, Nanomedicine, 2018, p. 385-395.

[88] G. DiVincenzo, C. Giordano and L. Schriever, *Biologic monitoring of workers exposed to silver,* vol. 56, Int. Arch. Occup. Environ. Health, 1985, p. 207-215.

Chapter 8

Silver Nanoparticles Impregnated Wound Dressings: Recent Progress and Future Challengcs

Atamjit Singh, Kirandeep Kaur, Komalpreet Kaur, Jaijeet Singh, Nitish Kumar, Neena Bedi and Preet Mohinder Singh Bedi

Abstract

Microbial infection remains all time and unresolved challenge in the management of burns and diabetic wounds. Especially in diabetic wounds infections are prominent reason of amputations. Microbial biofilms pose tough polymeric barrier that is difficult to cross by conventional antibiotics. Therefore, traditional approach of infection control using antibiotics is now failing at some extent that raised a need to shift this paradigm. Presently, silver nanoparticles incorporated scaffolds representing a new concept of nanoparticle dressings which is becoming popular in wound management. Recently developed silver nanoparticles functionalized wound dressings exhibited excellent profile in the management of wound infections and promotion of wound healing. This chapter throw light on the recent strategies used in the development of silver nanoparticles functionalized wound dressings and their outcomes along with potential benefits and future challenges in wound management.

Keywords: Silver nanoparticles, wound infections, biofilms, diabetic wounds, wound dressing

1. Introduction

A wound is any type of injury which can be in the form of a cut, bruise or contusion caused by some external force. Open wounds are susceptible of getting infected by microorganisms like bacteria, virus or fungi if left unattended. Infected wounds pose a major challenge for healthcare system due to its direct relation with mortality and morbidity of the affected patients [1]. When a wound is infected it causes stress (physiological and psychological) and slows down the wound healing process and in some cases worsens the situation [2]. Approximately 2% of all hospitalized patients worldwide have a chronic wound, and older adult patients are at highest risk, because aging impairs the healing process [3, 4]. As many as 70% of these wounds recur, and 34% are accompanied by infection [5]. A survey by Medicare beneficiaries in 2018 stated that approximately 8.2 million individuals were having some kind of wounds, with or without infection and they estimated the treatment cost to be near about 28.1 billion to 96.8 billion dollars making it a big economic challenge as well [6]. Infectious wounds like surgical wounds, diabetic foot and ulcer are more problematic and are reported more frequently than other infectious wounds [7]. Silver since time memorial is used for its good antimicrobial potential due to its interference with the thiol group of the microbial membrane. It also increases production of reactive oxygen species with in the microbe resulting in damage of DNA and bacterial proteins [8]. Due to its broad spectrum antimicrobial activity, silver nanoparticles are getting a great attention especially in the form of dressing for the infected wounds. This chapter will throw light on the various applications and recent trends in the health care system for treatment of infectious wounds using silver nanoparticles.

2. Wound and wound infections

In general, any form of breakage or harm to the surface of skin can be termed as a wound and broadly classified into two types i.e. acute wound and chronic wound. An acute wound is one in which the wound physiology tends to remain normal during the process of healing. Acute wounds are usually bites, minor burns, cuts and surgical wounds. It is seen that such kind of wounds heal within a predictable timespan depending upon the location, depth and type of wound. In case of chronic wounds the physiology of the wound is disturbed due to various endogenous mechanisms which results in damaging the integrity of the tissue. Examples of chronic wounds are ulcers, diabetic foot and pressure sores. These types of wounds aggravate by aging, malnutrition, diseases which results in immunosuppression within the patient like AIDS or in patients on immunosuppressant drugs [9, 10]. Wound is said to be infected when there is entry of a microbe from the breached skin, which slows down the healing process and results in appearance of signs and symptoms like pain, discoloration of the wounded area, edema, puss, abnormal smell, tenderness etc. [11]. Most of the hard to heal infected wounds are observed to possess biofilms in them [12]. Biofilms are surface linked microbial structural communities having sessile cells present in the matrix produced by the microbe itself, made of polysaccharides, deoxyribonucleic acid and other components which are an essential type of adaptation strategy for the survival of bacteria as it protects it from the harsh surroundings and several immune responses by the extracellular polymeric substance (EPS) [13]. Bacteria possessing biofilms has increased chance of gene transfer of antibiotic resistance gene to other bacterial species [14, 15]. Bacteria having biofilms are tough to treat even with higher doses as biofilms prevent antimicrobial agents to reach up to bacteria [16, 17]. Although there are many beliefs regarding the defensive mechanism of biofilms against the antimicrobial agents but till date its nature of defense is unknown [18].

3. Wound healing

Replacement of damaged tissue by newly produced tissue is termed as wound healing [19]. Skin (epidermis and dermis) acts as a protective layer against the harsh environment and when this barrier is breached a sequence of biochemical events takes place to repair the damage [19, 20]. The process of wound healing is generally described in four phases i.e. blood clotting, inflammation, tissue growth and tissue remodeling [21]. In the first phase with in some time platelets starts covering the area of the affected site. They release a particular chemical signal which promotes clotting resulting in activation of fibrin which produces a mesh to which platelets bind and forms a clot. This phase is also known as hemostasis [22, 23]. In second phase or inflammation stage clearing of debris and bacterial cells takes place via process of phagocytosis where white blood cells and macrophages engulf and destroys them [24]. In the next phase that is proliferation phase or tissue growth phase angiogenesis, collagen deposition, granulation tissue formation, epithelialization and wound contraction takes place [25]. At last the tissue enters the maturation or the remodeling phase in which collagen is realigned and the cells which are not needed are removed by apoptosis [26].

4. Silver as an antimicrobial agent

Since early times silver appeared in recorded history texts for its excellent antimicrobial action. The ancient Greek historian 'Herodotus' describes that at the times of war Persian kings among with his provisions used to take boiled water from silver jars [27, 28]. Raulin in 1869 was the first person who observed the antimicrobial activity of silver by observing that *Aspergillus niger* (most common type of fungus) was not able to grow in silver vessels [29, 30]. Carl Wilhelm von Nageli, a Swiss botanist came up with a term 'oligodynamics' (oligo means small; dynamics means power) which describes any metal having some antibacterial properties at very small concentration [31, 32]. Silver compounds are being used in the process of wound healing since 1970s, leading to discovery of silver sulphadiazine which has an effective, broad spectrum antibacterial activity [33]. The silver element in nature is inert and ionizes when it comes in contact with environment, producing Ag^+ ions which are believed to show antimicrobial activity [34]. Because of its strong antimicrobial activity, silver is a mostly used as an adjunct therapy in wound care. However, it also has the potential to delay the process of wound healing by producing toxic effects on keratinocytes and fibroblasts [35].

5. Silver nanoparticles

Any small particle which ranges between 1 to 100 nanometers (10^{-9}) in size is known as a nanoparticle. It cannot be detected with naked human eye and it exhibits different chemical and physical properties in comparison to their large material counterparts [36]. Among many inorganic nanoparticles, silver nanoparticles (AgNPs) have got researchers attention around the globe due its novel physical, chemical and biological properties as compared to their bulk form. They have particular chemical and physical properties like high electrical and thermal conductivity [37], surface enhanced Raman scattering effect [38], catalytic activity [39], chemical stability [40] and nonlinear optical behavior [41]. The above stated properties make it the material of choice to be used in electronics, and for medical use. Silver nanoparticles are also known for their antiviral, antibacterial and antifungal activity. Due to smaller size, surface area to volume ratio is increased which results in increase in the amount of atoms on the surface, that other forms. The net effect gives rise to unpredictable properties associated with nanoparticles [42]. Silver nanoparticles has shown proven antimicrobial activity in many in vivo and in vitro studies and have application in soaps, cosmetics, food packaging and wound dressing [43]. Due to epidermal keratin and phospholipids, and protein thiol groups, skin was thought to be impermeable to silver nanoparticles but some studies demonstrated, if any absorption beyond the stratum corneum [44, 45]. Silver wound dressings are in direct contact with damaged skin resulting in systemic absorption also associated with some toxicity [46].

6. Mechanistic insight of silver nanoparticles

In an average human concentration of silver in plasma is less than 2 μg/mL which is derived from inhalation of particulate matter and diet [44]. Silver can enter human body by inhalation, oral ingestion and dermal absorption [34]. Pinocytosis and endocytosis are believed to be two processes by which the silver nanoparticles may enter the body. It is seen that the particles that are of nanoscale penetrate much deeper than those of regular size leading to a novel delivery therapy [47, 48]. Till now exact mechanism of action of silver nanoparticles is not clear but several actions have been proposed by the scientists for its antimicrobial activity. Continual release of silver ion is considered to be the main reason for its antimicrobial activity [49]. Due to sulfur protein affinity and electrostatic attraction silver ions adhere to the wall of cells and cytoplasmic membrane which increases its permeability and penetrability into the cytoplasmic membrane leading to disruption of the bacterial cell wall [50]. When the silver ion enters the cell it can deactivate the respiratory enzymes and can generate reactive oxygen species [51]. Reactive oxygen species acts as a key component and a major reason for cell membrane disruption and DNA damage (by interacting with sulfur and phosphorus of DNA) causing problem in DNA replication, reproduction results in death of the microbes. Silver ions also inhibit the synthesis of proteins by denaturation of ribosomes and cause interruption the production of ATP [52]. After anchoring the surface of the cell silver nanoparticles gets accumulated in the pits of the cellular wall of microbe resulting in cell membrane denaturation [53]. Due to nanosize they easily penetrate cell membrane, leading to rupture of cell organelles and even lysis. They also affect the bacterial transduction process by interfering with the phosphorylation of protein substrates which can result in cell apoptosis and cell multiplication [53, 54]. Gram-negative bacterial strains are more sensitive towards the effect of silver nanoparticles because the cellular walls of these bacteria are narrower than the gram positive bacteria [55]. One drawback of silver nanoparticles is that they are not much effective in the case of bacteria having biofilms. Biofilms protects the membrane from both nanoparticles and silver ions by altering their transport due to its complicated structure [56]. The pathway of the nanoparticles penetration is highly obstructed if the size is grater that 50 nm [57]. It is also observed that adsorption and accumulation of the silver nanoparticles on the biofilm results in reduced diffusion of nanoparticles in bacteria [58].

7. Silver nanoparticles wound dressings

A dressing is a sterile material applied to a wound to promote healing and protect the wound from further harm [59, 60]. It has been designed in such a way that it is in direct contact with the wound, as distinguished from a bandage, which is most often used to hold a dressing in place. Silver dressings are used for both types of wounds (acute and chronic) and when there is risk of high level of bio burden or local infection for example in the case of burns [61]. Silver dressings helps in reducing the bioburden in infected or colonized wounds and also acts as a barrier to reduce any further chance of infection [62].

8. Synthesis of silver nanoparticles for wound dressing

There are four types of silver nanoparticles synthesis, namely chemical, irradiation, green and thermal. In chemical synthesis, two types of synthesis methods are used which are Brust-Schiffrin synthesis which is mainly used for golden nanoparticles and Turkevich method which is based on reduction of the boiling solution of silver salt with citrate salt solution [63, 64]. Irradiation synthesis is connected with radiation of precursors or intermediate products of reaction with electromagnetic radiation with different wavelengths [65–67]. In green synthesis plant, fungus or bacterial extract is mixed with silver ion usually silver nitrate, the bioactive molecules of extracts reduce silver ion to elementary silver and then it is precipitated in alcohol. Advantages of this synthesis are that cost involved in synthesis is low, environment friendly and plant extracts contain medicinal compounds which are used in conventional medicine [68–77]. Thermal synthesis is based on the principle of thermal reduction of silver salt. After the synthesis of the silver nanoparticles, they are incorporated in membrane or composite material, nanofibers, hydrogels, etc. and are used as a wound dressing.

8.1 Silver nanoparticles incorporated into membrane and composite material

Membrane and composite material-immobilized nanoparticles can have many functions including antimicrobial activity. Silver nanoparticles incorporated in membranes like polyethersulfone, acetate cellulose, polydopamine-coated poly(ether imide) etc. showed significant antimicrobial activity against diverse

Polymer used	Method of preparation of nanoparticles	Size (nm)	Result
Bacterial cellulose	Thermal method (thermal reduction at 80° C	10–30	Reduction seen in *Staphylococcus aureus*; 99%. Growth of cells observed with no cytotoxicity [79]
Chitin	Irradiation method, (gamma rays, ^{60}Co)	3–13	Significant bactericidal effect ($p < 0.01$) [80]
Bacterial cellulose	Green method (cellulose from *Acetobacter xylinum)*	50–150	Strong activity against *Escherichia coli* and *Staphylococcus aureus* [81]
Chitosan, Polyvinyl alcohol, Curcumine	Green method (chitosan)	16	Significant effects against *Escherichia coli, Staphylococcus aureus, Pseudomonas aeruginosa and Candida albicans* [82]
Chitosan and Chitin	Green method (*Camelia sinensis*)	60–150	Good healing activity [83]
Konjac Glucomannan	Green method (egg white)	8–32	Accelerates wound healing and fibroblast growth promotion [84]
Bacterial cellulose	Chemical method ($NaBH_4$)	3–17	Inhibiton of growth of *Escherichia coli and Staphylococcus aureus* [85]
Chitin	Chemical method (sodium citrate)	5	Inhibiton of growth of *Escherichia coli and Staphylococcus aureus* [86]
Poly vinyl pyrrolidine chitosan	Chemical method (sodium citrate)	10–30	Reduces growth of *Escherichia coli and Staphylococcus aureus* [87]
Chitosan	Chemical method ($NaBH_4$)	15	Facilitates cell proliferation and mitigate bacterial infection [88]
Bacterial cellulose	Chemical method ($NaBH_4$)	5–14	Inhibition of growth of *Escherichia coli and Staphylococcus aureus* [89]

Table 1.
Silver nanoparticles based membrane composites for wound management.

Material used for preparation of clothing and dressings	Method of preparation of nanoparticles	Size (nm)	Result
Cotton fabrics	Chemical method (alkali solution of starch)	22–24	Inhibition of growth of *Escherichia coli and Staphylococcus aureus* [90]
Dressing material	Chemical method ($NaBH_4$)	4–24	Reduction in wound inflammation and fibrogenic cytokines modulation [91]
Cotton fabrics	Green method (*Fusarium oxysporum*)	2	Reduction of growth on *Staphylococcus aureus* [92]
Silver nanoparticles incorporated wound dressing	Green method (*Aspergillus niger*)	200–800	Effective wound healing activity [93]

Table 2.
Silver nanoparticles incorporated clothing and dressings for wound management.

Polymer used for preparation of nanofibers	Method of preparation of nanoparticles	Size (nm)	Result
Collagen	Chemical method (sodium citrate)	25–55	Accelerated wound healing [96]
Poly vinyl pyrrolidine	Chemical method (N,N- DMF)	3–5	Effective antibacterial action [97]
Poly methyl methacrylate-co-dopamine	Chemical method (Silver ion dipped in PMMDM)	<20	Effective antibacterial and wound healing action [98]
Plumbagine	Chemical method (PBG reduction)	60	High antibacterial activity [99]
Gelatine	Chemical method (Silver nitrate reduced with gelatin powder)	11–20	High anti antibacterial activity against Gram positive bacteria [100]
Poly ethylene oxide poly caprolactone	Chemical method (Silver nitrate reduced with PEO and DMF)	13–17	Good antibacterial potential [101]
Alginate	Chemical method ($NaBH_4$)	5–17	Reduces the inflammatory phase and increased epidermal thickness [102]
Chitosan, glucose, Poly vinyl alcohol	Green method (chitosan, glucose)	10–30	Good antibacterial activity against gram negative bacteria [103]
Poly galacturonic acid	Green method (PGA, HA)	5–13	Maximum wound epithelization and collagen deposition [104]
Poly caprolactone	Green method (*P. nigrum*)	5–20	Excellent antibacterial activity against *Escherichia coli and Staphylococcus aureus* [105]
Poly vinyl alcohol	Irreadiation method	23–24	Significant antibacterial activity against *Staphylococcus aureus* and *Pseudomonas aeruginosa* [106]

Table 3.
Silver nanoparticles incorporated nanofibers for wound management.

range of microbes thus have capability to sterile wound environment and promote healing process [78], representative examples are summarized in **Table 1**.

8.2 Powdered silver nanoparticle and topical application

Powdered silver nanoparticles are used for incorporation into different types of clothing and dressings. Representative examples are summarized in **Table 2**.

8.3 Nanofibers

Nanofibers are emerged as an important structures with wide range of biological as well as physical applications like air filtration, immunoanalysis and as pseudo-enzymes etc. [81–84, 94, 95]. Apart from that active research is also ongoing for utilization of silver nanoparticles incorporated nanofibers for wound management. Examples of nanofibers are given in **Table 3**.

8.4 AgNPs-hydrogels

Hydrogels have excellent capacity to absorb wound exudates and at the same time maintain the moisture in wound environment to ensure proper healing. Hydrogels form impermeable physical barrier on wound surface and prevent bacterial invasion (**Figure 1**) and apart from that hydrogels also showcased its tendency to absorb wide range of metals [59, 96, 107, 108]. Some silver nanoparticles incorporated hydrogels showed excellent wound healing activity as shown in **Table 4**.

9. Future challenges

Silver nanoparticles functionalized wound dressings have significant antimicrobial activity and provide faster and effective tissue repair thus they are widely

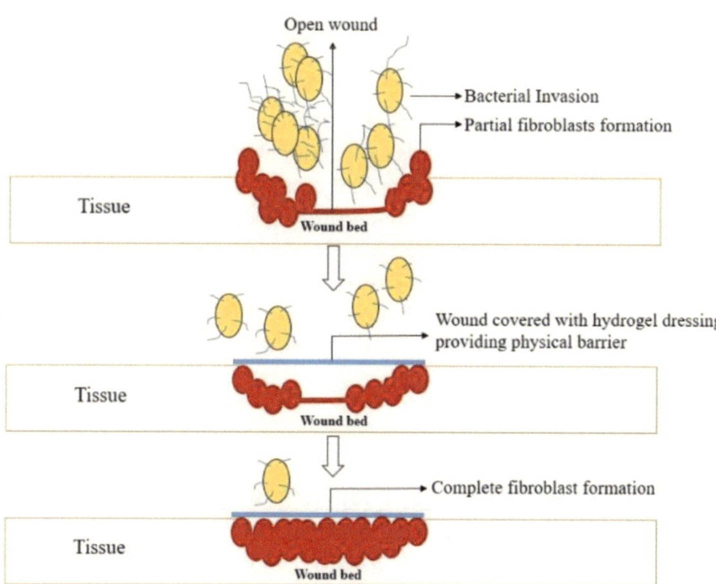

Figure 1.
Schematic layout of hydrogel membrane reducing bacterial invasion and accelerating wound healing process.

Polymer used for preparation of hydrogels	Method of preparation of nanoparticles	Size (nm)	Result
Poly acrylic acid and poly vinyl alcohol	Chemical method ($NaBH_4$)	2–3	Significant antibacterial activity against *Escherichia coli* [109]
Beta- chitin	Chemical method (sodium citrate)	4–8	It showed inhibitory effects on the bacteria growth [110]
Chitosan	Green method (sericin and chitosan)	240–970	Bactericidal action [111]
2-acrylamide-2-methylpropane sulphonic acid sodium salt	Irradiation method (UV radiation)	—	Used as a burn wound dressing due to its good antibacterial activity [112]
Poly vinyl alcohol	Irradiation method (UV radiation, gamma rays ^{60}Co)	90	Antimicrobial activity against *Escherichia coli, Staphylococcus aureus and Candida albicans* [113]
Collagen	Thermal method (reduction at 40°C	5–14	Antimicrobial activity against *Escherichia coli, Staphylococcus aureus and Pseudomonas aeruginosa* [114]
Carboxymethylcellulose	Thermal method (reduction at 70–100°C	7–21	Removes the exudates and prevents wound maceration [115]

Table 4.
Silver nanoparticles incorporated hydrogels for wound management.

explored for wound healing activity. Beside this they are far away from clinical practice as well as commercialization. More rigorous preclinical investigations are still required to validate their capability for tissue regeneration. Apart from that work is also needed to be done on industrial scale up techniques for commercialization.

10. Conclusion

Nanomaterial's now a days representing potential ways for combating microbial related diseases and disorders. Wounds especially chronic ones burdened with resistant microbes are posing serious challenge to healthcare system. For that purpose silver nanoparticles impregnated wound dressing due to their excellent antimicrobial potential are not less than a boon. Along with sterilization they are proven to fasten tissue repair in wounds. Presently more rigorous efforts are needed in their preclinical investigations to evaluate their efficacy verses safety ratio. They have capability to become potential wound dressing of future.

Acknowledgements

The authors are also thankful to Guru Nanak Dev University, Amritsar for providing various facilities to carry out the work.

Conflict of interest

The authors declare no conflict of interest.

Author details

Atamjit Singh*, Kirandeep Kaur, Komalpreet Kaur, Jaijeet Singh, Nitish Kumar, Neena Bedi and Preet Mohinder Singh Bedi
Department of Pharmaceutical Sciences, Guru Nanak Dev University, Amritsar, Punjab, India

*Address all correspondence to: atamjitpharma.rsh@gndu.ac.in

References

[1] Wilkinson LJ, White RJ, Chipman JK. Silver and nanoparticles of silver in wound dressings: a review of efficacy and safety. Journal of wound care. 2011 Nov;20(11):543-549

[2] Bowler PG, Duerden BI, Armstrong DG. Wound microbiology and associated approaches to wound management. Clinical microbiology reviews. 2001 Apr 1; 14(2):244-269.

[3] Powers JG, Higham C, Broussard K, Phillips TJ. Wound healing and treating wounds: chronic wound care and management. J Am Acad Dermatol 2016;74(4):607-625.

[4] Gottrup F, Holstein P, Jorgensen B, Lohmann M, Karlsmar T. A new concept of a multidisciplinary wound healing center and a national expert function of wound healing. Arch Surg 2001;136(7):765-772.

[5] Wong SY, Manikam R, Muniandy S. Prevalence and antibiotic susceptibility of bacteria from acute and chronic wounds in Malaysian subjects. J Infect Dev Ctries 2015;9(9):936-944.

[6] Nussbaum SR, Carter MJ, Fife CE, DaVanzo J, Haught R, Nusgart M, Cartwright D. An economic evaluation of the impact, cost, and medicare policy implications of chronic nonhealing wounds. Value in Health. 2018 Jan 1; 21(1):27-32.

[7] https://www.apma.org/diabetic woundcare

[8] Chaloupka, K., Malam, Y., Seifalian, A.M. Nanosilver as a new generation of nanoproduct in biomedical applications. Trends Biotechnol. 2010; 28: 11, 580-588.

[9] https://healthywa.wa.gov.au/Articles/U_Z/Wounds#:~:text=A%20wound%20is%20any%20damageon%20the%20legs%20and%20feet.

[10] Okur ME, Karantas ID, Şenyiğit Z, Okur NÜ, Siafaka PI. Recent trends on wound management: New therapeutic choices based on polymeric carriers. Asian Journal of Pharmaceutical Sciences. 2020 Jan 23.

[11] Aftab S, Tarik MM, Siddique MA, Yusuf MA. Clinical and microbiological aspect of wound infection: a Review Update. Bangladesh Journal of Infectious Diseases. 2014;1(2):32-37.

[12] Azeredo, J.; Azevedo, N.F.; Briandet, R.; Cerca, N.; Coenye, T.; Costa, A.R.; Desvaux, M.; Di Bonaventura, G.; Hébraud, M.; Jaglic, Z.; et al. Critical review on biofilm methods. Crit. Rev. Microbiol. 2017, 43, 313-351.

[13] Chew, S. C., & Yang, L. (2016). *Biofilms.* Encyclopedia of Food and Health, *407-415.*

[14] Galloway W.R.J.D., Hodgkinson J.T., Bowden S., Welch M., Spring D.R. Applications of small molecule activators and inhibitors of quorum sensing in Gram-negative bacteria. 2012; 20:449-458.

[15] Galloway W.R.J.D., Hodgkinson J.T., Bowden S.D., Welch M., Spring D.R. Quorum sensing in gram-negative Bacteria : small-molecule modulation of AHL and AI-2 quorum sensing pathways. Chem. Rev. 2011; 44:28-67.

[16] Abraham W.-R. Going beyond the control of quorum-sensing to combat biofilm infections. Antibiotics. 2016; 5:3-16.

[17] Hughes G., Webber M.A. Novel approaches to the treatment of bacterial biofilm infections. Br. J. Pharmacol. 2017; 174:2237-2246.

[18] McCarty S, Woods E, Percival SL. Biofilms: from Concept to Reality. InBiofilms in Infection Prevention and Control 2014 Jan 1 (pp. 143-163). Academic Press.

[19] Nguyen DT, Orgill DP, Murphy GF. The pathophysiologic basis for wound healing and cutaneous regeneration. InBiomaterials for treating skin loss 2009 Jan 1 (pp. 25-57). Woodhead Publishing.

[20] Rieger S, Zhao H, Martin P, Abe K, Lisse TS. The role of nuclear hormone receptors in cutaneous wound repair. Cell biochemistry and function. 2015 Jan;33(1):1-3.

[21] Stadelmann WK, Digenis AG, Tobin GR. Physiology and healing dynamics of chronic cutaneous wounds. The American Journal of Surgery. 1998 Aug 1;176(2):26S-38S.

[22] Rasche H. Haemostasis and thrombosis: an overview. European Heart Journal Supplements. 2001 Dec 1; 3(suppl_Q):Q3-7.

[23] Versteeg HH, Heemskerk JW, Levi M, Reitsma PH. New fundamentals in coagulation. Physiological Reviews. 2013 Jan; 93:327-358.

[24] Hirayama D, Iida T, Nakase H. The phagocytic function of macrophage-enforcing innate immunity and tissue homeostasis. International journal of molecular sciences. 2018 Jan;19(1):92.

[25] Chang HY, Sneddon JB, Alizadeh AA, Sood R, West RB, Montgomery K, Chi JT, Van De Rijn M, Botstein D, Brown PO. Gene expression signature of fibroblast serum response predicts human cancer progression: similarities between tumors and wounds. PLoS Biol. 2004 Jan 13; 2(2):e7.

[26] Medrado AR, Pugliese LS, Reis SR, Andrade ZA. Influence of low level laser therapy on wound healing and its biological action upon myofibroblasts. Lasers in surgery and medicine. 2003 Mar; 32(3):239-244.

[27] E.H. Blakeney (ed) (1945) "The History of Herodotus" translated by G. Rawlinson; Dent, London

[28] G. Sykes (1958) "Disinfection and Stedlizatiort'; Spon, London.

[29] J. Raulin (1869) Sci. Nat., 11, 93. Berk (3) Abstr. 1

[30] K.W. von Ngeli (1893) Denschr. schweiz natudorsch Ges. 33, 174. Berk (3) Abstr. 5.

[31] I.B. Romans (1968) in "Disinfection, Sterilization and Preservatiort' edited by C.A. Lawrence and S.S. Block; Lea and Febiger, London, pp372-400 and pp469-475.

[32] Berk RG. Abstracts of articles on oligodynamic sterilization. The Engineer Board, US Army Corps of Engineers [Project WS 768], Fort Belvoir, Virginia. 1947.

[33] Cooper, R., White, R.J.; Silver sulfadiazine: a review of the evidence. Wounds-UK; 2005; 1: 2, 51-61.

[34] Lansdown AB. A pharmacological and toxicological profile of silver as an antimicrobial agent in medical devices. Advances in pharmacological sciences. 2010 Oct;2010.

[35] Khansa I, Schoenbrunner AR, Kraft CT, Janis JE. Silver in wound care—friend or foe?: a comprehensive review. Plastic and Reconstructive Surgery Global Open. 2019 Aug;7(8).

[36] https://www.twi-global.com/technical-knowledge/faqs/what-are-nanoparticles

[37] Chan, K.L.; Mariatti, M.; Lockman, Z.; Sim, L.C. E_ects of the Size and Filler Loading on the Properties of

Copper- and Silver-Nanoparticle-Filled Epoxy Composites. J. Appl. Polym. Sci. 2011, 121, 3145-3152.

[38] Stamplecoskie, K.G.; Scaiano, J.C.; Tiwari, V.S.; Anis, H. Optimal size of silver nanoparticles for surface-enhanced raman spectroscopy. J. Phys. Chem. C 2011, 115, 1403-1409

[39] Zhang, P.; Shao, C.L.; Zhang, Z.Y.; Zhang, M.Y.; Mu, J.B.; Guo, Z.C.; Liu, Y.C. In Situ assembly of well-dispersed Ag nanoparticles (AgNPs) on electrospun carbon nanofibers (CNFs) for catalytic reduction of 4-nitrophenol. Nanoscale 2011, 3, 3357-3363.

[40] Chinnapongse, S.L.; MacCuspie, R.I.; Hackley, V.A. Persistence of singly dispersed silver nanoparticles in natural freshwaters, synthetic seawater, and simulated estuarine waters. Sci. Total Environ. 2011, 409, 2443-2450.

[41] Karimzadeh R, Mansour N. The effect of concentration on the thermo-optical properties of colloidal silver nanoparticles. Optics & Laser Technology. 2010 Jul 1;42(5):783-789.

[42] Jones, C.F., Grainger, D.W. *In vitro* assessments of nanomaterial toxicity. Adv Drug Deliv Rev. 2009; 61:438-456.

[43] Liao, C.Z.; Li, Y.C.; Tjong, S.C. Bactericidal and cytotoxic properties of silver nanoparticles. Int. J. Mol. Sci. 2019, 20, 449.

[44] Lansdown, A.B. Physiological and toxicological changes in the skin resulting from the action and interaction of metal ions. Crit Rev Toxicol.1995; 25: 5, 397-462.

[45] Stern, S.T., McNeil, S.E. Nanotechnology safety concerns revisited. Toxicol Sci. 2008; 101: 1, 4-21.

[46] Johnston, H.J., Hutchison, G., Christensen, F.M., et al. A review of the in vivo and in vitro toxicity of silver and gold particulates: Particle attributes and biological mechanisms responsible for the observed toxicity. Crit Rev Toxicol. 2010; 40: 4, 328-346.

[47] AshaRani, P.V., Low Kah Mun, G., Hande, M.P., Valiyaveettil, S. Cytotoxicity and genotoxicity of silver nanoparticles in human cells. ACS Nano. 2009; 24; 3: 2, 279-290.

[48] Liu, X., Lee, P.Y., Ho, C.M., et al. Silver nanoparticles mediate differential responses in keratinocytes and fibroblasts during skin wound healing. Chem Med Chem. 2010; 1; 3, 468-475.

[49] Bapat RA, Chaubal TV, Joshi CP, et al. An overview of application of silver nanoparticles for biomaterials in dentistry. Mater Sci Eng C. 2018;91:881-898. doi:10.1016/j.msec.2018.05.069

[50] Khorrami S, Zarrabi A, Khaleghi M, Danaei M, Mozafari M. Selective cytotoxicity of green synthesized silver nanoparticles against the MCF-7 tumor cell line and their enhanced antioxidant and antimicrobial properties. Int J Nanomedicine. 2018;13:8013-8024. doi:10.2147/IJN.S189295

[51] Ramkumar VS, Pugazhendhi A, Gopalakrishnan K, et al. Biofabrication and characterization of silver nanoparticles using aqueous extract of seaweed Enteromorpha compressa and its biomedical properties. Biotechnol Rep. 2017;14:1-7. doi:10.1016/j.btre.2017.02.001

[52] Durán N, Nakazato G, Seabra A. Antimicrobial activity of biogenic silver nanoparticles, and silver chloride nanoparticles: an overview and comments. Appl Microbiol Biotechnol. 2016;100(15):6555-6570. doi:10.1007/s00253-016-7657-7

[53] Liao C, Li Y, Tjong SC. Bactericidal and cytotoxic properties of silver nanoparticles. Int J Mol Sci.

2019;20(2):449. doi:10.3390/ijms20020449

[54] Li L, Li L, Zhou X, et al. Silver nanoparticles induce protective autophagy via Ca 2+/CaMKKβ/AMPK/mTOR pathway in SH-SY5Y cells and rat brains. Nanotoxicology. 2019;13(3):369-391. doi:10.1080/17435390.2018.1550226

[55] Meikle T, Dyett BP, Strachan JB, White J, Drummond CJ, Conn CE. Preparation, characterization, and antimicrobial activity of cubosome encapsulated metal nanocrystals. ACS Appl Mater Interfaces. 2020;12(6):6944-6954. doi:10.1021/acsami.9b21783

[56] Saravanan M, Arokiyaraj S, Lakshmi T, Pugazhendhi A. Synthesis of silver nanoparticles from Phenerochaete chrysosporium (MTCC-787) and their antibacterial activity against human pathogenic bacteria. Microb Pathog. 2018;117:68-72. doi:10.1016/j.micpath.2018.02.008

[57] Yin IX, Yu OY, Zhao IS, et al. Developing biocompatible silver nanoparticles using epigallocatechin gallate for dental use. Arch Oral Biol. 2019;102:106-112. doi:10.1016/j.archoralbio.2019.03.022

[58] Pugazhendhi A, Prabakar D, Jacob JM, Karuppusamy I, Saratale RG. Synthesis and characterization of silver nanoparticles using Gelidium amansii and its antimicrobial property against various pathogenic bacteria. Microb Pathog. 2018;114:41-45. doi:10.1016/j.micpath.2017.11.013

[59] First Aid Equipment, Supplies, Rescue, and Transportation". *Hospital Corpsman*. Naval Education and Training Command. 2003. pp. 3-1

[60] Zarrintaj, Payam; Moghaddam, Abolfazl Salehi; Manouchehri, Saeed; Atoufi, Zhaleh; Amiri, Anahita; Amirkhani, Mohammad Amir; Nilforoushzadeh, Mohammad Ali; Saeb, Mohammad Reza; Hamblin, Michael R; Mozafari, Masoud (October 2017). "Can regenerative medicine and nanotechnology combine to heal wounds? The search for the ideal wound dressing". Nanomedicine. 12 (19): 2403-2422. doi:10.2217/nnm-2017-0173

[61] International Wound Infection Institute (IWII) Wound infection in clinical practice. Wounds International 2016

[62] International consensus. Appropriate use of silver dressings in wounds. An expert working group consensus. London: Wounds International, 2012. Available to download from: www.woundsinternational.com

[63] Lee, N.-Y.; Ko, W.-C.; Hsueh, P.-R. Nanoparticles in the treatment of infections caused by multidrug-resistant organisms. Front. Pharmacol. 2019, 10.

[64] Chugh, H.; Sood, D.; Chandra, I.; Tomar, V.; Dhawan, G.; Chandra, R. Role of gold and silver nanoparticles in cancer nano-medicine. Artif. Cell. Nanomed. Biotechnol. 2018, 46, S1210–S1220.

[65] Jia, X.W.; Yao, Y.C.; Yu, G.F.; Qu, L.L.; Li, T.X.; Li, Z.J.; Xu, C.P. Synthesis of gold-silver nanoalloys under microwave-assisted irradiation by deposition of silver on gold nanoclusters/triple helix glucan and antifungal activity. Carbohydr. Polym. 2020, 238, 7

[66] Yu, Z.L.;Wang, W.; Dhital, R.; Kong, F.B.; Lin, M.S.; Mustaph, A. Antimicrobial effect and toxicity of cellulose nanofibril/silver nanoparticle nanocomposites prepared by an ultraviolet irradiation method. Colloid Surf. B Biointerfaces 2019, 180, 212-220.

[67] Zhao, X.M.; Li, N.; Jing, M.L.; Zhang, Y.F.; Wang, W.; Liu, L.S.; Xu, Z.W.; Liu, L.Y.; Li, F.Y.; Wu, N.

Monodispersed and spherical silver nanoparticles/graphene nanocomposites from gamma-ray assisted in-situ synthesis for nitrite electrochemical sensing. Electrochim. Acta 2019, 295, 434-443.

[68] Saha, J.; Begum, A.; Mukherjee, A.; Kumar, S. A novel green synthesis of silver nanoparticles and their catalytic action in reduction of Methylene Blue dye. Sustain. Environ. Res. 2017, 27, 245-250.

[69] Vorobyova, V.; Vasyliev, G.; Skiba, M. Eco-Friendly "green" synthesis of silver nanoparticles with the black currant pomace extract and its antibacterial, electrochemical, and antioxidant activity. Appl. Nanosci. 2020, 12.

[70] Bindhu, M.R.; Umadevi, M.; Esmail, G.A.; Al-Dhabi, N.A.; Arasu, M.V. Green synthesis and characterization of silver nanoparticles from Moringa oleifera flower and assessment of antimicrobial and sensing properties. J. Photochem. Photobiol. B Biol. 2020, 205, 7.

[71] Jyoti, K.; Baunthiyal, M.; Singh, A. Characterization of silver nanoparticles synthesized using Urtica dioic linn. leaves and their synergistic effects with antibiotics. J. Radiat. Res. Appl. Sci. 2016, 9, 217-227.

[72] Chinnasamy, G.; Chandrasekharan, S.; Bhatnagar, S. Biosynthesis of silver nanoparticles from Melia azedarach: Enhancement of antibacterial, wound healing, antidiabetic and antioxidant activities. Int. J. Nanomed. 2019, 14, 9823-9836.

[73] Barbosa, A.; Silva, L.P.C.; Ferraz, C.M.; Tobias, F.L.; de Araujo, J.V.; Loureiro, B.; Braga, G.; Veloso, F.B.R.; Soares, F.E.D.; Fronza, M.; et al. Nematicidal activity of silver nanoparticles from the fungus duddingtonia flagrans. Int. J. Nanomed. 2019, 14, 2341-2348.

[74] Hamedi, S.; Shojaosadati, S.; Shokrollahzadeh, S.; Hashemi-Najafabadi, S. Extracellular biosynthesis of silver nanoparticles using a novel and non-pathogenic fungus, neurospora intermedia: Controlled synthesis and antibacterial activity. World J. Microbiol. Biotechnol. 2014, 30, 693-704.

[75] Saravanan, M.; Arokiyaraj, S.; Lakshmi, T.; Pugazhendhi, A. Synthesis of silver nanoparticles from phenerochaete chrysosporium (MTCC-787) and their antibacterial activity against human pathogenic bacteria. Microb. Pathog. 2018, 117, 68-72.

[76] Li, F.S.; Weng, J.K. Demystifying traditional herbal medicine with modern approaches. Nat. Plants 2017, 3, 7. Pharmaceutics 2020, 12, 821
21 of 24

[77] Sehnal, K.; Hosnedlova, B.; Docekalova, M.; Stankova, M.; Uhlirova, D.; Tothova, Z.; Kepinska, M.; Milnerowicz, H.; Fernandez, C.; Ruttkay-Nedecky, B.; et al. An assessment of the effect of green synthesized silver nanoparticles using sage leaves (Salvia o cinalis L.) on germinated plants of maize (*Zea mays* L.). Nanomaterials 2019, 9, 1550.

[78] Liang, M.; Su, R.X.; Huang, R.L.; Qi, W.; Yu, Y.J.; Wang, L.B.; He, Z.M. Facile in situ synthesis of silver nanoparticles on procyanidin-grafted eggshell membrane and their catalytic properties. ACS Appl. Mater. Interfaces 2014, 6, 4638-4649.

[79] Wu, J.; Zheng, Y.; Song, W.; Luan, J.; Wen, X.; Wu, Z.; Chen, X.; Wang, Q.; Guo, S. In Situ synthesis of silver-nanoparticles/bacterial cellulose composites for slow-released antimicrobial wound dressing. Carbohydr. Polym. 2014, 102, 762-771.

[80] Singh, R.; Singh, D. Chitin membranes containing silver nanoparticles for wound dressing

application. Int. Wound J. 2014, 11, 264-268.

[81] Hu, W.; Chen, S.; Li, X.; Shi, S.; Shen, W.; Zhang, X.;Wang, H. In Situ synthesis of silver chloride nanoparticles into bacterial cellulose membranes. Mater. Sci. Eng. C 2009, 29, 1216-1219.

[82] Kanikireddy, V.; Yallapu, M.; Varaprasad, K.; Nagireddy, N.; Ravindra, S.; Neppalli, S.; Raju, K. Fabrication of Curcumin Encapsulated Chitosan-PVA Silver Nanocomposite Films for Improved Antimicrobial Activity. J. Biomater. Nanobiotechnol. 2011, 2, 55-64.

[83] Ahamed, M.I.; Sankar, S.; Kashif, P.M.; Basha, S.K.; Sastry, T.P. Evaluation of biomaterial containing regenerated cellulose and chitosan incorporated with silver nanoparticles. Int. J. Biol. Macromol. 2015, 72, 680-686.

[84] Chen, H.; Lan, G.; Ran, L.; Xiao, Y.; Yu, K.; Lu, B.; Dai, F.; Wu, D.; Lu, F. A novel wound dressing based on a Konjac glucomannan/silver nanoparticle composite sponge effectively kills bacteria and accelerates wound healing. Carbohydr. Polym. 2018, 183, 70-80.

[85] Maneerung, T.; Tokura, S.; Rujiravanit, R. Impregnation of silver nanoparticles into bacterial cellulose for antimicrobial wound dressing. Carbohydr. Polym. 2008, 72, 43-51.

[86] Madhumathi, K.; Sudheesh, K.P.T.; Abhilash, S.; Sreeja, V.; Tamura, H.; Manzoor, K.; Nair, S.V.; Jayakumar, R. Development of novel chitin/nanosilver composite scaffolds for wound dressing applications. J. Mater. Sci. Mater. Med. 2010, 21, 807-813.

[87] Archana, D.; Singh, B.K.; Dutta, J.; Dutta, P.K. Chitosan-PVP-nano silver oxide wound dressing: In Vitro and in vivo evaluation. Int. J. Biol. Macromol. 2015, 73, 49-57.

[88] Levi-Polyachenko, N.; Jacob, R.; Day, C.; Kuthirummal, N. Chitosan wound dressing with hexagonal silver nanoparticles for hyperthermia and enhanced delivery of small molecules. Colloids Surf. B Biointerfaces 2016, 142, 315-324.

[89] Yang, G.; Xie, J.; Hong, F.; Cao, Z.; Yang, X. Antimicrobial activity of silver nanoparticle impregnated bacterial cellulose membrane: Effect of fermentation carbon sources of bacterial cellulose. Carbohydr. Polym. 2012, 87, 839-845.

[90] Hebeish, A.; El-Rafie, M.H.; El-Sheikh, M.A.; Seleem, A.A.; El-Naggar, M.E. Antimicrobial wound dressing and anti-inflammatory efficacy of silver nanoparticles. Int. J. Biol. Macromol. 2014, 65, 509-515.

[91] Tian, J.; Wong, K.; Ho, C.-M.; Lok, C.-N.; Yu, W.-Y.; Che, C.-M.; Chiu, J.-F.; Tam, P. Topical Delivery of Silver Nanoparticles Promotes Wound Healing. Chem Med Chem 2007, 2, 129-136.

[92] Duran, N.; Marcato, P.; Souza, G.; Alves, O.; Esposito, E. Antibacterial Effect of Silver Nanoparticles Produced by Fungal Process on Textile Fabrics and Their Euent Treatment. J. Biomed. Nanotechnol.

[93] Sundaramoorthi, C.; Mathews, D.; Sivanandy, D.P.; Kalaiselvan, V.; Rajasekaran, A. Biosynthesis of silver nanoparticles from Aspergillus niger and evaluation of its wound healing activity in experimental rat model. Int. J. Pharm Tech Res. 2009, 1, 1523-1529.

[94] Yun, B.J.; Koh, W.G. Highly-Sensitive SERS-based immunoassay platform prepared on silver nanoparticle-decorated electrospun polymeric fibers. J. Ind. Eng. Chem. 2020, 82, 341-348.

[95] Shen, B.L.; Zhang, D.Y.; Wei, Y.J.; Zhao, Z.H.; Ma, X.F.; Zhao, X.D.; Wang,

S.; Yang, W.X. Preparation of Ag Doped Keratin/PA6 Nanofiber Membrane with Enhanced Air Filtration and Antimicrobial Properties. Polymers 2019, 11, 1511.

[96] Rath, G.; Hussain, T.; Chauhan, G.; Garg, T.; Goyal, A.K. Collagen nanofiber containing silver nanoparticles for improved wound-healing applications. J. Drug Target. 2016, 24, 520-529.

[97] Jin, W.-J.; Lee, H.; Jeong, E.; Park, W.H.; Youk, J. Preparation of Polymer Nanofibers Containing Silver Nanoparticles by Using Poly(N-vinylpyrrolidone). Macromol. Rapid Comm. 2005, 26, 1903-1907.

[98] Ghavaminejad, A.; Unnithan, R.A.; Ramachandra, K.S.A.; Samarikhalaj, M.; Thomas, R.; Jeong, Y.; Nasseri, S.; Murugesan, P.; Wu, D.; Park, C.; et al. Mussel-Inspired Electrospun Nanofibers Functionalized with Size Controlled Silver Nanoparticles for Wound Dressing Application. ACS Appl. Mater. Interf. 2015, 7.

[99] Natarajan, D.; Lakra, R.; Srivatsan, K.; Usha, R.; Korrapati, P.; Kiran, M. Plumbagin caged silver nanoparticle stabilized collagen sca_old for wound dressing. J. Mater. Chem. B 2014, 3.

[100] Rujitanaroj, P.-O.; Pimpha, N.; Supaphol, P. Wound-Dressing materials with antibacterial activity from electrospun gelatin fiber mats containing silver nanoparticles. Polymer 2008, 49, 4723-4732.

[101] Dubey, P.; Bhushan, B.; Sachdev, A.; Matai, I.; Kumar, U.; Packirisamy, G. Silver-Nanoparticle-Incorporated composite nanofibers for potential wound-dressing applications. J. Appl. Polymer Sci. 2015, 132.

[102] Neibert, K.; Gopishetty, V.; Grigoryev, A.; Tokarev, I.; Al-Hajaj, N.; Vorstenbosch, J.; Philip, A.; Minko, S.; Maysinger, D. Wound-Healing with mechanically robust and biodegradable hydrogel fibers loaded with silver nanoparticles. Adv. Healthc. Mater. 2012, 1, 621-630.

[103] Abdelgawad, A.M.; Hudson, S.M.; Rojas, O.J. Antimicrobial wound dressing nanofiber mats from multicomponent (chitosan/silver-NPs/ polyvinyl alcohol) systems. Carbohydr. Polym. 2014, 100, 166-178.

[104] El-Aassar, M.R.; Ibrahim, O.M.; Fouda, M.M.G.; El-Beheri, N.G.; Agwa, M.M. Wound healing of nanofiber comprising Polygalacturonic/ Hyaluronic acid embedded silver nanoparticles: In-Vitro and in-vivo studies. Carbohydr. Polym. 2020, 238, 11. Pharmaceutics 2020, 12, 821
23 of 24

[105] Augustine, R.; Kalarikkal, N.; Thomas, S. Electrospun PCL membranes incorporated with biosynthesized silver nanoparticles as antibacterial wound dressings. Appl. Nanosci. 2016, 6, 337-344.

[106] Uttayarat, P.; Jetawattana, S.; Suwanmala, P.; Eamsiri, J.; Tangthong, T.; Pongpat, S. Antimicrobial electrospun silk fibroin mats with silver nanoparticles for wound dressing application. Fiber. Polymer. 2012, 13, 999-1006.

[107] Dou, J.L.; Zhu, G.D.; Hu, B.; Yang, J.M.; Ge, Y.X.; Li, X.; Liu, J.Y. Wall thickness-tunable AgNPs-NCNTs for hydrogen peroxide sensing and oxygen reduction reaction. Electrochim. Acta 2019, 306, 466-476.

[108] Warren, David S.; Sutherland, Sam P. H.; Kao, Jacqueline Y.; Weal, Geoffrey R.; Mackay, Sean M. (2017-04-20). "The Preparation and Simple Analysis of a Clay Nanoparticle Composite Hydrogel". Journal of Chemical Education. 94 (11): 1772-1779. Bibcode: 2017JChEd..94.1772W. doi:10.1021/acs.jchemed.6b00389

[109] Varaprasad, K.; Yallapu, M.; Ravindra, S.; Nagireddy, N.; Kanikireddy, V.; Monika, K.; Bojja, S.; Raju, K. Hydrogel–Silver nanoparticle composites: A new generation of antimicrobials. J. Appl. Polymer Sci. 2010, 115, 1199-1207.

[110] Kumar, P.T.S.; Abhilash, S.; Manzoor, K.; Nair, S.V.; Tamura, H.; Jayakumar, R. Preparation and characterization of novel _-chitin/nanosilver composite sca_olds for wound dressing applications. Carbohydr. Polym. 2010, 80, 761-767.

[111] Verma, J.; Kanoujia, J.; Parashar, P.; Tripathi, C.B.; Saraf, S.A. Wound healing applications of sericin/chitosan-capped silver nanoparticles incorporated hydrogel. Drug Deliv. Translat. Res. 2017, 7, 77-88.

[112] Boonkaew, B.; Kempf, M.; Kimble, R.; Supaphol, P.; Cuttle, L. Antimicrobial efficacy of a novel silver hydrogel dressing compared to two common silver burn wound dressings: Acticoat. and PolyMem Silver®. Burns 2014, 40, 89-96.

[113] Oliveira, R.N.; Rouze, R.; Quilty, B.; Alves, G.G.; Soares, G.D.; Thire, R.M.; McGuinness, G.B. Mechanical properties and in vitro characterization of polyvinyl alcohol-nano-silver hydrogel wound dressings. Interface Focus 2014, 4.

[114] Rattanaruengsrikul, V.; Pimpha, N.; Supaphol, P. In Vitro e_cacy and toxicology evaluation of silver nanoparticle-loaded gelatin hydrogel pads as antibacterial wound dressings. J. Appl. Polymer. Sci. 2012, 124.

[115] Das, A.; Kumar, A.; Patil, N.B.; Viswanathan, C.; Ghosh, D. Preparation and characterization of silver nanoparticle loaded amorphous hydrogel of carboxymethylcellulose for infected wounds. Carbohydr. Polym. 2015, 130, 254-261.

Chapter 9

Silver Nanoparticles for Photocatalysis and Biomedical Applications

William Leonardo da Silva,
Daniel Moro Druzian, Leandro Rodrigues Oviedo,
Pâmela Cristine Ladwig Muraro and Vinícius Rodrigues Oviedo

Abstract

The present chapter aims to overview the application of silver nanoparticles (AgNPs) in photocatalysis and biomedical field. Firstly, the relevance of AgNPs will be addressed. Then, the discussion about the photocatalytic activity of the AgNPs (either in suspension or impregnation), and correlation with your properties and its potential application to organic pollutants degradation under UV and visible/solar radiation will be described. Thus, applications of the AgNPs as antimicrobial agents, such as *Escherichia coli, Schizophyllum commune, Staphylococcus aureus, Pseudomonas aeruginosa, Klebsiella pneumoniae, Haemophilus influenzae, Bacillus subtilis, Bacillus cereus* and *Enterobactor faecalis*, and in the development of biosensors will be discussed. Therefore, the present work will be important to contextualize different scenarios to AgNPs mainly to wastewater treatment and diagnosis/therapeutic applications.

Keywords: nanotechnology, metallic nanoparticles, heterogeneous photocatalysis, antimicrobial properties, biosensors

1. Introduction

Nanotechnology involves the manipulation of materials at nanometric scale (10^{-9} m) and have evolved to novel solutions for water/wastewater treatment as well as biomedical applications [1]. These applications fields are possible due to the unique properties of nanomaterials, such as high surface area, high reactivity and considerable porosity and morphological, electrical, magnetic and/or optical properties, which turn them into useful materials in catalysis, adsorption, sensing and optic-electronic applications [2]. Succinctly, nanomaterials can be divided into 2D, 1D, 0D, according to the number if dimensions the electrons are confined [3]. Metallic nanoparticles are 0D materials, that is, they have the three dimensions within the nanoscale [4]. Among metallic nanoparticles, silver nanoparticles (AgNPs) are largely investigated due to versatility in synthesis, easy processing, fast kinetic reaction rate, high thermal and chemical stability and so forth [5]. Both related to water/wastewater treatment and biomedical applications, AgNPs features allows them to control the interaction with bacteria and, in the case of wastewater

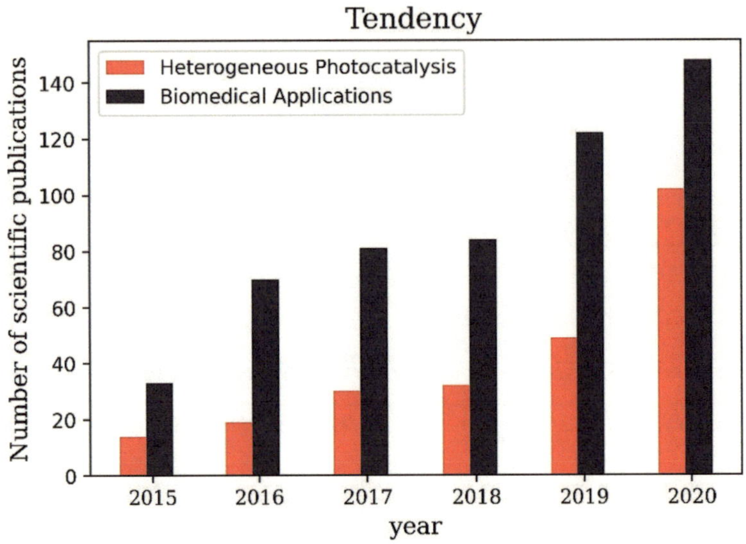

Figure 1.
Number of published papers involving AgNPS in photocatalysis and biomedical applications [7, 8].

treatment, this control can be applied to several pollutants (dyes, hydrocarbons, pesticides, pathogens, and so forth) [6]. Based on that, AgNPs can be applied to photocatalysis and to biomedical applications, such as antimicrobial agents and biosensing [7, 8]. **Figure 1** shows the number of published papers involving AgNPs applied to photocatalysis and biomedical applications.

According to the **Figure 1**, it is possible to notice an increasing tendency of works related to application of AgNPs in the last 5 five years, mainly due to great efficiency of these nanoparticles in sensing applications.

2. Photocatalysis applications with AgNPs

2.1 Water quality deterioration

Mainly due to industrial expansion, climate change, population growth and anthropogenic activities, wastewater quality deterioration has been increased along the years [9]. Wastewater contaminants can be either organic or inorganic [10]. Therefore, the application of an adequate wastewater treatment technology has become a need for minimizing the pollution and the adverse environmental impact as well as for preserving the environment and attending to legal policies of water management [11].

Some of the inorganic ones are well known since ancient times such as chromium, copper, lead, nickel, cadmium, arsenic, mercury and others heavy metals. Theses pollutants pose a serious threat to human health and to environment due to their high toxicity [12]. Meanwhile, organic contaminants, such as benzene, toluene, xylene and natural organic matter (measured by the dissolved or total organic carbon content), are of great concern wastewater management [13]. Usually, the natural organic matter removal from wastewater is challenging and plays a crucial role in defining the treatment technology to be used [14].

In addition, emergent pollutants had been identified in wastewaters all around the world. Most of these contaminants have low biodegradability, high chemical stability, water solubility and are resistant against the conventional wastewater treatment processes [15]. Pharmaceuticals and Personal Care Products (PPCPs) and organic dyes belong to this class of contaminants [16].

Similarly, bacteria commonly pose a threat to humans and to the ecosystem. *Escherichia coli, Enterococcus faecalis*, and *Fusarium solani* are the most contaminants found in wastewater and are of great public concern [17]. Even though they are inactivated by conventional technologies (eg.: chlorination), secondary toxic pollutants are generally found after the treatment, which demonstrate the relatively low-efficiency of the traditional methods of disinfection [18]. Therefore, sophisticated technologies for their inactivation are required, and at same time, it is expected that they do not result in secondary pollutants generation [19].

For these reasons, emergent pollutants (such as pharmaceuticals and dyes), as well as heavy metals, are difficult to remove from wastewater due to either the cost of sophisticated technologies or low efficiency in removing them [20]. In this view, nanotechnology-enable processes seem to be promising to solve the wastewater deterioration quality problem [21]. Therefore, the use of nanoparticles (iron, silver, titanium and zinc oxides) in wastewater applications has been increased due to the unique features of these nanomaterials, requiring treatments with relative low cost and reduction on labor time [22].

Advanced Oxidation Processes (AOPs) have proved to be highly efficient in the degradation of various contaminants like pharmaceuticals, dyes, hydrocarbons, pesticides, and pathogen [23–25]. AOPs are divided into two systems, being [26]: (a) homogeneous, where there are no solid catalysts and the degradation of the organic compound can happen with or without irradiation (direct photolysis), using strong oxidizers (such as hydrogen peroxide, H_2O_2 and ozone, O_3) with or without irradiation, such as photocatalytic ozonolysis and photo-Fenton and, (b) heterogeneous that are characterized by the presence of semiconductor catalysts with or without irradiation, such as heterogeneous photocatalysis and electro-Fenton. Among them, there are processes which are either based on the use of ozone or hydrogen peroxide, decomposition of hydrogen peroxide in acidic media or semiconductors such as titanium dioxide (TiO_2) [27]. The latter one is referred to as a heterogeneous process due to the ease of operation and to sustainable character. Also, it is promising in treating wastewater containing resistant contaminants, once a high percentage of refractory organic contaminants degradation can be achieved [28].

2.2 Heterogeneous photocatalysis

Heterogeneous photocatalysis is an Advanced Oxidation Process (AOPs) useful in the degradation of various contaminants, such as dyes, pharmaceuticals, pesticides, herbicides, hydrocarbons and microorganism inactivation. It can use either UV radiation or visible light to activate a metal-based photocatalyst (generally a semiconductor), promoting oxi-reduction reactions on the catalytic surface with considerable application in wastewater treatment [29]. In this initial step, an electron moves from valence band (VB) to conduction band (CB) of the semiconductor and an electron–hole pair is generated (e^-_{CB} and h^+_{VB}) [30]. Moreover, in this process the water molecules as well as the dissolved oxygen are used as precursors of the reactive oxygen species (ROS) generation [31].

Therefore, ROS (HO• and O_2–•, mainly) tend to react non-selectively with the organic matter and to mineralize it altogether [32]. Thus, heterogeneous photocatalysis is a light-induced catalytic process that reduces or oxidizes organic molecules through redox reactions, which are activated through the electron–hole pairs generated on the surface of the catalyst beyond band gap light irradiation, according to the Eqs. (1)–(8) [33]:

$$Supported\ AgNPs + h\nu \rightarrow e^-_{CB} + h^+_{VB} \tag{1}$$

$$O_2 + e^-_{CB} \rightarrow O_2^- \bullet \tag{2}$$

$$h^+_{VB} + H_2O \rightarrow HO\bullet + H^+ \tag{3}$$

$$e^-_{CB} + h^+_{VB} \rightarrow heat\ loss \tag{4}$$

$$HO\bullet + HO\bullet \rightarrow H_2O_2 \tag{5}$$

$$O_2^- \bullet + H_2O_2 \rightarrow HO\bullet + OH^- + O_2 \tag{6}$$

$$O_2^- \bullet + H^+ \rightarrow HOO\bullet \tag{7}$$

$$HO\bullet + organic\ matter\ O_2 \rightarrow CO_2 + H_2O \tag{8}$$

Thus, the Eq. (1) represents the metal-based photocatalyst (semiconductor) activation yielding to electron–hole pair. Eqs. (2), (3), (6), and (7) are related to ROS generation, which are responsible for organic matter degradation. Eqs. (4) and (5) are undesirable recombination that takes place in the process. Meanwhile, Eq. (4) represents the spontaneous reaction of electron–hole pair recombination, which reduces the photocatalytic degradation efficiency, and Eq. (5) indicates hydrogen peroxide production. Eq. (8) holds for degradation of the organic matter by hydroxyl radicals. Thus, the great advantage of AOPs is that, during the treatment of organic compounds, they are destroyed and not just transferred from one phase to another, as occurs in some conventional homogeneous treatment processes.

Low human toxicity, high stability and relatively low cost are some of the advantages of the heterogeneous photocatalysis process [34]. Moreover, there is a possibility of the complete mineralization of organic pollutants, generating CO_2 and H_2O [35]. In this view, heterogeneous photocatalysis can be applied to degradation of a great number of recalcitrant and non-biodegradable organic pollutants [36, 37].

Moreover, this process can be combined with other processes (either pre-treatment or post-treatment), being performed *in situ*, making use of high oxidative potential agents with fast kinetic reaction and without to need for after-treatment or disposal [38]. In addition, high-quality organoleptic characteristics of water can be achieved, with less energy consumption and low-cost of operation. It is important to mention that the absorbed energy by the semiconductor (and in its photoactivation) is related to the catalyst photocatalytic efficiency, which depends on the competition between the removal of the electron from the semiconductor surface and the recombination of electron–hole pair [39].

2.3 AgNPs-based supported photocatalysts

Individually applied to wastewater treatment, AgNPs proved to be extremely toxic to humans and aquatic life [40]. Moreover, AgNPs can access various organs which most part of the substances cannot [34]. For this reason, AgNPs have been associated to catalytic supports used in heterogeneous photocatalysis [41]. In fact, AgNPs possess considerable photocatalytic activity and when they are impregnated in a less active material, labeled catalytic support, toxicity issues can be fixed [42].

Material	Pollutant	Comment	Reference
Ag@MGO-TA/Fe^{3+}	Methylene Blue (MB)	Excellent stability of the photocatalyst in the aqueous media, high reduction rate (0.054 s^{-1}) for MB, under the dosage of 0.05 mg mL^{-1}. 100% inactivation of *E. coli* using 20 μg.mL^{-1} of photocatalyst	[43]
$nAgFeO_2$	Imidaclopride (IMI)	80% degradation Imidaclopride under UV radiation	[44]
Ag/wood	Methylene Blue and oil separation	The filter can separate selectively the oil from water and dye (about 99%). AgNPs incorporated to filter wood can improve the photocatalytic activity for MB degradation (94.03%)	[45]
Ag@AgBr/Bi_2MoO_6	Reactive Blue-19	Excellent photodegradation of Reactive Blue 19 (98.7%) under visible light after 120 minutes	[46]
C@$CoFe_2O_4$@Ag	Red and Methyl Orange	Fast kinetic of degradation (7 min) for Red Orange and for Methyl Orange (10 min)	[47]
Ag/ZnO/PMMA	Methylene Blue, Paracetamol and Sodium Lauryl Sulfate	90% degradation for the target pollutants after 4 hours under UV radiation	[48]
Ag@CAF and Ag@TiO_2	4-nitrophenol (4-NP), 2-nitrophenol (2-NP), 2-nitroaninile (2-NA), trinitrophenol (TNP), Rhodamine B (RhB) and Methyl Orange (MO)	Nanomaterials exhibited high photocatalytic activity (95%) efficiency was achieved after 10 min	[49]
glass/Ag	Textile wastewater	The coat glass (with AgNPs) yields to about 95% of dye removal after 5 h. In addition, the reusability was studied, targeting microbes found in wastewater. After 2 h, 80% of microbes were inactivated	[50]
Chitosan/Ag	Sodium Fluoresceine	48% dye degradation under anaerobic condition after 2,5 hours. About 51% degradation was achieved under aerobic condition. Chitosan/AgNPs nanocomposite showed antimicrobial activity for gram-positive (*E. coli*) and gram-negative bacteria (*G. bacillus*).	[51]
Fe_3O_4@PPy-MAA/Ag	4-nitrophenol (4-NP), Methylene Blue and Methyl Orange	Excellent catalytic activity towards 4-Nitrophenol, Methyl Orang and Methylene Blue. Good reusability of the nanophotocatalyst	[52]
rGO-Ag	MB and RhB	Degradation of 95% is observed, which is significantly greater that the pristine nanophotocatalyst AgNPs (78%) and rGO (55%)	[53–55]

Table 1.
Paper that apply supported AgNPs as heterogeneous photocatalyst.

It is noteworthy that the main catalytic supports are metallic oxides, since they have favorable characteristics for applicability in heterogeneous photocatalysis, such as photoactivity within a UV–vis radiation range, redox potential of the positive conduction band high enough to promote the mineralization of the organic pollutant, high physical–chemical stability and efficiency in the oxygen reduction reaction. **Table 1** shows some scientific papers found during the 2015–2020, which apply supported AgNPs as heterogeneous photocatalysis.

In addition, the main drawbacks of the metallic nanoparticles, such as nanoparticle agglomeration, can be solved with the application of the Ag-based supported photocatalysts [56]. Thus, impregnation of AgNPs onto ZnO photocatalyst results in better photocatalytic activity, when compared to the use of unsupported ZnO for degradation of Methylene Blue dye. In addition, associations of AgNPs to AuNPs in a core-shell structure leads to a photocatalytic activity comparable with the commonly used TiO_2 photocatalyst. As can be seen, about 80% Methylene Blue degradation can be achieved, when AgNPs supported onto bismuth vanadate ($BiVO_4$) are used [57].

Moreover, with respect to inactivation of bacteria, efficient inactivation degrees for *E. Coli*, *F. Solani*, and *E. faecalis* are reported by means of heterogeneous photocatalysis [58]. Additionally, the efficiency of supported AgNPs for inactivating some pathogens in real wastewaters has been confirmed, resulting in about 80% inactivation [59].

Meanwhile, the utilization of AgNPs as for potentializing the metal-based catalyst and others materials applied to the degradation of organic pollutants or the discoloration of the wastewater is considered an efficient technology, accounting for up to 90% after 180 minutes under UV radiation.

3. Biomedical applications with AgNPs

Silver nanoparticles have been the most investigated, among other metallic nanoparticles for biomedical applications, such as antimicrobial agents and biosensing due their unique physicochemical and biological properties [60]. However, the effective application of AgNPs in the biomedical field is strictly related to their morphology, that is, size and shape [61].

Moreover, one of the main applications of AgNPs, within the biomedical area, consists of acting as an antimicrobial agent, capable of inhibiting the growth of pathogenic microorganisms, being indicated for the treatment of bacterial infections [62]. There are several proposed mechanisms that explain the antimicrobial activity of AgNPs, and all of them lead to applications in wound healing [63], bone tissue [64], and surface coatings [65]. Besides that, optical, electrical and plasmonic properties of AgNPs turn them into interesting nanostructured materials to be used in chemical and biological sensing [66].

Despite being effective against several pathogens and showing promising potential in biosensing, safety concerns are yet a challenge nowadays [67]. In spite of their outstanding properties to biomedical applications, it is known that AgNPs can be toxic to humans depending on the concentration [68] or due to toxic chemicals involved during the synthesis process [69]. To overcome this, nanotechnology has been used together with green chemistry [70], leading to the synthesis of AgNPs by using alternative sources/materials, such as plant extracts, biopolymers and microorganisms (e.g. bacteria and fungi) [71]. Thus, particular interest in evaluating the biocompatibility of green-synthesis derived AgNPs have also been attracting the attention of the scientific community [72]. In the following subsections, both AgNPs applications as antimicrobial agents and as biosensors are discussed.

3.1 Silver nanoparticles as antimicrobial agents

The pathogenic microorganisms are constantly evolving, with a wide genetic diversification, being capable of eliciting several diseases [73]. Although, there are various antimicrobial therapies commercially available, the use of conventional therapies led to the gain of resistance by these pathogenic bacteria [74].

The emerging resistance of bacteria against the conventional antibiotics and metallic ions increased the research in the field of applied nanotechnology, with the AgNPs being among the most potent compounds due to their high specific surface area and number of atoms available to interact with the surroundings, resulting in exhibiting unique properties (electronic, bactericide, magnetic, and optical), since these favorable properties favor the generation of reactive oxygen species (ROS), which they cause changes in the structure of proteins and nucleic acids, and in the permeability of the cell wall, culminating in the lysis of the bacterial cell [75].

The biosynthesis (both intracellular and extracellular) uses microorganisms and show advantages compared to chemical processes: (i) easy strain manipulation, supporting the synthesis process, (ii) easy scaling-up, and (iii) no generation of toxic substances to the environment. However, as the main disadvantages of the biosynthesis using microorganisms is the need to use of the ultrasound to unbind the AgNPs [76, 77].

It is known that the AgNPs have great potential against several gram-negative, gram-positive and antibiotics-resistant bacterial strains [74]. The antimicrobial activity depends on the nanoparticle size and concentration, where low particle sizes with low concentrations can kill bacterial strains and, in the case of green-synthesis derived nanoparticles, this is allied to the advantage of showing lower toxicity to human health and to the environment, leading to a high interest in developing them to combat pathogenic bacteria [78].

The antimicrobial activity of AgNPs against pathogenic bacteria follows two action mechanisms: according to the first one (i), the nanoparticles adhere to the cellular membrane and penetrate the bacteria, promoting alterations on their cellular membrane (due to interactions of silver ions with proteins, sulfur, and phosphorous within the cell, avoiding the electron transport) and, then, resulting in bacterial growth suppression [79]; the other mechanism (ii) involves the silver ions releasing and the production of reactive oxygen species which generates an antimicrobial effect [80].

Moreover, the bacteria are generally unable to develop resistance against AgNPs, which are specially formulated due to their particle size and that allows them to attack a wide range of targets present in the microorganisms, such as proteins, thiol groups and cellular walls [81]. In fact, AgNPs have a huge potential to be used as antimicrobial agents depending on the physicochemical characteristics of these nanoparticles. Therefore, the synergistic effect of these properties, associated with low production cost, good thermal and radiation stability (UV and visible), doing AgNPs effective against pathogenic microorganisms, being promising in biomedical applications to reduce infections. **Table 2** shows the different results of AgNPs against several pathogenic bacteria.

3.2 Silver nanoparticles applied to biosensing

AgNPs have been investigated for chemical and biological sensing. It was already found in literature that the AgNPs present better results than gold nanoparticles (AuNPs) when related to biosensor sensitivity, despite the AuNPs being more investigated for biosensing applications [87]. In addition, AgNPs are plasmonic nanostructures, which means that they can absorb and scatter light [88].

Nanoparticle	Antimicrobial agent	Comments	Reference
Green AgNPs	*Escherichia coli* and *Schizophyllum commune*	Green synthesis using *Eucalyptus camaldulensis* (EC). Moreover, High antimicrobial activity due to the high surface are, which supports better cellular interactions with pathogenic microbes	[82]
Green AgNPs	*Escherichia coli, Staphylococcus aureus, Pseudomonas aeruginosa, Klebsiella pneumoniae* and *Haemophilus influenzae*	Green synthesis using *Artemisia vulgaris* leaves extract (AVLE). In addition, antimicrobial activity with mechanisms involving penetration in the bacteria and silver ions releasing, leading to bacterial growth suppression.	[83]
Green AgNPs	*Bacillus subtilis, Bacillus cereus* and *Staphylococcus aureus*	Green synthesis using *Acorus calamus* rhizome extract. Moreover, nanoparticles exhibited antimicrobial activity when adhering to the bacterial cellular membrane, inhibiting the cellular growth	[84]
Green AgNPs	*Escherichia coli* and *Staphylococcus aureus*	Green synthesis using *Vitex negundo L.* AgNPs showed antimicrobial activity against both gram-positive and gram-negative bacteria	[85]
AgNPs	*Escherichia coli* and *Staphylococcus aureus*	Synthesis by chemical reduction and by using *Petroselinum crispum* plant, and showed antimicrobial activity by penetrating the cellular wall of bacteria	[86]

Table 2.
Results of antimicrobial activity of AgNPs against pathogenic bacteria.

Thus, AgNPs could be used to provide a colorimetric/plasmonic method to detect several biomolecules (including eye-naked verification) due to their light absorption bands - around 400–500 nm [89] - being within the visible range of the electromagnetic spectrum. It is worth pointing out that there are two types of colorimetric/plasmonic biosensors: the aggregation-based and the LSPR-based [90] - some of them are summarized in **Table 3**. Nevertheless, control over morphology during the synthesis of AgNPs is crucial, as anisotropic AgNPs can display various light absorption bands rather than just one [100].

Moreover, Localized Surface Plasmon Resonance (LSPR) phenomenon of AgNPs have turned them into interesting nanomaterials for applications, which involve interaction with light [101]. Thus, when a metallic nanoparticle is irradiated, superficial electrons oscillate collectively, and these generate an electromagnetic field around the nanostructure, called Surface Plasmon Resonance (SPR) [102]. If an external electromagnetic field is applied, in such a way that it is in resonance with the generated electromagnetic field around the metallic nanoparticle, LSPR phenomenon takes place [103]. Thus, the LSPR is possible due to the confinement of the resulting electromagnetic field within the metallic nanoparticle [104, 105].

Moreover, the aggregation-based AgNPs biosensors are considered low-cost, and high-sensitivity biosensing devices, as the aggregation of AgNPs depending on induced changes on the solution medium can be applied to detect DNA molecules, proteins (recognizing) [106]. In this case, aggregation phenomenon and chemical instability of AgNPs is desirable, once it is the working principle of the

Biosensor	Application	Biosensor type	Comment	Reference
AgNPs-based SPR biosensor	MicroRNA (miRNA) detection	LSPR-based	Good sensitivity and selectivity. Excellent reliability. No need for modification procedures to amplify biosensing	[91]
AuNPs/AgNPs biosensor	Cyclin A2	Aggregation-based	Simplicity, high sensitivity, and selectivity. Eye-naked verification allied to quantitative detection. No need for functionalization of the AuNPs/AgNPS. Suitable for peptide-based protein detection. Detection limit: 30 nM.	[92]
Citrate capped silver nanoparticles (Cit-AgNPs)	*Acinetobacter baumannii* detection	Aggregation-based	Accurate and quick detection (about 2 min); low detection limit of quantification (LLOQ) of 1zM.	[93]
Carbon quantum dots (CQDs)/AgNPs nanocomposite	Melamine detection	Aggregation-based	High sensitivity, simple method of detection, and eye-naked verification allied to quantitative detection. Detection limit: 65.3 $pmol.L^{-1}$.	[94]
Glutathione-coated AgNPs	Vitamin B1 (thiamine) detection	LSPR-based	Provides both qualitative (colorimetric) and quantitative sensing. High sensitivity and selectivity; low-cost, quick, and simple detection. Worked well with real samples, such as blood and urine.	[95]
Grown-AgNPs on AuNS	Alkaline phosphatase (ALP) detection	LSPR-based	Provides both qualitative (colorimetric) and quantitative sensing. Could be extended to a general device for immunosensors/aptasensors designs.	[96]
SiO_x/AgNPs/Graphene	DNA hybridization detection	LSPR-based	Other applications may involve enzyme detection, medical diagnostics, food safety, and environmental monitoring. Sensitivity improvement of 304.60% compared to pure AgNPs.	[97]
Ag@AgCl nanotubes loaded onto reduced graphene oxide (RGO)	Ochratoxin A (OTA) detection	LSPR-based	Good accuracy, high sensitivity, and good reproducibility. High stability and photocurrent response under visible light irradiation. Range of detection: 0.05 to 300 n mol L^{-1}, limit of detection (LOD): 0.01 n mol L^{-1} (4.0 pg. mL^{-1}).	[98]
AgNPs-based aptasensor	Adenosine detection	LSPR-based	High sensitivity and selectivity. Simple and specific design, low-cost, and quick detection. Linear range of detection: 200 n mol L^{-1} to 200 μ mol L^{-1}, detection limit: 48 n mol L^{-1}.	[99]

Table 3.
AgNPs-based biosensors for biomedical applications.

biosensor [107]. LSPR-based ones are established on changes in the occurring refractive index now that photons are directed to the nanoparticles, leading them to oscillation [108], being used in biomolecules detection [109]. It is also important to mention that AgNPs are normally functionalized before applying them as biosensors to overcome chemical stability and toxicity aspects [110]. Therefore, coating AgNPs with organic or inorganic materials are the common approaches. Furthermore, polymeric coatings are also used to functionalize AgNPs by using either synthetic polymers, such as (poly)-ethylene glycol (PEG) [111], (poly)-vinyl alcohol (PVA) [112] and (Poly)-vinylpyrrolidone (PVP) [113], or natural polymers such as starch [114], sodium alginate [115] and chitosan [116]. Functionalization with polymeric blends that uses both synthetic and natural polymers is also an interesting approach (e.g. PVA/Chitosan-coated AgNPs [117]). The inorganic coating involves the functionalization of AgNPs with silicon dioxide [118], while organic coating involves citrates mainly [119]. Furthermore, there are plenty of electrochemical-based AgNPs biosensors [120, 121], however, they will not be covered here as the focus is on the plasmonic ones. Another possible application of AgNPs is the surface enhanced Raman Spectroscopy (SERS), which involves the adsorption of molecules on the AgNPs to achieve a high-quality spectroscopy technique. The applications of SERS focus on disease diagnosis caused by microbial infections or cancer [122].

Therefore, AgNPs-based biosensors are good alternatives against conventional sensing devices, as nanostructured biosensors show greater sensibility, reliability, wide limits of detection, precision, speed and provides eye-naked colorimetric assays together with quantitative analysis [123], among other unique characteristics that are shown in the papers summarized in **Table 3.**

4. Conclusion

Regarding the use AgNPs in heterogeneous photocatalysis, it can be proved highly efficient in the degradation a large amount of organic pollutants and inactivation of bacteria and pathogens, under either UV radiation or visible light. Moreover, when supported AgNPs-based nanophotocatalysts are used in wastewater not only the photocatalytic activity is enhanced, but also some operational problems (nanoparticles agglomeration) can be fixed. With respect to the use of AgNPs as antimicrobial agents, it is a current alternative against common pathogens and multi-resistant bacteria due to the toxicity to microorganisms compared to antibiotics and conventional approaches. In addition, AgNPs-based biosensors are resulting in high sensitivity and selectivity aligned to wide detection limits, which turns them suitable for clinical practice. It is worth to point out that the green synthesis of AgNPs is increasing along the years, and when combined with photocatalytic and biomedical applications, contributes to sustainable development and biocompatibility aspects.

Acknowledgements

This work received financial support from the Foundation for Research of the State of Rio Grande do Sul (FAPERGS – Project 19/2551-0001362-0), so all thanks.

Conflict of interest

The authors declare no competing interests.

Author details

William Leonardo da Silva*, Daniel Moro Druzian, Leandro Rodrigues Oviedo, Pâmela Cristine Ladwig Muraro and Vinícius Rodrigues Oviedo
Nanoscience Graduate Program – Universidade Franciscana, Santa Maria - RS, Brazil

*Address all correspondence to: w.silva@ufn.edu.br

References

[1] Wen M, Li G, Liu H, Chen J, An T, Yamashita H. Metal–organic framework-based nanomaterials for adsorption and photocatalytic degradation of gaseous pollutants: recent progress and challenges. Environmental Science: Nano. 2019;6(4):1006-1025. DOI:10.1039/c8en01167b

[2] Das R, Ali ME, Hamid SBA, Ramakrishna S, Chowdhury ZZ. Carbon nanotube membranes for water purification: A bright future in water desalination. Desalination. 2014;336:97-109. DOI:10.1016/j.desal.2013.12.026

[3] Zhang B, Sun JY, Ruan MY, Gao PX. Tailoring two-dimensional nanomaterials by structural engineering for chemical and biological sensing. Sensors and Actuators Reports. 2020;2(1):1-12. DOI:10.1016/j.snr.2020.100024

[4] Varshney S, Kumar M, Gowda A, Kumar S. Soft discotic matrix with 0-D silver nanoparticles: Impact on molecular ordering and conductivity. Journal of Molecular Liquids. 2017;238:290-295. DOI:10.1016/j.molliq.2017.05.008

[5] Nagaraju G, Udayabhanu, Shivaraj, Prashanth SA, Shastri M, Yathish KV, Anupama, C; Rangappa, D, Nagaraju, G. Electro-chemical heavy metal detection, photocatalytic, photoluminescence, biodiesel production and antibacterial activities of Ag–ZnO nanomaterial. Materials Research Bulletin. 2017;94:54-63. DOI:10.1016/j.materresbull.2017.05.043

[6] Qu X, Brame J, Li Q, Alvarez PJJ. Nanotechnology for a Safe and Sustainable Water Supply: Enabling Integrated Water Treatment and Reuse. Accounts of Chemical Research. 2012;46(3):834-843. DOI:10.1021/ar300029v

[7] Kamarudin NS, Jusoh R, Setiabudi HD, Sukor NF, Shariffuddin JH. Potential nanomaterials application in wastewater treatment: Physical, chemical and biological approaches. Materials Today: Proceedings. 2020:1-13. DOI:10.1016/j.matpr.2020.10.221

[8] Shen T, Wang J, Xia Z, Dai X, Li B, Feng Y. Ultraviolet sensing characteristics of Ag-doped ZnO micro-nano fiber. Sensors and Actuators A: Physical. 2020;307:1-8. DOI:10.1016/j.sna.2020.111989

[9] Zhang S, Liu Y, Gu P, Ma R, Wen T, Zhao G, Li L, Ai Y, Hu C, Wang X. Enhanced photodegradation of toxic organic pollutants using dual-oxygen-doped porous g-C3N4: Mechanism exploration from both experimental and DFT studies. Applied Catalysis B: Environmental. 2019;248:1-10. DOI:10.1016/j.apcatb.2019.02.008

[10] Tahir MB, Nawaz T, Nabi G, Sagir M, Khan MI, Malik N. Role of nanophotocatalysts for the treatment of hazardous organic and inorganic pollutants in wastewater. International Journal of Environmental Analytical Chemistry. 2020:1-25. DOI:10.1080/03067319.2020.1723570

[11] Hernández-Chover V, Castellet-Viciano L, Hernández-Sancho F. Preventive maintenance versus cost of repairs in asset management: An efficiency analysis in wastewater treatment plants. Process Safety and Environmental Protection. 2020;141:215-221. DOI:10.1016/j.psep.2020.04.035

[12] Ali I. New Generation Adsorbents for Water Treatment. Chemical Reviews. 2012;112(10):5073-5091. DOI:10.1021/cr300133d

[13] Mello JMM, Brandao HL, Valério A, de Souza AAU, de Oliveira D,

da Silva A, et al. Biodegradation of BTEX compounds from petrochemical wastewater: Kinetic and toxicity. Journal of Water Process Engineering. 2019;32:1-7. DOI:10.1016/j.jwpe.2019.100914

[14] Matilainen A, Gjessing ET, Lahtinen T, Hed L, Bhatnagar A, Sillanpaää M. An overview of the methods used in the characterization of natural organic matter (NOM) in relation to drinking water treatment. Chemosphere. 2011;83(11):1431-1442. DOI:10.1016/j.chemosphere.2011.01.018

[15] Bernabeu A, Vercher RF, Santos-Juanes L, Simon PJ, Lardíın C, Martíınez MA, Vicente JA, González R, Llosá C, Amant AM. Solar photocatalysis as a tertiary treatment to remove emerging pollutants from wastewater treatment plant effluents. Catalysis Today. 2011;161(1):235-240. DOI:10.1016/j.cattod.2010.09.025

[16] Mohapatra DP, Brar SK, Tyagi RD, Picard P, Surampalli RY. Analysis and advanced oxidation treatment of a persistent pharmaceutical compound in wastewater and wastewater sludge-carbamazepine. Science of the Total Environment. 2014;470-471:58-75. DOI:10.1016/j.scitotenv.2013.09.034

[17] Yan Y, Zhou X, Yu P, Li Z, Zheng T. Characteristics, mechanisms and bacteria behavior of photocatalysis with a solid Z-scheme Ag/AgBr/g-C_3N_4 nanosheet in water disinfection. Applied Catalysis A: General. 2020;590:1-27. DOI:10.1016/j.apcata.2019.117282

[18] Meng Y, Wang Y, Han Q, Xue N, Sun Y, Gao B, Li Q. Trihalomethane (THM) formation from synergic disinfection of biologically treated municipal wastewater: Effect of ultraviolet (UV) irradiation and titanium dioxide photocatalysis on dissolve organic matter fractions. Chemical Engineering Journal. 2016;303:252-260. DOI:10.1016/j.cej.2016.05.141

[19] Madhura L, Singh S, Kanchi S, Sabela M, Bisetty K, Inamuddin. Nanotechnology-based water quality management for wastewater treatment. Environmental Chemistry Letters. 2018;17(1):65-121. DOI:10.1007/s10311-018-0778-8

[20] Shah AI, Dar MUD, Bhat RA, Singh JP, Singh K, Bhat SA. Perspectives and challenges of wastewater treatment technologies to combat contaminants of emerging concerns. Ecological Engineering. 2020;152:105882. DOI:10.1016/j.ecoleng.2020.105882

[21] Theron J, Walker JA, Cloete TE. Nanotechnology and Water Treatment: Applications and Emerging Opportunities. Critical Reviews in Microbiology. 2008;34(1):43-69. DOI:10.1080/10408410701710442

[22] Kanchi S. Nanotechnology for Water Treatment. Journal of Environmental Analytical Chemistry. 2014;01(02):1-3. DOI:10.4172/JREAC.1000e102

[23] Kosera VS, Cruz TM, Chaves ES, Tiburtius ERL. Triclosan degradation by heterogeneous photocatalysis using ZnO immobilized in biopolymer as catalyst. Journal of Photochemistry and Photobiology A: Chemistry. 2017;344:184-191. DOI:10.1016/j.jphotochem.2017.05.014

[24] Nobre FX, Mariano FAF, Santos FEP, Rocco MLM, Manzato L, de Matos JME, Couceiro P, Brito W. Heterogeneous photocatalysis of Tordon 2, 4-D herbicide using the phase mixture of TiO_2. Journal of Environmental Chemical Engineering. 2019;7(6):1-46. DOI:10.1016/j.jece.2019.103501

[25] Castaneda-Juáarez M, Martínez-Miranda V, Almazan-Sánchez PT, Linares-Hernández I, Santoyo-Tepole F,

Vázquez-Mejía G. Synthesis of TiO_2 catalysts doped with Cu, Fe, and Fe/Cu supported on clinoptilolite zeolite by an electrochemical-thermal method for the degradation of diclofenac by heterogeneous photocatalysis. Journal of Photochemistry and Photobiology A: Chemistry. 2019;380:1-10. DOI:10.1016/j.jphotochem.2019.04.045

[26] Nie G, Xiao L. New insight into wastewater treatment by activation of sulfite with photosensitive organic dyes under visible light irradiation. Chemical Engineering Journal. 2020;389:1-9. DOI:10.1016/j.cej.2019.123446

[27] Pandoli O, Rosso TD, Santos VM, de Siqueira Rezende R, Marinkovic BA. Prototyping of photocatalytic microreactor and testing of photodegradation of organic dye. Química Nova. 2015;38(6):859-863. DOI:10.5935/0100-4042.20150079

[28] Pereira VJ, Fernandes D, Carvalho G, Benoliel MJ, Romão MVS, Crespo MTB. Assessment of the presence and dynamics of fungi in drinking water sources using cultural and molecular methods. Water Research. 2010;44(17):4850-4859. DOI:10.1016/j.watres.2010.07.018

[29] Díez AM, Moreira FC, Marinho BA, Espíndola JCA, Paulista LO, Sanroman MA, Pazos M, Boaventura RAR, Vilar VJP. A step forward in heterogeneous photocatalysis: Process intensification by using a static mixer as catalyst support. Chemical Engineering Journal. 2018;343:597-606. DOI:10.1016/j.cej.2018.03.041

[30] Gonçalves NPF, Paganini MC, Armillotta P, Cerrato E, Calza P. The effect of cobalt doping on the efficiency of semiconductor oxides in the photocatalytic water remediation. Journal of Environmental Chemical Engineering. 2019;7(6):1-16. DOI:10.1016/j.jece.2019.103475

[31] Cambrussi ANCO, de Sena Neto LR, da Silva Filho EC, Osajima JAF, Ribeiro AB. Heterogeneous photocatalysis using TiO_2 in suspension applied to antioxidant activity assays. Materials Today: Proceedings. 2019;14:648-655. DOI:10.1016/j.matpr.2019.02.002

[32] Zerjav G, Albreht A, Vovk I, Pintar A. Revisiting terephthalic acid and coumarin as probes for photoluminescent determination of hydroxyl radical formation rate in heterogeneous photocatalysis. Applied Catalysis A: General. 2020;598:1-7. DOI:10.1016/j.apcata.2020.117566

[33] Gaya UI, Abdullah AH. Heterogeneousphotocatalyticdegradation of organic contaminants over titanium dioxide: A review of fundamentals, progress and problems. Journal of Photochemistry and Photobiology C: Photochemistry Reviews. 2008;9(1):1-12. DOI:10.1016/j.jphotochemrev.2007.12.003

[34] Kunduru KR, Nazarkovsky M, Farah S, Pawar RP, Basu A, Domb AJ. Nanotechnology for water purification: applications of nanotechnology methods in wastewater treatment. In: Grumezescu AM, editor. Water Purification. Amsterdan: Elsevier; 2017. 33-74. DOI:https://doi.org/10.1016/B978-0-12-804300- 4.00002-2

[35] Patel SG, Yadav NR, Patel SK. Evaluation of Degradation Characteristics of Reactive Dyes by UV/Fenton, UV/Fenton/Activated Charcoal, and UV/Fenton/TiO_2 Processes: A Comparative Study. Separation Science and Technology. 2013;48(12):1788-1800. DOI:10.1080/01496395.2012.756035

[36] Da Silva WL, Lansarin MA, Stedlie FC, Dos Santos JH. The potential of chemical industrial and academic wastes as a source of supported photocatalysts. Journal of Molecular

Catalysis. A, Chemical. 2014; 393: 125-133. DOI: 10.1016/j.molcata.2014.05.040

[37] Apostila do Curso da Escola Piloto: Técnicas de Controle Ambiental em efluentes líquidos – Processos Oxidativos Avançados. Rio de Janeiro: Universidade Federal do Rio de Janeiro; 1993

[38] Gehrke I, Geiser A, Somborn-Schulz A. Innovations in nanotechnology for water treatment. Nanotechnology, Science and Applications. 2015;8:1-17. DOI:10.2147/NSA.S43773

[39] Lazar M, Varghese S, Nair S. Photocatalytic Water Treatment by Titanium Dioxide: Recent Updates. Catalysts. 2012;2(4):572-601. DOI: https://doi.org/10.3390/catal2040572

[40] Aani SA, Gomez V, Wright CJ, Hilal N. Fabrication of antibacterial mixed matrix nanocomposite membranes using hybrid nanostructure of silver coated multi-walled carbon nanotubes. Chemical Engineering Journal. 2017;326:721-736. DOI: https://doi.org/10.1016/j.cej.2017.06.029

[41] Zabihi-Mobarakeh H, Nezamzadeh-Ejhieh A. Application of supported TiO_2 onto Iranian clinoptilolite nanoparticles in the photodegradation of mixture of aniline and 2, 4-dinitroaniline aqueous solution. Journal of Industrial and Engineering Chemistry. 2015;26:315-321. DOI:10.1016/j.jiec.2014.12.003

[42] Zou Y, Huang H, Li S, Wang J, Zhang Y. Synthesis of supported Ag/AgCl composite materials and their photocatalytic activity. Journal of Photochemistry and Photobiology A: Chemistry. 2019;376:43-53. DOI:10.1016/j.jphotochem.2019.03.00

[43] Yang W, Hu W, Zhang J, Wang W, Cai R, Pan M, Huang C, Chen X, Yan B, Zeng H. Tannic acid/Fe^{3+}functionalized magnetic graphene oxide nanocomposite with high loading of silver nanoparticles as ultra-efficient catalyst and disinfectant for wastewater treatment. Chemical Engineering Journal. 2021;405:1-10. DOI:10.1016/j.cej.2020.126629

[44] Kan Q, Lu K, Dong S, Shen D, Huang Q, Tong Y, Wu W, Gao S, Mao L. Transformation and removal of imidacloprid mediated by silver ferrite nanoparticle facilitated peroxymonosulfate activation in water: Reaction rates, products, and pathways. Environmental Pollution. 2020;267:1-13. DOI:10.1016/j.envpol.2020.115438

[45] Cheng Z, Guan H, Meng J, Wang X. Dual-Functional Porous Wood Filter for Simultaneous Oil/Water Separation and Organic Pollutant Removal. ACS Omega. 2020;5(23):14096-14103. DOI:10.1021/acsomega.0c01606

[46] Yang R, Zhao Q, Liu B. Two-step method to prepare the direct Z-scheme heterojunction hierarchical flowerlike $Ag@AgBr/Bi_2MoO_6$ microsphere photocatalysts for wastewater treatment under visible light. Journal of Materials Science: Materials in Electronics. 2020;31(7):5054-5067. DOI:10.1007/s10854-020-03040-3

[47] Bodaghi Z, Pakpour F, Ghanbari D. $Carbon@CoFe_2O_4@Ag$ and hollow $CoFe_2O_4@Ag$ nanocomposite: green synthesis of a photocatalyst and magnetic adsorbent for antibiotic removal from aqueous solutions. Journal of Materials Science: Materials in Electronics. 2020. DOI:10.1007/s10854-020-04439-8

[48] Mauro AD, Farrugia C, Abela S, Refalo P, Grech M, Falqui L, Nicotra G, Sfuncia G, Mio A, Buccheri MA, Rappazzo G, Brundo MV, Scalisi EM, Pecoraro RP, Iaria C, Privitera V, Impellizzeri G. Ag/ZnO/PMMA Nanocomposites for Efficient Water Reuse. ACS Applied Bio

Materials. 2020;3(7):4417-4426. DOI:10.1021/acsabm.0c00409

[49] Albukhari SM, Ismail M, Akhtar K, Danish EY. Catalytic reduction of nitrophenols and dyes using silver nanoparticles @ cellulose polymer paper for the resolution of wastewater treatment challenges. Colloids and Surfaces A: Physicochemical and Engineering Aspects. 2019;577:548-561. DOI:10.1016/j.colsurfa.2019.05.058

[50] Mazumder JA, Perwez M, Noori R, Sardar M. Development of sustainable and reusable silver nanoparticlecoated glass for the treatment of contaminated water. Environmental Science and Pollution Research. 2019;26(22):23070-23081. DOI:10.1007/s11356-019-05647-4

[51] Al-Sherbini ASA, Ghannam HEA, El-Ghanam GMA, El-Ella AA, Youssef AM. Utilization of chitosan/Ag bionanocomposites as eco-friendly photocatalytic reactor for Bactericidal effect and heavy metals removal. Heliyon. 2019;5(6):e01980. DOI:10.1016/j.heliyon.2019.e01980

[52] Das R, Sypu VS, Paumo HK, Bhaumik M, Maharaj V, Maity A. Silver decorated magnetic nanocomposite (Fe_3O_4@PPy-MAA/Ag) as highly active catalyst towards reduction of 4-nitrophenol and toxic organic dyes. Applied Catalysis B: Environmental. 2019;244:546-558. DOI:10.1016/j.apcatb.2018.11.073

[53] Pratheesya T, Harish S, M N, Sohila S, Ramesh R. Enhanced antibacterial and photocatalytic activities of silver nanoparticles anchored reduced graphene oxide nanostructure. Materials Research Express. 2019;6(7):1-25. DOI:10.1088/2053-1591/ab1567

[54] Rasool K, Lee DS. Inhibitory Effects of silver nanoparticles on removal of organic pollutants and sulfate in an anaerobic biological wastewater treatment process. Journal of Nanoscience and Nanotechnology. 2016;16(5):4456-4463. DOI:10.1166/jnn.2016.10984

[55] Casa M, Sarno M, Cirillo C, Ciambelli P. Reduced graphene oxide-based silver nanoparticle-containing natural hydrogel as highly efficient catalysts for nitrile wastewater treatment. Chemical Engineering Transactions. 2016;47:307-312. DOI:10.3303/CET1647052

[56] Chen X, Cen C, Tang Z, Zeng W, Chen D, Fang P, Chen Z. The key role of pH value in the synthesis of titanate nanotubes-loaded manganese oxides as a superior catalyst for the selective catalytic reduction of NO with NH_3. Journal of Nanomaterials. 2013;2013:1-7. DOI:10.1155/2013/871528

[57] Mahardika T, Putri NA, Putri AE, Fauzia V, Roza L, Sugihartono I, Herbani Y. Rapid and low temperature synthesis of Ag nanoparticles on the ZnO nanorods for photocatalytic activity improvement. Results in Physics. 2019;13:1-9. DOI:10.1016/j.rinp.2019.102209

[58] Booshehri AY, Goh SCK, Hong J, Jiang R, Xu R. Effect of depositing silver nanoparticles on BiVO4in enhancing visible light photocatalytic inactivation of bacteria in water. J Mater Chem A. 2014;2(17):6209-6217. DOI:10.1039/c3ta15392d

[59] Radhu S. Photocatalytic degradation of textile dye molecules by Ag@Au core–shell nanoparticles. Materials Today: Proceedings. 2020;25:285-288. DOI:10.1016/j.matpr.2020.01.413

[60] Burdusel AC, Gherasim O, Grumezescu AM, Mogoanta L, Ficai A, Andronescu E. Biomedical Applications of Silver Nanoparticles: An Up-to-Date Overview. Nanomaterials.

2018;8(9):681. DOI:10.3390/nano8090681

[61] Nedelcu IA, Ficai A, Sonmez M, Ficai D, Oprea O, Andronescu E. Silver Based Materials for Biomedical Applications. Current Organic Chemistry. 2014;18(2):173-184. DOI:10.2174/13852728113176660141

[62] Ivask A, Juganson K, Bondarenko O, Mortimer M, Aruoja V, Kasemets K, et al. Mechanisms of toxic action of Ag, ZnO and CuO nanoparticles to selected ecotoxicological test organisms and mammalian cells in vitro: A comparative review. Nanotoxicology. 2013;8(sup1):57-71. DOI:10.3109/17435390.2013.855831

[63] Kumar SSD, Rajendran NK, Houreld NN, Abrahamse H. Recent advances on silver nanoparticle and biopolymer-based biomaterials for wound healing applications. International Journal of Biological Macromolecules. 2018;115:165-175. DOI:10.1016/j.ijbiomac.2018.04.003

[64] Brennan SA, Fhoghlú CN, Devitt BM, O'Mahony FJ, Brabazon D, Walsh A. Silver nanoparticles and their orthopaedic applications. The Bone & Joint Journal. 2015;97-B(5):582-589. DOI:10.1302/0301- 620x.97b5.33336

[65] Marassi V, Cristo LD, Smith SGJ, Ortelli S, Blosi M, Costa AL, Reschiglian P, Volkov Y, Prina-Melo A. Silver nanoparticles as a medical device in healthcare settings: a five-step approach for candidate screening of coating agents. Royal Society Open Science. 2018;5(1):1-21. DOI:10.1098/rsos.171113

[66] Pala R, Zeng Y, Pattnaik S, Busi S, Alomari N, Nauli SM, Liu G. Functionalized silver nanoparticles for sensing, molecular imaging and therapeutic applications. Current Nanomedicine. 2019;8(3):234-250. DOI:10.2174/2468187308666180508144919

[67] Mao BH, Chen ZY, Wang YJ, Yan SJ. Silver nanoparticles have lethal and sublethal adverse effects on development and longevity by inducing ROS-mediated stress responses. Scientific Reports. 2018;8(1). DOI:10.1038/s41598-018-20728-z

[68] Pauksch L, Hartmann S, Rohnke M, Szalay G, Alt V, Schnettler R, Lips KS. Biocompatibility of silver nanoparticles and silver ions in primary human mesenchymal stem cells and osteoblasts. Acta Biomaterialia. 2014;10(1):439-449. DOI:10.1016/j.actbio.2013.09.037

[69] Ibrahim EH, Kilany M, Ghramh HA, Khan KA, Ul Islam S. Cellular proliferation/cytotoxicity and antimicrobial potentials of green synthesized silver nanoparticles (AgNPs) using Juniperus procera. Saudi Journal of Biological Sciences. 2019;26(7):1689-1694. DOI:10.1016/j.sjbs.2018.08.014

[70] Elia P, Zach R, Hazan S, Kolusheva S, Porat Z, Zeiri Y. Green synthesis of gold nanoparticles using plant extracts as reducing agents. International journal of nanomedicine. 2014;9:4007-4021. DOI:10.2147/IJN.S57343

[71] Tarannum N, Divya D, Gautam YK. Facile green synthesis and applications of silver nanoparticles: a state-of-the-art review. RSC Advances. 2019;9(60):34926-34948. DOI:10.1039/c9ra04164h

[72] Srikar SK, Giri DD, Pal DB, Mishra PK, Upadhyay SN. Green synthesis of silver nanoparticles: A review. Green and Sustainable Chemistry. 2016;06(01):34-56. DOI:10.4236/gsc.2016.61004

[73] Kyriacou SV, Brownlow WJ, Xu XHN. Using nanoparticle optics assay for direct observation of the function of antimicrobial agents in single live bacterial cells. Biochemistry.

2004;43(1):140-147. DOI:10.1021/bi0351110

[74] Kim JS, Kuk E, Yu KN, Kim JH, Park SJ, Lee HJ, Kim SH, Park YK, Park, YH, Hwang C, Kim Y, Lee Y, Jeong H, Cho M. Antimicrobial effects of silver nanoparticles. Nanomedicine: Nanotechnology. Biology and Medicine. 2007;3(1):95-101. DOI:10.1016/j.nano.2006.12.001

[75] Sahar MO. Some Nanoparticles effects on proteus sp. and klebsiella sp. isolated from water. American Journal of Infectious Diseases and Microbiology. 2014;2(1):4-10. DOI:10.12691/ajidm-2-1-2

[76] Jain D, Daima H, Kachhwala S, Kothari S. Synthesis of plant-mediated silver nanoparticles using papaya fruit extract and evaluation of their Anti Microbial Activities. Digest Journal of Nanomaterials and Biostructures. 2009;4:557-563

[77] Kvítek L, Panácek A, Soukupová J, Kolár M, Vecerovpa R, Prucek R, Holecová M, Zbořil R. Effect of surfactants and polymers on stability and antibacterial activity of silver nanoparticles (NPs). The Journal of Physical Chemistry C. 2008;112(15):5825-5834. DOI:10.1021/jp711616v

[78] Khan SU, Anjum SI, Ansari MJ, Khan MHU, Kamal S, Rahman K, Shoaib M, Man S, Khan AJ, Khan SU, Khan D. Antimicrobial potentials of medicinal plant's extract and their derived silver nanoparticles: A focus on honey bee pathogen. Saudi Journal of Biological Sciences. 2019;26(7):1815-1834. DOI:10.1016/j.sjbs.2018.02.010

[79] Sanjivkumar M, Vaishnavi R, Neelakannan M, Kannan D, Silambarasan T, Immanuel G. Investigation on characterization and biomedical properties of silver nanoparticles synthesized by an actinobacterium Streptomyces olivaceus (MSU3). Biocatalysis and Agricultural Biotechnology. 2019;17:151-159. DOI:10.1016/j.bcab.2018.11.014

[80] Scheringer M, MacLeod M, Behra R, Sigg L, Hungerbühler K. Environmental risks associated with nanoparticulate silver used as biocide. Household Pers Care Today. 2010;1:34-37

[81] MubarakAli D, Thajuddin N, Jeganathan K, Gunasekaran M. Plant extract mediated synthesis of silver and gold nanoparticles and its antibacterial activity against clinically isolated pathogens. Colloids and Surfaces B: Biointerfaces. 2011;85(2):360-365. DOI:10.1016/j.colsurfb.2011.03.009

[82] Munir H, Mumtaz A, Rashid R, Najeeb J, Zubair MT, Munir S, Bilal M, Cheng H. Eucalyptus camaldulensis gum as a green matrix to fabrication of zinc and silver nanoparticles: Characterization and novel prospects as antimicrobial and dye-degrading agents. Journal of Materials Research and Technology. 2020;9(6):15513-15524. DOI:10.1016/j.jmrt.2020.11.026

[83] Rasheed T, Bilal M, Iqbal HMN, Li C. Green biosynthesis of silver nanoparticles using leaves extract of Artemisia vulgaris and their potential biomedical applications. Colloids and Surfaces B: Biointerfaces. 2017;158:408-415. DOI:10.1016/j.colsurfb.2017.07.020

[84] Nakkala JR, Mata R, Gupta AK, Sadras SR. Biological activities of green silver nanoparticles synthesized with Acorous calamus rhizome extract. European Journal of Medicinal Chemistry. 2014;85:784-794. DOI:10.1016/j.ejmech.2014.08.024

[85] Zargar M, Hamid AA, Bakar FA, Shamsudin MN, Shameli K, Jahanshiri F, Farahani F. Green Synthesis and antibacterial effect of silver nanoparticles using vitex

negundo L. Molecules. 2011;16(8):6667-6676.doi:10.3390/molecules16086667

[86] Khan ZUH, Shah NS, Iqbal J, Khan AU, Imran M, Alshehri SM, et al. Biomedical and photocatalytic applications of biosynthesized silver nanoparticles: Ecotoxicology study of brilliant green dye and its mechanistic degradation pathways. Journal of Molecular Liquids. 2020;319:114114. DOI:10.1016/j.molliq.2020.114114

[87] Unser S, Bruzas I, He J, Sagle L. Localized Surface Plasmon Resonance Biosensing: Current Challenges and Approaches. Sensors. 2015;15(7):15684-15716. DOI:10.3390/s150715684

[88] Cao Z, He P, Huang T, Yang S, Han S, Wang X, Ding G. Plasmonic Coupling of AgNPs near Graphene Edges: A Cross-Section Strategy for High-Performance SERS Sensing. Chemistry of Materials. 2020;32(9):3813-3822. DOI:10.1021/acs.chemmater.9b05293

[89] Ashraf JM, Ansari MA, Khan HM, Alzohairy MA, Choi I. Green synthesis of silver nanoparticles and characterization of their inhibitory effects on AGEs formation using biophysical techniques. Scientific Reports. 2016;6(1). DOI:10.1038/srep20414

[90] Loiseau A, Asila V, Boitel-Aullen G, Lam M, Salmain M, Boujday S. Silver-Based Plasmonic Nanoparticles for and Their Use in Biosensing. Biosensors. 2019;9(2):78. DOI:10.3390/bios9020078

[91] Wang X, Hou T, Lin H, Lv W, Li H, Li F. In situ template generation of silver nanoparticles as amplification tags for ultrasensitive surface plasmon resonance biosensing of microRNA. Biosensors and Bioelectronics. 2019;137:82-87. DOI:10.1016/j.bios.2019.05.006

[92] Wang X, Wu L, Ren J, Miyoshi D, Sugimoto N, Qu X. Label-free colorimetric and quantitative detection of cancer marker protein using noncrosslinking aggregation of Au/Ag nanoparticles induced by target-specific peptide probe. Biosensors and Bioelectronics. 2011;26(12):4804-4809. DOI:10.1016/j.bios.2011.06.012

[93] Bahavarnia F, Pashazadeh-Panahi P, Hasanzadeh M, Razmi N. DNA based biosensing of Acinetobacter baumannii using nanoparticles aggregation method. Heliyon. 2020;6(7):1-6. DOI:10.1016/j.heliyon.2020.e04474

[94] Wang WF, Qiang Y, Meng XH, Yang JL, Shi YP. Ultrasensitive colorimetric assay melamine based on in situ reduction to formation of CQDs-silver nanocomposite. Sensors and Actuators B: Chemical. 2018;260:808-815. DOI:10.1016/j.snb.2018.01.108

[95] Rajamanikandan R, Ilanchelian M. Simple and visual approach for highly selective biosensing of vitamin B1 based on glutathione coated silver nanoparticles as a colorimetric probe. Sensors and Actuators B: Chemical. 2017;244:380-386. DOI:10.1016/j.snb.2016.12.129

[96] Guo Y, Wu J, Li J, Ju H. A plasmonic colorimetric strategy for biosensing through enzyme guided growth of silver nanoparticles on gold nanostars. Biosensors and Bioelectronics. 2016;78:267-273. DOI:10.1016/j.bios.2015.11.056

[97] Barghouti ME, Akjouj A, Mir A. Design of silver nanoparticles with graphene coatings layers used for LSPR biosensor applications. Vacuum. 2020;180:1-23. DOI:10.1016/j.vacuum.2020.109497

[98] Tang J, Xiong P, Cheng Y, Chen Y, Peng S, Zhu ZQ. Enzymatic oxydate-triggered AgNPs etching: A novel signal-on photoelectrochemical

immunosensing platform based on Ag@AgCl nanocubes loaded RGO plasmonic heterostructure. Biosensors and Bioelectronics. 2019;130:125-131. DOI:10.1016/j.bios.2019.01.014

[99] Wang Y, Li Z, Li H, Vuki M, Xu D, Chen HY. A novel aptasensor based on silver nanoparticle enhanced fluorescence. Biosensors and Bioelectronics. 2012;32(1):76-81. DOI:10.1016/j.bios.2011.11.030

[100] Sangappa Y, Latha S, Asha S, Sindhu P, Parushuram N, Shilpa M, Byrappa K, Narayana B. Synthesis of anisotropic silver nanoparticles using silk fibroin: characterization and antimicrobial properties. Materials Research Innovations. 2017;23(2):79-85. DOI:10.1080/14328917.2017.1383680

[101] Wu Q, Si M, Zhang B, Zhang K, Li H, Mi L, et al. Strong damping of the localized surface plasmon resonance of Ag nanoparticles by Ag2O. Nanotechnology. 2018;29(29):1-18. DOI:10.1088/1361- 6528/aac031

[102] Mlalila N, Swai H, Hilonga A, Kadam D. Antimicrobial dependence of silver nanoparticles on surface plasmon resonance bands against Escherichia coli. Nanotechnology, Science and Applications. 201610:1-9. DOI:10.2147/nsa.s123681

[103] Ribeiro MS, de Melo LSA, Farooq S, Baptista A, Kato IT, Núñez SC, et al. Photodynamic inactivation assisted by localized surface plasmon resonance of silver nanoparticles: In vitro evaluation on *Escherichia coli* and *Streptococcus mutans*. Photodiagnosis and Photodynamic Therapy. 2018;22:191-196. DOI:10.1016/j.pdpdt.2018.04.007

[104] Louis C, Pluchery O. Gold Nanoparticles for Physics, Chemistry and Biology. 1st ed. London: Imperial College Press; 2012. 308 p. DOI:10.1142/p815

[105] Ahn H, Song H, ryul Choi J, Kim K. A localized surface plasmon resonance sensor using double-metal-complex nanostructures and a review of recent approaches. Sensors. 2017;18(2):1-20. DOI:10.3390/s18010098

[106] Caro C, M P, Klippstein R, Pozo D, P A. Silver Nanoparticles: Sensing and Imaging Applications. In: Perez DP, editor. Silver Nanoparticles. London: InTech; 2010. p. 201-223. DOI:10.5772/8513

[107] Liu B, Zhuang J, Wei G. Recent advances in the design of colorimetric sensors for environmental monitoring. Environmental Science: Nano. 2020;7(8):2195-2213. DOI:10.1039/d0en00449a

[108] Anker JN, Hall WP, Lyandres O, Shah NC, Zhao J, Van Duyne RP. Biosensing with plasmonic nanosensors. In: Rodgers P, editor. Nanoscience and Technology: A Collection of Reviews from Nature Journals. CoPublished with Macmillan Publishers Ltd, UK; 2010. p. 308-319. DOI:10.1142/9789814287005 0032

[109] Oh SY, Heo NS, Bajpai VK, Jang SC, Ok G, Cho Y, et al. Development of a Cuvette-Based LSPR Sensor Chip Using a Plasmonically Active Transparent Strip. Frontiers in Bioengineering and Biotechnology. 2019;7:1-11. DOI:10.3389/fbioe.2019.00299

[110] Bastos V, de Oliveira JMPF, Brown D, Jonhston H, Malheiro E, da Silva ALD, et al. The influence of Citrate or PEG coating on silver nanoparticle toxicity to a human keratinocyte cell line. Toxicology Letters. 2016;249:29-41. DOI:10.1016/j.toxlet.2016.03.005

[111] Pinzaru I, Coricovac D, Dehelean C, Moacă EA, Mioc M, Baderca F, et al. Stable PEG-coated silver nanoparticles – A comprehensive toxicological profile. Food and

Chemical Toxicology. 2018;111:546-556. DOI:10.1016/j.fct.2017.11.051

[112] Devi JM, Umadevi M. Synthesis and Characterization of Silver–PVA Nanocomposite for Sensor and Antibacterial Applications. Journal of Cluster Science. 2013;25(2):639-650. DOI:10.1007/s10876-013- 0660-6

[113] Haberl N, Hirn S, Wenk A, Diendorf J, Epple M, Johnston BD, et al. Cytotoxic and proinflammatory effects of PVP-coated silver nanoparticles after intratracheal instillation in rats. Beilstein Journal of Nanotechnology. 2013;4:933-940. DOI:10.3762/bjnano.4.105

[114] Ban DK, Paul S. Rapid colorimetric and spectroscopy based sensing of heavy metal and cellular free oxygen radical by surface functionalized silver nanoparticles. Applied Surface Science. 2018;458:245-251. DOI:10.1016/j.apsusc.2018.07.069

[115] Faghiri F, Ghorbani F. Colorimetric and naked eye detection of trace Hg^{2+} ions in the environmental water samples based on plasmonic response of sodium alginate impregnated by silver nanoparticles. Journal of Hazardous Materials. 2019;374:329-340. DOI:10.1016/j.jhazmat.2019.04.052

[116] Ranjani B, Pandian K, Kumar GA, Gopinath SCB. D-glucosamine chitosan base molecule-assisted synthesis of different shape and sized silver nanoparticles by a single pot method: A greener approach for sensor and microbial applications. International Journal of Biological Macromolecules. 2019;133:1280-1287. DOI:10.1016/j.ijbiomac.2019.04.196

[117] Rezaei ZB, Rastegarzadeh S, Kiasat A. In-situ decorated silver nanoparticles on electrospun poly (vinyl alcohol)/chitosan nanofibers as a plasmonic sensor for azathioprine determination. Colloids and Surfaces A: Physicochemical and Engineering Aspects. 2018;559:266-274. DOI:10.1016/j.colsurfa.2018.09.047

[118] Yilmaz H, Cobandede Z, Yilmaz D, Cinkilic A, Culha M, Demiralay EC. Monitoring forced degradation of drugs using silica coated AgNPs with surface-enhanced Raman scattering. Talanta. 2020;214:1-8. DOI:10.1016/j.talanta.2020.120828

[119] Sharma VK, Yngard RA, Lin Y. Silver nanoparticles: Green synthesis and their antimicrobial activities. Advances in Colloid and Interface Science. 2009;145(1-2):83-96. DOI:10.1016/j.cis.2008.09.002

[120] Kurbanoglu S, Ozkan SA, Merkoc¸i A. Nanomaterials-based enzyme electrochemical biosensors operating through inhibition for biosensing applications. Biosensors and Bioelectronics. 2017;89:886-898. DOI:10.1016/j.bios.2016.09.102

[121] Meng F, Sun H, Huang Y, Tang Y, Chen Q, Miao P. Peptide cleavage-based electrochemical biosensor coupling graphene oxide and silver nanoparticles. Analytica Chimica Acta. 2019;1047:45-51. DOI:10.1016/j.aca.2018.09.053

[122] Chisanga M, Muhamadali H, Ellis D, Goodacre R. Enhancing Disease Diagnosis: Biomedical Applications of Surface-Enhanced Raman Scattering. Applied Sciences. 2019;9(6):1163. DOI:10.3390/app9061163

[123] Kucherenko IS, Soldatkin OO, Kucherenko DY, Soldatkina OV, Dzyadevych SV. Advances in nanomaterial application in enzyme-based electrochemical biosensors: a review. Nanoscale Advances. 2019;1(12):4560-4577. DOI:10.1039/c9na00491b

Section 3

Silver Nanostructures for SERS and Photoelectrochemical Applications

Chapter 10

Theoretical and Experimental Study on the Functionalization Effect on the SERS Enhancement Factor of SiO_2-Ag Composite Films

Paulina De León Portilla, Ana Lilia González Ronquillo and Enrique Sánchez Mora

Abstract

Herein we addressed a study to determine the enhancement factor (EF) of the Raman signal reached by composite films with two main components, Ag nanoparticles and SiO_2 spheres. The study involves the synthesis, structural composition and optical response by using experimental techniques and theoretical-numerical modeling. A colloid with single NPs and agglomerates of them, with a tannic acid layer on its surface, was produced. Separately, porous SiO_2 spheres were obtained. A mixture of both, Ag NPs and SiO_2 particles was used to produce the films by solvent evaporation method. It is shown that single or agglomerated Ag NPs are preferentially located at the interstices of the SiO_2 spheres. Using discrete dipole approximation, the SERS EF has been estimated considering the agglomeration and tannic acid layer. Both, the dielectric spheres and tannic acid layer diminish the electric field intensity and therefore the SERS EF. When a Ag NP with/without a dielectric shell is touching a SiO_2 sphere, the EF is as high as 1×10^3, the zones where this value is reached are smaller when the dielectric layer is present. With a cluster of 3 nude Ag NPs surrounded by SiO_2 spheres an EF of 2.4×10^3 is obtained.

Keywords: SiO_2 spheres, silver nanoparticles, composite film, tannic acid, functionalization, near field intensity, discrete dipole approximation

1. Introduction

Research around the optical properties of Ag nanoparticles (NPs) has been a topic of high importance for diverse applications as coating, plasmonic antennas, for drug delivery, as components of molecular detectors, among others [1–6]. The convenience of using Ag NPs lies in their intense plasmonic response with a surface plasmon resonance (SPR) in the visible optical range. The SPR is a phenomenon that occurs when a metal NP is excited by an electromagnetic wave, the conduction electrons respond to the electromagnetic field in such a way that oscillations of the electron charge distribution occur in the vicinity or on the surface of the particle. Ag NPs have their SPR in the range from 350 to 600 nm, depending on their size and morphology [7, 8].

IntechOpen

To improve the colloidal stability and surface properties of Ag NPs the functionalization is a feasible option. This chemical process modifies the NP's surface charge using a specific chemical compound that successfully encapsulates the NP, attributing the surface with a neutral, negative or positive character [9]. Once NPs are functionalized can be used for different purposes, for example, to selectively adsorb drugs to later release them in a controlled way [3], to inhibit the growth of bacteria [5], to prevent agglomeration in a colloid [10], among others.

Tannic acid is commonly used in synthesis methods as a stabilizer, it is a typical hydrolyzable tannin derived from plants with various biological properties such as antioxidant, antitumor, antimutagenic, antimicrobial and anti-inflammatory, besides its ability to interact with proteins [11]. The structure of tannic acid is made up of tannins that contain digalloyl ester groups connected to a central glucose nucleus. These groups can interact ionically with others, also can interact through hydrogen bonding with alike or different molecules, and moreover bond to a metal [12].

Silicon dioxide, SiO_2, is considered a useful material for coating because of its physical and chemical properties such as chemical stability, optical transparency, biocompatibility, inert-material character and reactivity with various coupling agents [13, 14]. Specifically, amorphous SiO_2 substrates have the advantage of a very high specific surface area. In addition, they promote the dispersion and homogeneous distribution of metallic NPs on it [14]. The last, seems promising for Surface Enhanced Raman Scattering (SERS) based-metal substrates purposes. The SERS enhancement factor (EF) that can be reached is directly related to the SPR of the used NPs through the $|\boldsymbol{E}|^4$ approximation.

To determine the electric field intensity on the NP's surface and away from it, exist various options, some of the most common are based on finite elements methods, finite difference time domain or volume integral equations. In particular, Discrete Dipole Approximation (DDA) is a flexible integral equation method that allows to study the optical response of targets with a size smaller or similar to that of the incident wavelength. With this method is possible to study isolated or periodic targets with arbitrary shapes [15, 16]. On the other hand, this method has been widely used to study the absorbed and scattered light by several systems as nanoparticles, bacteria, films, red blood cells, among others [17, 18].

Along this chapter we present a study of films composed by Ag NPs and SiO_2 spheres. The films are proposed as SERS substrates with the characteristic of Ag NPs very well located at the interstices of the silica spheres. It is worth mentioning that the NPs have a tannic acid layer on its surface. Through a combination of experimental techniques and theoretical-numerical modeling a study of structural and optical properties has been realized, in addition the SERS EF has been estimated. The content of the chapter is as follows. In section 2 the methodology followed for the synthesis of Ag NPs, SiO_2 spheres and composite films is addressed. In sections 3 and 4, the determination of size and morphology of Ag NPs is presented, and the presence of a layer of tannic acid on its surface is also shown. Subsequently, in section 5 and 6 the structural composition of the SiO_2 spheres and composite films is shown, respectively. Section 7 contains the details of the methodology followed to estimate the SERS EF of systems made of Ag NPs and SiO_2 spheres. Finally, the conclusions are stated.

2. Synthesis methodology

2.1 Ag NPs synthesis

Ag NPs were synthesized using the seed growth method reported by Bastús et al. [19], they reported an experimental methodology para obtainer Ag NP de 4 a

200 nm de diameter. The methodology describes different conditions to obtain a specific size. In essence, the required size is achieved from the number of growth steps. Here, Ag seeds of approximately 35 nm in diameter were obtained as follows. 50 ml of 0.1 mM tannic acid (Sigma Aldrich, 98%) and 50 ml of 5 mM sodium citrate (J. T. Baker, 96%) were mixed and then the solution was heated up to 90 ° C. Subsequently, 1 ml of 25 mM silver nitrate (Riedel-de Haën, 98%) was added and the solution was kept stirring for 30 minutes. At this point, to eliminate secondary products, the solution was centrifuged at 20,000 rpm, for 15 minutes. The supernatant solution was separated and subsequently the Ag seeds were redispersed in 20 ml of deionized water. The elimination of secondary products was repeated 3 times. The final solution is named the seed solution.

In order to obtain Ag NPs of 100 nm of diameter, a 4-stage growth procedure was followed. The first stage consisted of diluting 20 ml of the seed solution (0.21 M) with 16 ml of deionized water and then heated up at 90 °C. Subsequently, 500 μl of 25 mM sodium citrate, 1.5 ml of 2.5 mM tannic acid and 1 ml of 25 mM silver nitrate were added. This stage ends after 30 minutes of reaction maintaining the temperature at 90 °C. As a second stage, the same amounts of sodium citrate, tannic acid and silver nitrate were added keeping the same reaction conditions. Third and fourth stages were alike than the second one. The elimination process of secondary products was carried out according to the one indicated above, with the difference that a centrifugation velocity of 18,000 rpm was used.

2.2 SiO_2 spheres synthesis

A common method to synthesize SiO_2 spheres is the Stöber method [20], which consists of the hydrolysis of tetraethylorthosilicate (TEOS) with an alcohol as solvent, and an hydrolysis catalyst. With this method polydisperse SiO_2 colloids are obtained. On the other hand, Razo and collaborators [21] obtained monodispersed SiO_2 microspheres with characteristic sizes between 150 and 600 nm in diameter. In their work they reported the relation between the sphere diameter of SiO_2 as a function of the moles of TEOS. During all the study the number of moles of NH_3/H_2O/C_2H_5OH was fixed. Using 0.46 moles of NH_3, 2.89 moles of H_2O and 2.15 moles of C_2H_5OH (relation A) they obtained microspheres with size from 150 nm to 600 nm in diameter. Whereas using 0.41 moles of NH_3, 2.20 moles of H_2O and 2.15 moles of C_2H_5OH (relation B) microspheres from 350 nm to 600 nm were produce.

Herein we synthesize SiO_2 spheres of 300 nm based on relation A. Separately, two solutions were prepared: the first one consisted of 9 ml of TEOS (Sigma Aldrich, 98%) and 125 ml of ethanol (Sigma Aldrich, 80%). The second solution was formed from 52 ml of deionized water and 30 ml of ammonium hydroxide (J. T. Baker, 64%). Later, both solutions were mixed and the reaction was maintained for 3 hours, under constant stirring at room temperature. For the purification of the SiO_2 spheres, the same washing procedure (elimination of the secondary products) used for the Ag NPs was followed.

2.3 SiO_2-Ag composite film deposition

To obtain the films of the SiO_2-Ag composites, the solvent evaporation method was used. For this, 25 ml of solution of SiO_2 spheres (5 mM) dispersed in ethanol were mixed with 5 ml of Ag NPs solution (3 mM) dispersed in water.

On the other hand, the glass substrate was cleaned with neutral soap in a sonic bath, dried with pressurized air and immersed for 1 hour in 100 ml of piranha solution (3: 1 of sulfuric acid and hydrogen peroxide). After this time, the substrate was rinsed with deionized water and dried with compressed air. The substrate was introduced into the colloid formed by microspheres of SiO_2 and Ag NPs. Subsequently, it was heated in a muffle at 70 °C for 24 hours.

3. Determination of the size and shape of Ag NPs

Figure 1 shows the Absorbance of the Ag NPs colloidal solution measured from the ultraviolet to the visible region. The maximum of the spectrum is located at 421 nm and the shape-line is asymmetrical, with a shoulder covering from 480 nm to 550 nm. To have an insight about the origin of the maximum and the shoulder we proceed to calculate the optical extinction efficiency (Q_{ext}) of a spherical Ag NP of various diameters. The Absorbance and extinction efficiency are directly related through the next expression:

$$A(\lambda) = CQ_{ext}(\lambda)\pi r^2 L/\ln(10), \quad (1)$$

where C is the concentration of Ag NPs (in particles/cm^3), L is the length of the sample (1 cm), r the radius of the NP and ln is the natural logarithm function. The linear relation between A and Q_{ext} indicates that both of them have the same spectral shape-line but different intensities. The Q_{ext} was calculated using the well-known Mie theory [22].

As we can see in the **Figure 1**, the Q_{ext} spectra that better fit to the Absorbance are those of a Ag NP with a diameter of 63 nm and other of 100 nm. This indicates that in the colloidal solution predominates the presence of Ag NPs of 63 nm and 100 nm. According to the intensity of the spectrum one can expect that the number of NPs with a size of 63 nm is larger than that with a size of 100 nm.

Apart, the size distribution was also determined using Dynamic Light Scattering (DLS), a technique that provides information of the hydrodynamic diameter of the entities in the solution. Under optimal particle concentration, that is, sufficiently low to disesteem particle-particle interaction and sufficiently high to have a strong signal, the hydrodynamic diameter of spherical particles matches to the physical diameter. From the size distribution a main value around 66 nm is observed, also

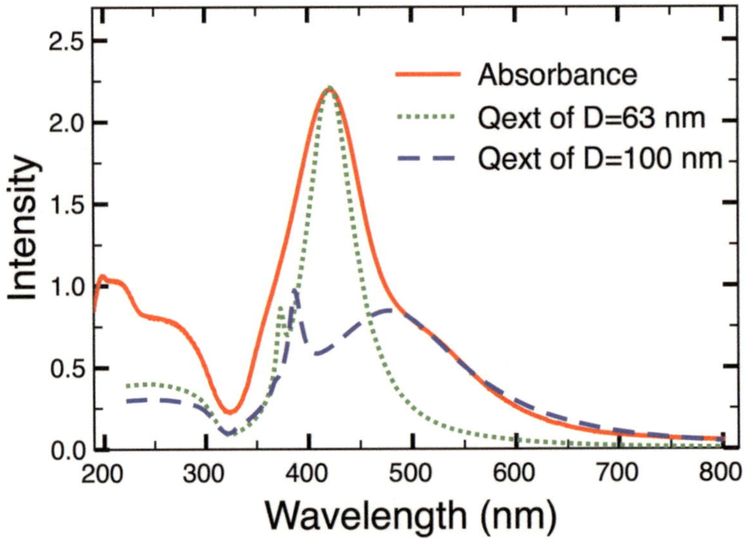

Figure 1.
Measured and calculated optical response of the Ag colloid. The calculated spectra were obtained using Mie theory for a spherical NP with a diameter D = 63 nm and other with a D = 100 nm.

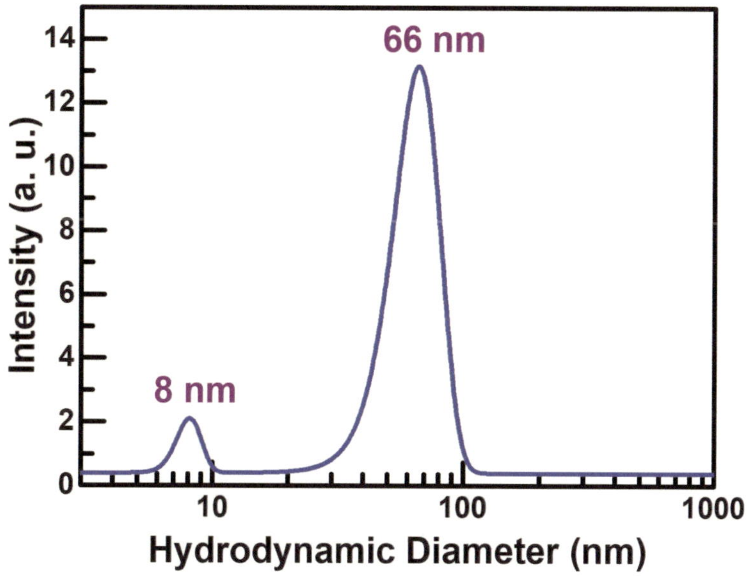

Figure 2.
Size distribution of the Ag NPs in solution, measured with DLS technique.

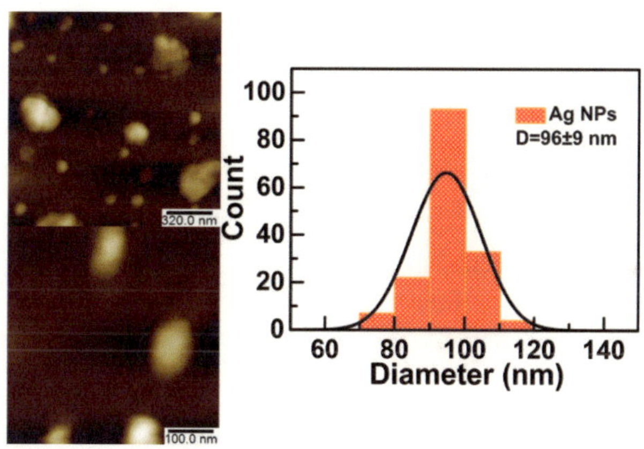

Figure 3.
Images by AFM showing single Ag NPs and agglomerates on a crystalline silicon wafer, the estimated mean value is 96 ± 9 nm.

the presence of NPs with a diameter of 100 nm is detected, see **Figure 2**. Therefore, there is a good agreement with the predicted calculations obtained with Mie theory and shown in **Figure 1**.

Small and big Ag NPs were also observed by using an Atomic Force Microscope (AFM), see **Figure 3**. However, the "big NPs" are actually quasi spherical agglomerates of Ag NPs with an average size of 96±9 nm, the statistic was done over 166 particles. This fact clarifies the deductions made from DLS and spectroscopy techniques, and Mie theory, the solution contains single Ag NPs and agglomerates of them.

4. Functionalization of Ag NPs

In **Figure 4** the zeta potential profile of the Ag NPs is shown, there a negative superficial charge of −52.5 mV is detected. To expose the origin of the negative

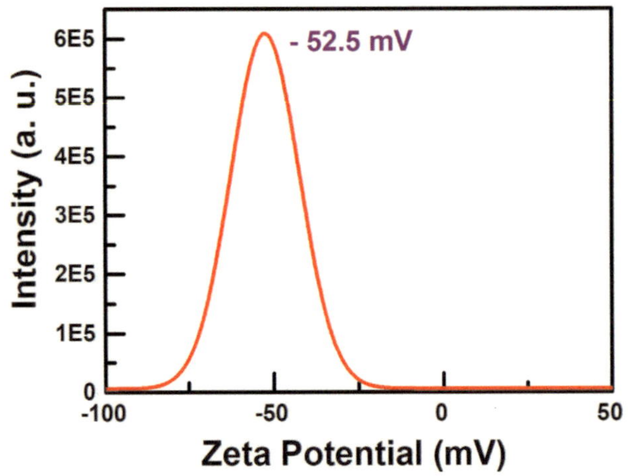

Figure 4.
Zeta potential of the Ag NPs in solution.

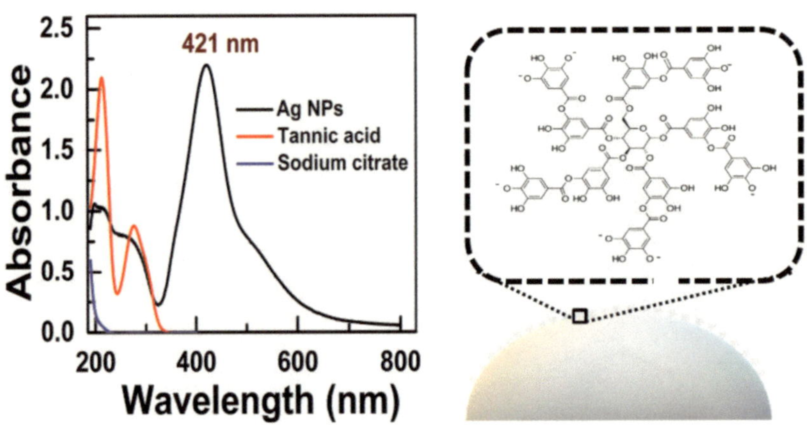

Figure 5.
Measured absorbance of the Ag NPs (black line), tannic acid (red line) and sodium citrate (blue line). To the right a scheme of the adsorption of tannic acid molecules on the NP surface is depicted.

superficial charge, a comparison among the optical spectrum of the Ag NPs, the tannic acid and the sodium citrate is presented in **Figure 5**. The last two were used as reducing and stabilizing agents during the synthesis procedure of the NPs. The band of the sodium citrate is out of the measured interval, whereas the optical response of the tannic acid is characterized by one band located at 214 and other at 268 nm. These two bands are well superimposed to the shape line of the Ag NPs spectrum in the interval of ultraviolet light. Therefore, the tannic acid is present in the colloid even after the several washing times. The negative charge of the NPs is explained as follows. Assuming the tannic acid molecules are adsorbed on the surface of the NPs, the negative charge detected is probably because of a partial deprotonation of the OH groups bonded to aromatic ring (see scheme in **Figure 5**). It is worth mentioning the fact that the electronic interband transitions of Ag take place at wavelengths lower than 320 nm, therefore they also contribute to the Absorbance in the UV range [23].

5. Determination of size and shape of SiO_2 particles

A transmission electron microscopy (TEM) image of the SiO_2 spheres is shown in **Figure 6A**. It is observed that the spheres are amorphous because of their porosity. In the inset of **Figure 6A** we illustrate the structure of one pore. The formation of the pores takes place when some periodic chains of Si-O were not carried out during the polymerization reaction, in addition, some OH groups may stay on the surface [13].

The optical UV–Vis spectrum is shown in **Figure 6B**. An absorption edge starting at approximately 300 nm and a maximum in 290 nm are observed. The width of absorbance spectrum corresponds to the band gap (Eg) energy of amorphous SiO_2. The Eg was estimated by extrapolating the absorption edge to the photon energy axis through a linear fit, then the Eg value in the direct transition of SiO_2 turned out to be 3.98 eV. This value is in agreement with the reported Eg of 3.8 eV in the case of SiO_2 nanostructures [13, 24]. This value is lower than that reported for crystalline SiO_2 (quartz), which is approximately 9 eV [25] and is a consequence of the defects mentioned above.

The zeta potential of the SiO_2 colloid is shown in **Figure 7**, a negative surface charge of −44.1 mV is detected. The minus sign is attributed to the oxygens of the OH groups located on the surface or in the pores.

The spherical shape of the SiO_2 particles is corroborated by the AFM images in **Figure 8**. A large concentration of SiO_2 spheres covering a glass substrate is

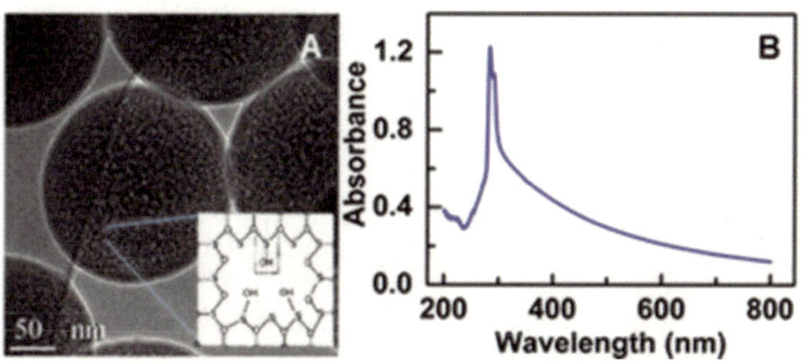

Figure 6.
(A) a TEM image of SiO_2 spheres. (B) UV–vis spectrum of the colloidal solution of SiO_2 particles.

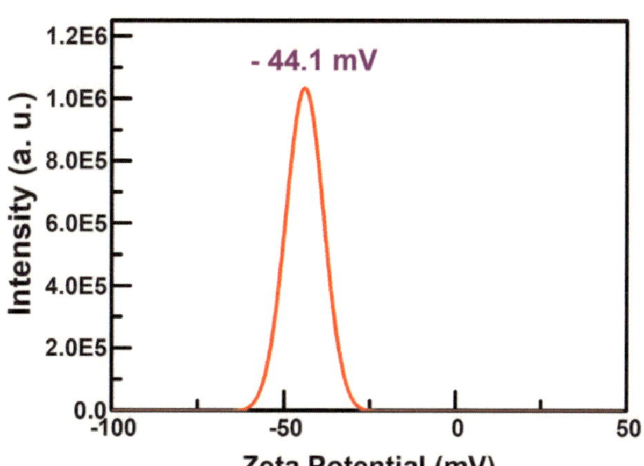

Figure 7.
Zeta potential of the SiO_2 particles in solution.

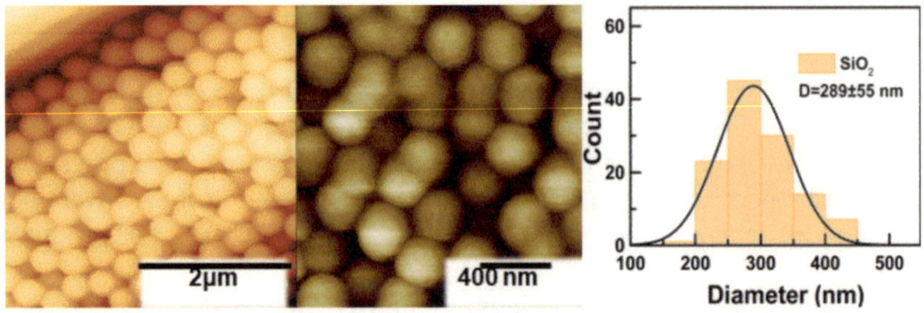

Figure 8.
Images obtained by AFM showing the shape and size of the SiO_2 particles.

observed. The main feature is a short-range order with spheres forming hexagonal arrays and multilayers. From a statistical analysis over 120 particles, an average size is estimated in 289 ± 55 nm.

6. Structural composition of the SiO_2-Ag composite films

AFM technique was also used to analyze the structure of the SiO_2-Ag composite films, the micrographs are shown in **Figure 9**. In different areas of the film it is observed that the spheres are arranged in a short-range hexagonal packing, a similar spatial distribution than the observed in the SiO_2 films. Because of the absence of SiO_2 and Ag NPs some voids are created. The single or agglomerates of Ag NPs are located in the interstices among SiO_2 spheres. The fact that the Ag NPs do not surround the SiO_2 spheres is because of the surface charge, of both Ag NPs and SiO_2 spheres, is negative (−52.5 mV and − 44.1 mV, respectively).

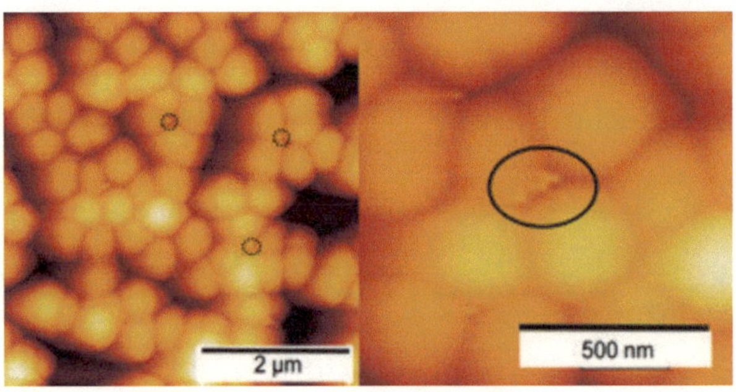

Figure 9.
AFM micrographs of the SiO_2-Ag composite films at different amplifications. The right panel shows that Ag NPs are found in the interstices.

7. Estimation of the SERS enhancement factor

As Raman signal of molecules is very weak for detecting them at very low concentrations various proposals for different condition are required. One way to enhance the signal is with the use of metallic surfaces, because of their excitation of the surface plasmon. The magnitude of the Raman signal enhancement is influenced by some factors:

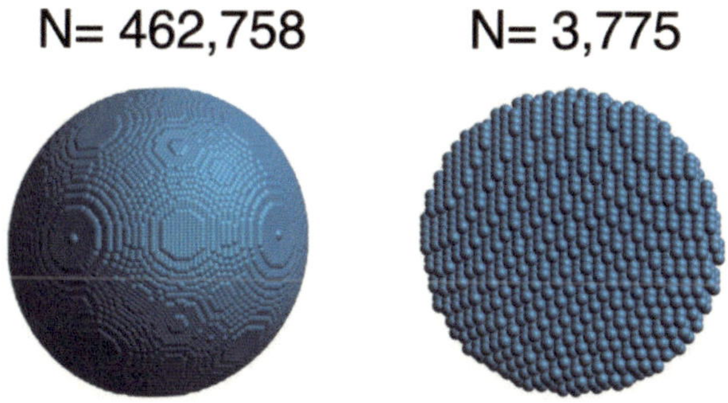

Figure 10.
A particle with a spherical shape represented by an array of N point dipoles (tiny spheres) in DDA approximation.

- Laser excitation: wavelength, angle of incidence, power, interval of time.
- Plasmonic Surface: chemical composition, refractive index, shape, surface roughness.
- Analyte: Intrinsic properties such as polarizability tensor or optical cross section, concentration, compatibility to be adsorbed to the metal surface.

To quantify the enhancement of the Raman signal factor (EF) the Single-Molecule SERS EF or the Average SERS EF are defined. The first one quantifies the amplification of the Raman signal of one molecule at a particular position, whereas the second one is the average of the intensity when considering random positions along an area of the substrate. Here we are focused on the Single-Molecule SERS EF (SMEF) expressed as

$$SMEF \approx \frac{|E_{loc}(\omega_{exc})|^2}{|E_{inc}(\omega)|^2} \frac{|E_{loc}(\omega_{Raman})|^2}{|E_{inc}(\omega)|^2}, \quad (2)$$

where $E_{inc}(\omega)$ is the incident field, $E_{loc}(\omega_{exc})$ and $E_{loc}(\omega_{Raman})$ are the local electric field at the excitation frequency and Raman frequency, respectively. When the excitation frequency is similar to the Raman frequency then

$$SMEF \approx \frac{|E_{loc}(\omega_{exc})|^4}{|E_{inc}(\omega)|^4}, \quad (3)$$

this expression is called the $|E|^4$ approximation.

To calculate the local field, the DDA and its numerical implementation DDSCAT code have been used [26]. The fundamentals of DDA are widely detailed elsewhere [27]. The basic idea is the representation of the target by a discrete array of N point dipoles. Each dipole is at the site of a cubic lattice with a lattice parameter d_0, being this smaller than the wavelength of the incident electric field. The chosen N value determines the convergence and numerical error in the calculations, in **Figure 10** a particle with a spherical shape is represented by N = 462,758 and N = 3,775,

DDA assumes that the target is illuminated by an external source of monochromatic radiation represented by a plane wave, $\vec{E}_{inc}(\omega) = \vec{E}_0 e^{i(\vec{k}\cdot\vec{r}-\omega t)}$. In **Figure 11** a plane wave with a) y-polarization and b) z-polarization state is depicted. The electric plane wave induces a dipole moment on each dipole (see **Figure 11**). Because of the dipole–dipole interaction an induced electric field is originated given by:

$$\vec{E}_{ind,i} = -\sum_{j}{}' \mathbb{A}_{ij} \cdot \vec{p}_j \quad i = 1, \ldots, N, \tag{4}$$

where $\mathbb{A}_{ij}$ represents the dipole–dipole interaction and depends on the relative position between the dipole i and j ($\vec{r}_{ij} = \vec{r}_j - \vec{r}_i$), $\sum_j$' indicates that $i \neq j$, $\vec{p}_j$ is the dipole moment of the j dipole. Conventionally $\mathbb{A}_{jj} = 0$. Considering that $\vec{p}_i = \boldsymbol{\alpha}_i \vec{E}_{T,i}$, being $\boldsymbol{\alpha}_i$ the polarizability tensor and $\vec{E}_{T,i}$ the total field at $\vec{r}_i$ and that

$$\vec{E}_{T,i} = \vec{E}_{inc,i} + \vec{E}_{ind,i}, \tag{5}$$

is possible to get a system of 3 N complex equations that in matrix notation is expressed as:

$$\hat{\mathbb{B}}\vec{\mathrm{P}} = \vec{E}_{inc}, \tag{6}$$

with $\vec{\mathrm{P}} = \left(\vec{p}_1, \vec{p}_{2,\ldots,}\vec{p}_N\right)$, $\vec{E}_{inc} = \left(\vec{E}_{inc,1}, \ldots, \vec{E}_{inc,N}\right)$ and $\hat{\mathbb{B}}$ a $N \times N$ symmetric matrix [27].

Once the Eq. (6) is solved, that is, $\vec{p}_1, \vec{p}_{2,\ldots,}\vec{p}_N$ are known, then the electric field in, on and outside the target can be calculated. To solve Eq. (6) we used the software DDSCAT, which is a numerical implementation of DDA. Because we are interested in the SERS EF, we focused on the electric field on and outside the particle. DDSCAT code allows to calculate the electric field on a point with coordinates (x, y, z):

$$\vec{E}(x,y,z) = \vec{E}_{inc}(x,y,z) + \vec{E}_{scat}(x,y,z), \tag{7}$$

where an incident field of magnitude 1 is specified, that is, $\left|\vec{E}_{inc}\right| = E_{inc} = 1$, and $\vec{E}_{scat}$ is the scattered field of the N radiating dipoles representing the target [28].

In addition, we used the Lattice Dispersion Relation (an option in the DDSCAT code) to describe the polarizability of each dipole [29].

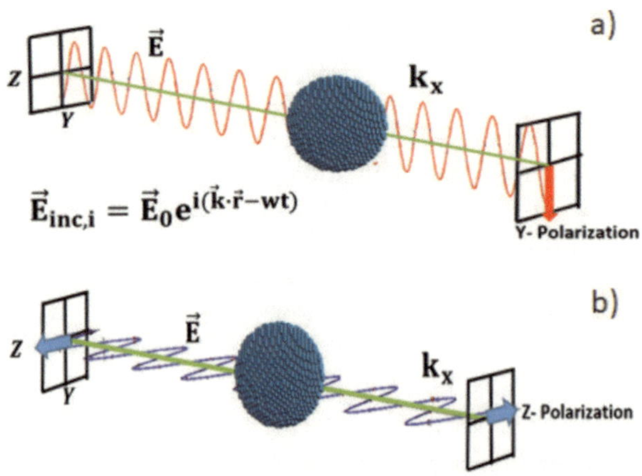

Figure 11.
Scheme of an incident plane wave on a spherical target with (a) y-polarization and (b) z-polarization. The electric field induces a dipole moment on each dipole (tiny spheres).

$$\alpha_i = \frac{\alpha_i^{nr}}{1 - 2i\,k^3\alpha_i^{nr}/3} \quad \mathrm{i} = 1, 2, ..\mathrm{N}, \tag{8}$$

where $\alpha_i^{nr} = \frac{\alpha^{CM}}{1+\alpha^{CM}k^2[b_1+m^2b_2+m^2b_3S]/d}$, being k the wavenumber of the incident plane wave, α^{CM} the well-known Clausius- Mossotti polarizability, m the refractive index of the material, S contains information of the polarization state and direction of propagation, b_1, b_2 and b_3 are constants, and d is the separation distance between two adjacent dipoles.

The polarizability depends on the dielectric function through the refractive index and the Clausius- Mossotti polarizability. To describe properly the polarizability an analysis of the size-dependent dielectric function was carried out.

A size correction in the bulk dielectric function of a material was firstly proposed by Kreigib [30]. He proposed whether the dimensions of the material are smaller than a critical diameter D_c, then quantum effects should be considered. As the diameter of the particle becomes of the order or greater than D_c then the electronic levels become a finite set of energy levels forming the conduction band, as it occurs in a bulk material.

Finite size correction (FSC) theory is based in the assumption that the free electrons have instant collisions with the lattice ions and also with the surface of the particle, the last acts as a scatterer of the electrons reflecting them diffusely. Then, the damping constant is expressed as

$$\Gamma = \Gamma_{bulk} + \frac{V_F}{l}, \tag{9}$$

where Γ_{bulk} is the bulk damping constant, V_F is the Fermi velocity and l is the mean free path.

On the other hand, the bulk dielectric function is expressed as

$$\epsilon_{bulk}(\omega) = \epsilon_{inter}(\omega) + \epsilon_{intra}(\omega), \tag{10}$$

where the intraband contribution is because of free electron transitions in the conduction band, whereas the interband contribution is because of band-to-band transitions of the core electrons. The interband transitions become important at smaller wavelengths than 320 nm for Ag [23]. Because interband electrons are not notably affected by the surface, $\epsilon_{inter}(\omega)$ can be assumed size independent. Therefore, considering that intraband electrons behave as Drude electrons with a damping constant as Eq. (9), the dielectric function of a particle with radius a is [31].

$$\epsilon(\omega, a) = \epsilon_{bulk}(\omega) - \epsilon_{intra} + \left\{1 - \frac{\omega_p^2}{\omega[\omega + i\Gamma_{bulk} + iV_F/a]}\right\}, \tag{11}$$

$\Gamma_{bulk} = 1/\tau$, with τ the relaxation time and ω_p the plasma frequency. τ is also related to the mean free path of the electrons, l. For Ag $\hbar\omega_p = 9.6$ eV, $1/\omega_p\tau = 0.00188$, V_F= 1.39×10^6 m/s and l =52 nm [32].

We explore the FSC in the dielectric function in an interval of diameters between 20 nm and 100 nm. In **Figure 12** we present the real and imaginary part of the dielectric function for only the extreme values of the studied interval, the curves of other diameters are in between. The optical response when the size of the material is 100 nm matches with that of the bulk Ag.

The SPR of a Ag NP in water and with a diameter (D) in the interval between 20 nm and 100 nm is shown in **Figure 13** a). For this interval, as the size increases,

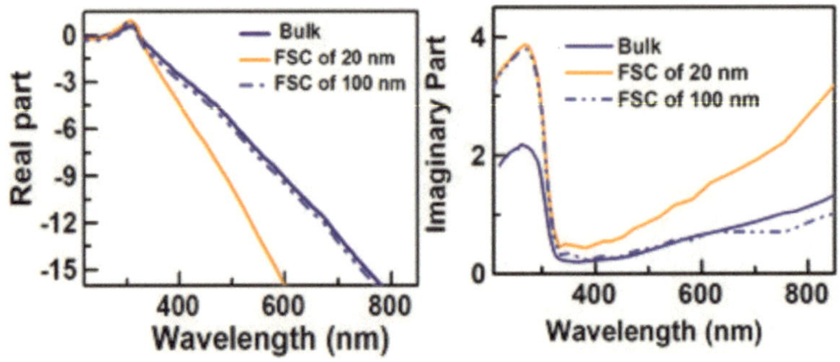

Figure 12.
Real and imaginary part of the dielectric function of silver in bulk (blue line) and taking into account FSC with a diameter of 20 nm (yellow line) and 100 nm (dashed blue line).

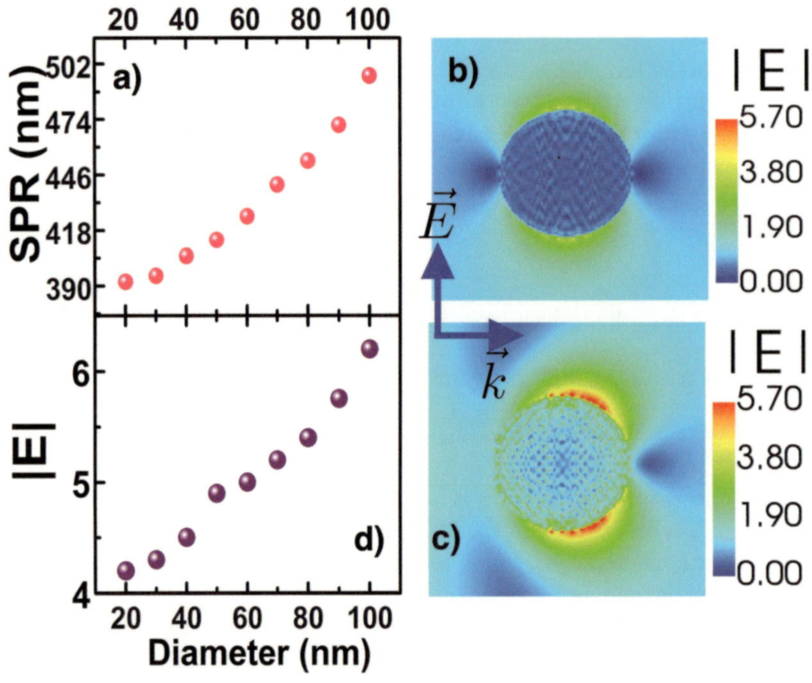

Figure 13.
The SPR vs *diameter (D) of a spherical Ag NP is shown in panel (a). The magnitude of the electric field of a Ag NP with a D = 20 nm and D = 100 nm is shown in panel (b) and (c), respectively, the incident plane wave has a wavelength of 632 nm. In panel (d) |E| at a point in the zenith of the NP* vs *D is presented. The NP is in water (refractive index of 1.33).*

the SRP is red shifted. In addition, the magnitude of the complex electric field vector on a plane crossing the center of the NP is shown in panel b) and c) for a D = 20 nm and D = 100 nm, respectively. The plane wave has a wave vector $\vec{k}$ in x-direction, a linear polarization state in y-direction and a wavelength of 632 nm. Along the xy plane the small NP has the characteristic field intensity distribution of a dipole plasmon. Whereas, for the large NP the intensity distribution is slightly distorted from that of a dipole surface plasmon. To visualize the relation $\left|\vec{E}\right|$ *vs* D a specific hot spot in the zenith of the sphere was chosen, its electric field magnitude

$\left|\vec{E}\right|$ (not the intensity or irradiance) *vs* D is shown in panel d). For the sizes studied here the magnitude of the electric field varies from 4 to 6, therefore the SERS EF of a molecule located at the zenith of the sphere is SERS EF $= \left|\vec{E}\right|^4 = 1.3x10^3$ for a Ag NP with a D = 100 nm.

Because a common laser to study SERS signal is that with a 632 nm wavelength, and based on the results shown in **Figure 13**, the Ag NP with a SPR wavelength close to the laser has a diameter of 100 nm. Therefore, in the following sections we have considered a Ag NP with that size and an excitation wavelength of 632 nm.

7.1 Electric field magnitude of the composite thin film

From the experimental techniques we observed that the Ag NPs have a preferred location at the interstices of the SiO_2 spheres when in substrate, particularly creating clusters or agglomerates. Moreover, the Ag NPs are covered by a layer of tannic acid.

To estimate the SERS EF of the composite material the layer of the tannic acid was taken into account and a study of the agglomerated NPs was carried out. **Figure 14** shows the electric field magnitude of a single SiO_2 sphere with a D = 290 nm (see panel a)), of a single Ag NP with a D = 100 nm (see panel b)), same NP as in b) with a shell of 1 nm of tannic acid (see panel c)), a SiO_2 sphere and a Ag NP touching (see panel d)), and same array as in d) with the tannic acid shell of the NP (see panel e)). For all the cases mentioned, the plane wave with the same characteristics as in **Figure 13** were considered. The refractive index is 1.46 for SiO_2 [33], 1.704 for tannic acid [34], and 1.0003 for the surrounding medium (the composites are in air). Clearly, the electric field on and away from the surface of the dielectric SiO_2 sphere is less intense than that of the Ag NP. The presence of the dielectric tannic acid layer affects the plasmonic response of the NP and the result is a less intense electric field with a spatially symmetric distribution. There is a "screening" effect because of the presence of the dielectric shell, this is observed in the presence and absence of the SiO_2 sphere. When a Ag NP with/without a tannic acid shell is touching a SiO_2 sphere a SERS EF can be as high as 1×10^3, the zones where this value can be reached are smaller when the dielectric layer is present.

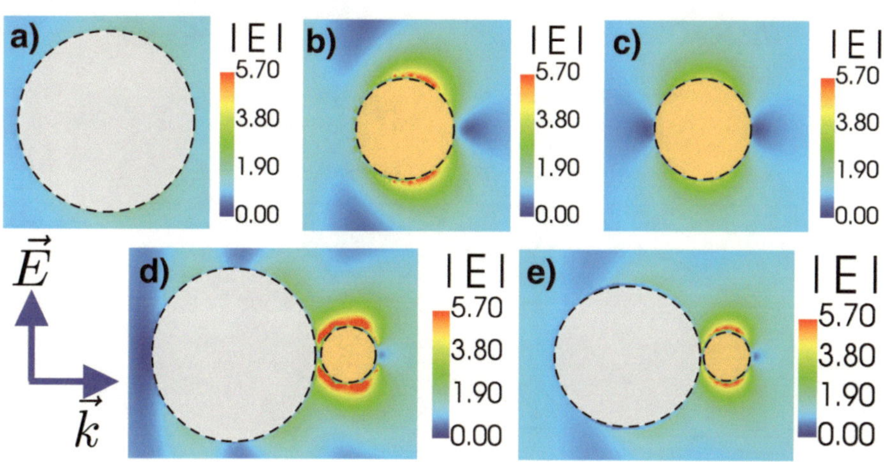

Figure 14.
Local electric field of (a) a single SiO_2 sphere, (b) a single nude Ag NP, (c) a single Ag NP with a tannic acid shell of 1 nm thick, (d) nude Ag NP touching a SiO_2 sphere, and (e) Ag NP-tannic acid shell touching a SiO_2 sphere.

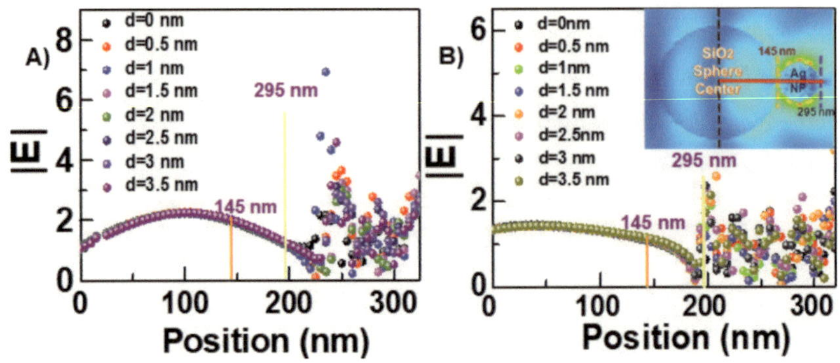

Figure 15.
Magnitude of the electric field along a center-to-center path, see red line in the inset of panel (B). When the sphere and NP are touching d = 0 and a point on the surface of SiO_2 sphere is at a position of 145 nm (D = 290 nm of a SiO_2 sphere). Panel (A) shows |E| values as d increases and the Ag NP is nude, whereas in panel (B) the tannic acid shell has been considered.

Because SiO_2 spheres and Ag NPs have a negative superficial charge, −44.1 mV and − 52.5 mV, respectively, we deduce there is not surface-to-surface contact. We explore surface-to-surface separation distances (d)between 0.5 nm and 3.5 nm and how this modifies the electric field.

In **Figure 15** is shown the magnitude of the electric field along a straight path starting at the center of the SiO_2 sphere and ending at the center of the Ag NP, this is illustrated as the red line shown in the inset of panel B). The size of the SiO_2 sphere is D = 290 nm and of the Ag NP D = 100 nm, therefore when $d = 0$, that is, the sphere and Ag NP are in contact, the contact point is at a position of 145 nm. Panel (A) and (B) shows |E| values as d increases when the Ag NP is nude and with

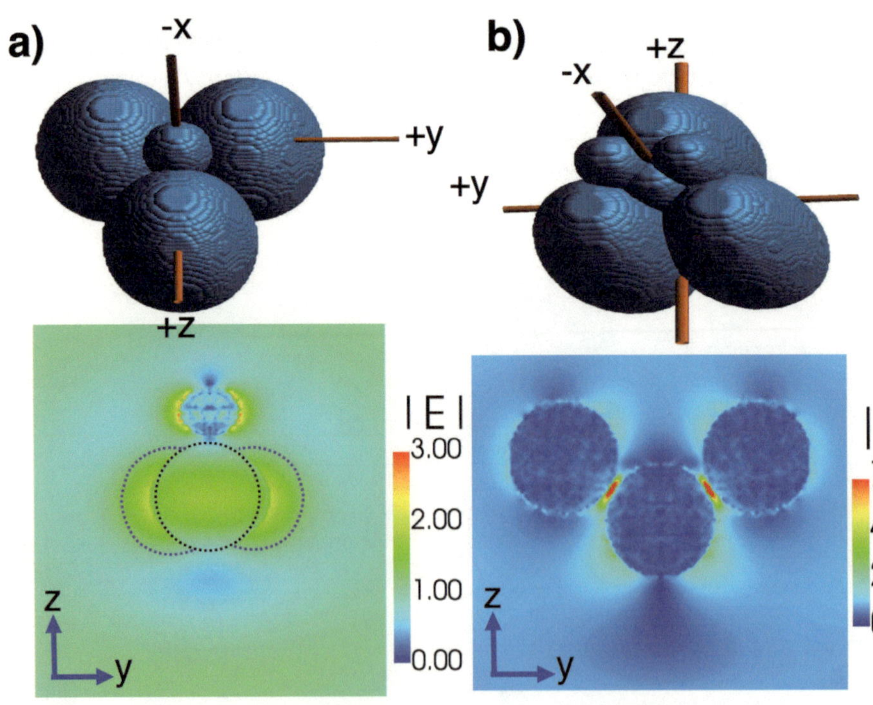

Figure 16.
Dipoles array and |E| values along the yz-plane of an array of 3 SiO_2 spheres with (a) one Ag NP, and (b) 3 Ag NPs with no coplanar centers, at the interstice. For both cases, the incident field has a wave vector along x, y-polarization and a wavelength of 632 nm.

the tannic acid shell, respectively. For both cases, the magnitude of the electric field presents inappreciable changes.

To inquire about the electric field intensities in the composite material, we explore the case with one and three Ag NPs at the interstice of three SiO_2 spheres. Again, the size of the SiO_2 sphere is D = 290 nm and of the Ag NP D = 100 nm. The refractive index of the medium is air and the incident electromagnetic field travels in the positive x-direction with a y-polarization, its wavelength is 632 nm. It is worth mentioning that in these cases the tannic acid layer was not considered. To illustrate the systems under study, their respective array of dipoles is shown in **Figure 16**. Because of the complexity of the 3D arrays we chose only the yz-plane to exemplify the field magnitude. As expected, the electric field of the Ag NP in presence of three SiO_2 spheres is weak because of the screening effect promoted by the dielectric material, whereas, in the array of three Ag NPs, relative hot spots are present in the space among them. The hot spots may provide a SERS EF of the order of 2.4×10^3.

8. Conclusions

In this chapter we present the synthesis, structural composition and optical response of composite films with two main components, Ag nanoparticles and SiO_2 spheres. The independent production of Ag NP and SiO_2 sphere colloids were the basis to fabricate the composite films by evaporation solvent method. According to the analysis made to Atomic Force microscopy images, measurements of Absorbance and Dynamic light scattering, and numerical calculations supported on Mie theory, we conclude that single NPs and also agglomerates of Ag NPs were produced with a NP mean diameter of 96 ± 9 nm. In addition, the presence of a tannic acid layer covering the surface of the NP is deduced. With the synthesis method followed to produce the SiO_2 particles it was possible to obtain porous spherical particles with a mean diameter about 289 ± 55 nm.

The composite films are characterized by a short-range order with local hexagonal arrays of SiO_2 spheres with Ag NPs, single or agglomerated, located at the interstices. Some voids are also observed, that is spaces without the presence of NPs and/or spheres. We attribute the preferred location of the NPs to the negative surface charge of both, Ag NPs and SiO_2 spheres. Another aspect deduced from the same sign of the surface charge is that the NP and sphere are not touching, that is, there is no surface-to-surface contact. To have an insight about how this affects the magnitude of the electric field $|\vec{E}|$, and therefore, the SERS EF, DDA calculations were carried out.

When the separation distance between the surface of a NP and that of a SiO_2 sphere is varied from 0 to 3.5 nm the intensity of the field is practically the same. The last was observed with a nude NP and with a NP covered by the tannic acid shell. Furthermore, the dielectric tannic acid layer generates a screening effect on the field intensity, besides, a different intensity distribution is observed when compared to that of a nude NP. With both a nude or covered Ag NP when in presence of the SiO_2 sphere, a maximum SERS EF estimated is of the order of 1×10^3. The difference is that this EF can be reached in smaller regions when the NP is covered compared to that when the NP is nude. Finally, the SERS EF of an agglomerate of three Ag NPs surrounded by three SiO_2 spheres was estimated. Even with the presence of the dielectric SiO_2 spheres that screens the electric field intensity, relative hot spots are observed in the regions where the nude NPs are very close one each other, giving place to a maximum SERS EF of the order of 2.4×10^3.

Despite this order of magnitude is not as large to the usually reported with plasmonic NPs, this composite films have the advantage of being prepared by inexpensive methods, moreover, the NPs are located at specific positions, a fact that can be taken advantage of for SERS applications, as proposed here, or many others.

Acknowledgements

The authors thank to Dr. Francisco Flores for the facilities and guide with the AFM characterization. P D-Leon P. thank CONACYT for the grand 330633. This project has been financially supported by VIEP-BUAP 100504244-VIEP2019.

Conflict of interest

The authors declare no conflict of interest.

Appendices and nomenclature

SERS	Surface Enhanced Raman Scattering
NPs	Nanoparticles
DDA	Discrete Dipole Approximation
SPR	Surface Plasmon Resonance
DLS	Dynamic Light Scattering
AFM	Atomic Force Microscope
TEM	Transmission Electron Microscopy
EF	Enhancement Factor
SMEF	Single-Molecule SERS Enhancement Factor
FSC	Finite Size Correction

Author details

Paulina De León Portilla, Ana Lilia González Ronquillo and Enrique Sánchez Mora*
Institute of Physics, Meritorious Autonomous University of Puebla, Puebla, Mexico

*Address all correspondence to: esanchez@ifuap.buap.mx

References

[1] Li WR, Xie XB, Shi QS, Zeng HY, Ou-Yang YS, Chen YB. Antibacterial activity and mechanism of silver nanoparticles on Escherichia coli. Applied Microbial and Cell Physiology. 2010; 85; 1115–1122. DOI: 10.1007/s00253-009-2159-5

[2] Hoang TB, Akselrod GM, Argyropoulos C. Ultrafast spontaneous emission source using plasmonic nanoantennas. Nature Communications. 2015; 6; 1–7. DOI: 10.1038/ncomms8788

[3] Ivanova N, Gugleva V, Dobreva M, Pehlivanov I, Stefanov S, Andonova V. Silver Nanoparticles as Multi-Functional Drug Delivery Systems. In: Muhammad Akhyar F, editor. Nanomedicines IntechOpen, p. 71–92. DOI: 10.5772/intechopen.80238.

[4] Beckett R, Jue Z, Giddings JC. 1987 Determination of molecular weight distribution of fulvic and humic acids using flow field-flow fractionation. Environmental Science & Technology. 1987; 21; 289–295. DOI:10.1021/es00157

[5] Keziban A, Münteha Ö, Mahmut Ö. Investigation of antibacterial properties of novel papain immobilized on tannic acid modified Ag/CuFe2O4 magnetic nanoparticles. International Journal of Biological Macromolecules. 2017;109; 720–731. DOI: 10.1016/j.ijbiomac.2017.12.066

[6] Shafer-Peltier K, Haynes CL, Van Duyne RP. Toward a Glucose Biosensor Based on Surface-Enhanced Raman Scattering. 2003; 125; 588–593. DOI: 10.1021/ja028255v

[7] Sijo F, Siby J, Ebey K, Beena M. Microwave assisted green synthesis of silver nanoparticles using leaf extract of elephantopus scaber and its environmental and biological applications. Artificial Cells, Nanomedicine, and Biotechnology. 2017; 46 ;795–804. DOI:10.1080/21691401.2017.1345921

[8] Alesenkov A, Pilipavicius J, Beganskiene A. Nonlinear properties of silver nanoparticles explored by a femtosecond z-scan technique. Lithuanian Journal of Physics. 2015; 55; 100–109. DOI: 10.3952/physics.v55i2.3100

[9] Le Trequesser Q, Seznec H, Delville MH. Functionalized nanomaterials: their use as contrast agents in bioimaging: mono- and multimodal approaches. Nanotechnology Reviews. 2013; 2(2); 1–44. DOI:10.1515/ntrev-2012-0080

[10] Liu Y, Li Y, Li XM, He T. Kinetics of (3-Aminopropyl) triethoxylsilane (APTES) Silanization of Superparamagnetic Iron Oxide Nanoparticles. Langmuir. 2013; 29; 15275–15282. DOI: 10.1021/la403269u

[11] Kumari MM, Aromal SA, Philip D. Synthesis of monodispersed palladium nanoparticles using tannic acid and its optical non-linearity. Spectrochimica Acta Part A: Molecular and Biomolecular Spectroscopy. 2013; 103; 130–133. DOI: 10.1016/j.saa.2012.11.020

[12] Opes L, Brito LM., Bezerra TT, Gomes KN, Carvalho FA. Silver and gold nanoparticles from tannic acid: synthesis, characterization and evaluation of antileishmanial and cytotoxic activities. Anais da Academia Brasileira de Ciências, 2018; *90*(3); 2679–2689. DOI: 10.1590/0001-3765201820170598

[13] Carmona-Carmona AJ, Palomino MA, Hernández O, Sánchez E, Toledo M. Synthesis and characterization of magnetic opal/Fe3O4 colloidal crystal. Journal of Crystal Growth. 2017; 462; 6–11. DOI: 10.1016/j.jcrysgro.2016.12.105

[14] Munnik P, de Jongh PE, de Jong KP. Recent Developments in the Synthesis of Supported Catalysts. Chemical Reviews. 2015; 115(14); 6687–6718. DOI: 10.1021/cr500486u

[15] González AL, Santos A, Toledo M. Metallo-Dielectric Colloidal Films as SERS Substrate. In: Morteza Sasani Ghamsari, Soumen Dhara, editors. Nanorods and Nanocomposites. IntechOpen 90313. 2019. 1-17 p. DOI: 10.5772/intechopen.90313

[16] Santos A, González AL. Far Field Optical Proprieties of a Monolayer of SiO_2 Spheres and Small Au Nanoparticles. Material Research Society Advances. 2018; 3; 3917–3923. DOI: 10.1557/adv.2019.128

[17] Santos A, González AL. Far field optical properties of a monolayer of SiO_2 spheres and small Au nanoparticles. MRS Advances. 2018;3(64):3917–3923. DOI: 10.1557/adv.2019.128

[18] Méndez PA, Soto L, Sánchez E, González AL, Romo JM, Gervacio JJ, Martínez LJ. Damage on Escherichia coli and Staphylococcus aureus using White light photoactivation of Au and Ag nanoparticles. Journal of Applied Physics. 2019; 125; 213102. DOI: 10.1063/1.5090273

[19] Bastús G, Merkoçi F, Piella J. Synthesis of Highly Monodisperse Citrate-Stabilized Silver Nanoparticles of up to 200 nm: Kinetic Control and Catalytic Properties. Chemistry of Materials. 2014; 26; 2836–2846. DOI: 10.1021/cm500316k

[20] Stöber W, Bohn FA. Controlled growth of monodisperse silica spheres in the micron size range. Journal Colloid Interface Science. 1968; 26; 62–69. DOI: 10.1016/0021-9797(68)90272-5

[21] Razo D, Pallavidino L, Garrone E, Geobaldo F. A version of Stöber synthesis enabling the facile prediction of silica nanospheres size for the fabrication of opal photonic crystals. Journal of Nanoparticle Research. 2008; 10; 1225–1229. DOI: 10.1007 / s11051-008–9373-4

[22] Mei KK. MEI Method. In: K. Chang, editor. Encyclopedia of RF and Microwave Engineering. 1st ed. Chichester: Wiley; 2005. DOI: 10.1002/0471654507.eme570

[23] González AL, Reyes-Esqueda JA, Noguez C. Optical Properties of Elongated Noble Metal Nanoparticles. Journal Physical Chemistry C. 2008; 112; 7356–7362. DOI: 10.1021/jp800432q

[24] Bachan N, Asha A, Jeyarani WJ, Kumar DA, Shyla JM. A Comparative Investigation on the Structural, Optical and Electrical Properties of SiO_2–Fe3O4 Core–Shell Nanostructures with Their Single Components. Acta Metallurgica Sinica. 2015; 28; 1317–1325. DOI: 10.1007/s40195-015-0328-3

[25] Weinberg ZA, Rubloff GW, Bassous E. Transmission, photoconductivity, and the experimental band gap of thermally grown SiO_2 films. Physical Review B. 1979; 19 (3107); 3107–3117. DOI: 10.1021/la403269u

[26] Draine BT. The discrete-dipole approximation and its application to interstellar graphite grains. Astrophysical Journal. 1988; 333; 848–872. DOI: 10.1086/166795

[27] Purcell EM, Pennypacker CR. Scattering and absorption of light by nonspherical dielectric grains. Astrophysical Journal. 1973; 186; 705–714. DOI:10.1086/152538

[28] Jackson JD. Classical Electrodynamics. 3rd ed. New York: John Wiley and Sons Inc; 1999. 832 p.

[29] Draine BT, Goodman J. Beyond Clausius-Mossotti: Wave propagation on

a polarizable point lattice and the discrete dipole approximation. Astrophysical Journal. 1993; 405; 685–697. DOI: 10.1086/172396

[30] Kreibig U. Electronic properties of small silver particles: the optical constants and their temperature dependence. Journal of Physics F: Metal Physics. 1974; 4; 999–1014. DOI: 10.1088/0305-4608/4/7/007

[31] González AL, Noguez C, Ortiz GP, Rodríguez -Gattorno G. Optical Absorbance of Colloidal Suspensions of Silver Polyhedral Nanoparticles. Journal Physical Chemistry B. 2005; 109; 17512–17517. DOI: 10.1021/jp0533832

[32] Ashcroft NW, Mermin ND. Solid State Physics. 7th ed. Philadelphia: Saunders college; 1976. 826 p.

[33] Gao L, Lemarchand F, Lequime M. Refractive index determination of SiO_2 layer in the UV/Vis/NIR range: spectrophotometric reverse engineering on single and bi-layer designs. Journal of European Optical Society Rapid Publications. 2013; 8; 13010. DOI: 10.2971 / jeos.2013.13010

[34] CAS. Refractive index Tannic Acid [Internet]. Available from: http://www.chemnet.com/cas/en/1401-55-4/tannic-acid.html [Accessed: 2019-05-20]

Chapter 11

Silicon-Silver Dendritic Nanostructures Enabled Photoelectrochemical Solar Water Splitting for Energy Applications

Uday Dadwal and Rajendra Singh

Abstract

Photoelectrochemical (PEC) splitting of natural water was studied using silicon nanowires decorated with silver dendrites (dendritic nanostructures) as working electrode. A metal assisted wet chemical etching method has been used for the synthesis of dendritic heteronanostructures. Measured photocurrent density 1.7 mA/cm^2 under white light illumination exhibits the efficient decomposition of natural water. The decomposition of water is primarily ascribed to the enhancement in the working electrode surface and water effective interface and the decrease in the recombination of light induced (photoexcited) carriers in the existence of silver dendritic nanostructures. Enhancement in photoinduced charge carriers separation caused due to the existence of Schottky barrier between the silicon and silver dendritic nanostructures. The light induced carriers (holes) in silicon are transferred to the metal (Ag) dendritic nanostructures that work as a charge basin to effectively carry out the oxidation reaction of water during PEC measurement. The solar-to-hydrogen (STH) conversion efficiency of about 4.5% was reported, indicating the efficient PEC solar water (pH 7) splitting. A cost-effective and efficient method for the PEC solar water splitting is presented in order to enhance the STH efficiency for the production of clean and renewable fuel.

Keywords: photoelectrochemical, water splitting, dendritic nanostructures, metal-assisted wet chemical etching

1. Introduction

Water splitting is the chemical method which involves the decomposition or splitting of water molecules into hydrogen and oxygen gases [1, 2]. The splitting of water molecules in the presence of external bias (such as sunlight) and working electrode is called as photoelectrochemical (PEC) water splitting method [1–3]. In this method, suitable electrodes immersed in an aqueous solution which split water molecules under light [3, 5, 6]. A form of PEC water splitting occurs in nature is the photosynthesis process used by the plants and other organisms in order to convert the sunlight into chemical energy with the release of oxygen. The working electrode used for the PEC splitting of water can act as photoanode or photocathode depending upon the PEC reactions [1, 4–6]. If the PEC oxidation of water occurs at the working electrode, then working electrode acts as photoanode (oxygen release at the working electrode). On the other hand, with the PEC reduction of water at the working electrode is referred to as photocathode (release of hydrogen). The requirement for the PEC water splitting process is that the band gap of the working electrode should be properly positioned with respect to the normal hydrogen electrode (NHE) in order to split the water under light illumination. This means that the top of the valence band must be more positive than the O_2/H_2O Redox potential (1.23 V) and the bottom of the conduction band must be more negative than the H^+/H_2 Redox potential (0 V) [1, 6, 7].

In the presence of external light photon with energy more than the working electrode band gap, light is absorbed by the photogenerated electron. This excited electron jumps into the conduction band from the valance band of the working electrode. This leaves behind an oxidizing hole in the valence band. The excited electron goes to reduce the proton while the hole accepts electrons from the water [1, 6]. The presence of the noble metal nanomaterials such as silver or gold nanoparticles on the surface of the working electrode (silver dendrites on the surface of silicon nanowires in this case) can enhance the PEC reactions [1]. The metal nanoparticles act as a sink for photoexcited electrons from the working electrode. This increases the charge carrier density and improves the electronic conductivity near the surface region of the working electrode. It provides the efficient overall charge transfer process at the working electrode and water interface [1, 2, 6]. This eventually increases the water splitting process for the generation of alternate source of clean and renewable fuel using sunlight. The production of hydrogen using water splitting method provides a renewable and sustainable energy that presents a number of advantages including overcoming the global warming and greenhouse gases from fossil fuels [1–6]. Thus, the use of hydrogen in replacing conventional fossil fuels for renewable and sustainable energy applications presents a number of advantages [1–4].

Semiconductor nanostructures have shown an excellent photocatalytic activity for solar water splitting. These nanostructures act as nano-photocatalysts for achieving an optimum solar-to-hydrogen efficiency [6–8]. Hydrogen produced by the photoelectrochemical method is of specific attention to the case where photocatalyst decomposes water molecules in the existence of solar radiation without the association of any substance or external biasing. This makes the hydrogen production cost-effective and simple as compare with previously reported water splitting methods. There is an increase in concerns over rising global energy demand, environmental sustainability, and development of new energy solutions with minimal impact on the environment [1–4]. In the recent few years, significant attention has been given to hydrogen production via PEC water splitting which could be one of the prominent energy sources [1, 5, 6].

It has been shown that the solar water splitting uses both oxidization and reduction nature of semiconductor nanostructures as photocatalysts [8, 9]. Water splitting occurs spontaneously only when hydrogen and oxygen reactions lie between the conduction band minimum and a valence band maximum of the semiconductor. Investigations have been going on for the detail understanding, optimization and development of the photoelectrochemical method using various semiconductor nanostructures [10–12]. Semiconductor nanostructures have shown to be excellent nano-photocatalysts for the photoelectrochemical solar water splitting. Based on recent research work, various photocatalysts such as CdS, TiO_2 and SiC for solar-to-hydrogen production has been investigated [13, 14]. A photocatalyst with non-destructive, environmentally pleasant, plentiful and cost effective in nature is promising for the photoelectrochemical solar water splitting. The nanostructures of additional semiconductors such as third-fifth nitrides, cobalt oxide (CoO) and silicon are also shown to be important for photoelectrochemical solar water splitting [3, 12, 15]. Several methods have been reported for the synthesis of semiconductor nanostructures. Among these, metal assisted wet chemical etching method has been reported investigated for the synthesis of semiconductor nanostructures [16, 17]. Vertical standing silicon nanowires have been prepared using a standard aqueous solution of HF and $AgNO_3$ for different applications. It includes the fabrication of the solar cell and related devices [18–21]. Optical properties of a vertical standing array of silicon nanowires have been also explored for the fundamental understanding of various concepts. Due to the simple process flow of metal assisted chemical etching method, synthesis of metal assisted porous silicon has been investigated [20, 21]. Semiconductor nanostructures have been studied from the photoelectrochemical-induced solar water splitting. Other materials such as ruthenium polypyridyl complexes have been investigated as visible light-induced photo-sensitizers for water splitting applications [22]. Synthesis of carbon nanotube anode and tin oxide nanoribbons has been carried out for various applications [23]. In particular, such materials and others have been investigated for the hydrogen production by photocatalytic decomposition of solar water using a water splitting method.

In the context of the current status of the energy availability, energy sources should be clean and environmentally friendly. The renewable energy sources are of top priority in this context. One of the prominent alternate sources of renewable and clean energy emerged out to be hydrogen production by phoelectrolysis of water using solar energy. Hydrogen production by such a way is the promising fuel, which can be used in fuel cells for the generation of electricity with high efficiency. In addition, many chemical reactions and petroleum refining require hydrogen as a reactant. Therefore, in the recent past, a number of semiconductor nanostructures have been proposed for the development of solar water splitting methods. However, solar-to-hydrogen conversion energy efficiency is reportedly lower.

The aim of this report is to investigate the metal assisted wet chemical etching of semiconductor (silicon), and semiconductor nanostructures-induced solar water splitting for achieving the optimum solar-to-hydrogen efficiency. From the existing fundamental understanding, various parameters/conditions of the metal assisted wet chemical etching and photoelectrochemical solar water splitting will be optimized for the progress of hydrogen production as an alternate fuel in the context of the current status. It has been found that metal assisted chemical etching of silicon can result in the synthesis of tree branches-like silicon nanowires. Such nanostructures can be the promising nano-photocatalysts for the hydrogen generation as an alternate fuel using photoelectrochemical solar water splitting. The solar water splitting using photoelectrochemical method has been explored by reviewing the current density and impedance measurement of the dendritic nanostructures (working electrode) and water interface using electrochemical method. The electrical properties of the dendritic nanostructures as working electrode have been studied using Mott-Schottky (M-S) measurements. The oxygen gas evolution is also considered during the PEC process.

This study reports that the PEC solar water splitting using silicon silver dendrites nanostructures as working electrode can provide effective way for lower cost collection of sunlight for the renewable fuel development.

2. Experimental

N-type Czochralski (CZ) grown (100) silicon wafers of diameter 2-inch were cut into small size samples (3×3 cm^2). The samples were then exposed to a typical silicon cleaning procedure [24, 25]. The cleaning was carried out using isopropanol

(IPA), de-ionized wafer (DI) and boiling in a piranha solution of H_2SO_4 and H_2O_2 (1: 1 by volume) for 15 minutes [24, 25]. This follows the DI wafer rinse and 5% HF dip for 30 seconds in order to take away the native oxide from the Si substrate. The schematic of galvanic cell for the metal assisted chemical etching of semiconductor is shown in **Figure 1**. The cleaned silicon wafer pieces were then etched by a metal assisted wet chemical etching method for the synthesis of nanostructures. Metal assisted wet chemical etching was carried out using an optimized process consisting of an aqueous solution of 5 M HF (49%): 0.02 M $AgNO_3$ (1: 1 by volume) (see **Figure 1**) [1]. The metal assisted chemical etching of the silicon sample was carried out for 30 min at room temperature (RT). The details of the metal assisted wet chemical etching method can be found from the earlier reports [26–28]. Topological study of the etched silicon has been done using (FEI QUAN TA 3D FEG) field emission scanning electron microscope (FE-SEM). The structural quality of the dendritic nanostructures has been evaluated using X-ray diffraction method (XRD) in bulk mode in the range between 40 and 80 degree using Rigaku Ultima IV thin film X-ray diffractometer. A Cu $K\alpha_1$ radiation source of wavelength 1.54 Å was used in the XRD measurements. The band gap of the nanostructures was studied at room temperature using ultraviolet–visible–near infrared spectrometer [1]. The PEC work station from Zahner Instruments has been employed for studying the natural water splitting [1]. PEC work station consists of a high power light emitting diode as a white light irradiation source (430–730 nm) and a three electrode assembly photoelectrochemical cell. The heteronanostructures were submerged in water filled within the PEC cell. Natural water (pH 7) was taken as an electrolyte for all PEC water splitting experiment. Thus, a photoelectrochemical cell was consisting of dendritic heteronanostructures as photoanodes (working electrodes), a platinum wire as counter electrode and a double junction Ag/AgCl that acted as a reference electrode.

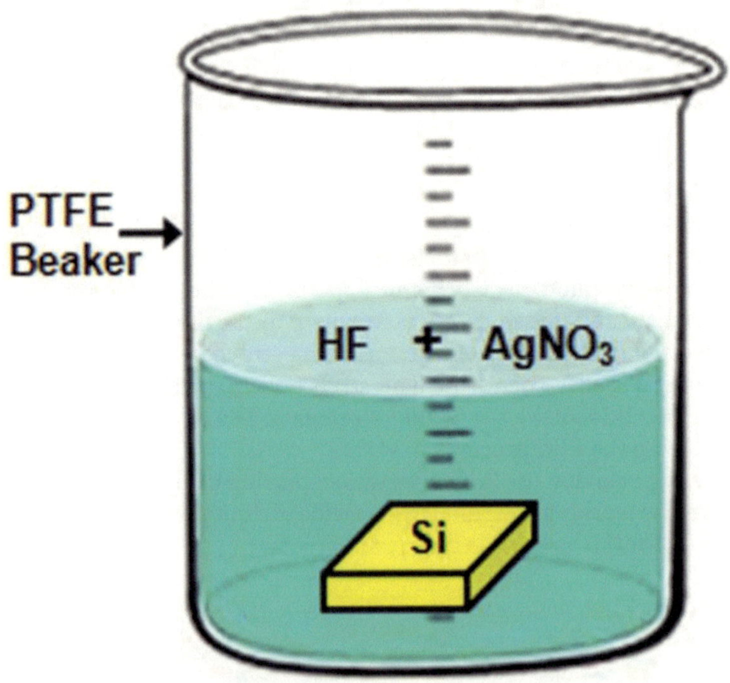

Figure 1.
Schematic of the galvanic cell consisting of Si immersed in HF and $AgNO_3$ solution. A polytetrafluoroethylene (PTFE) beaker was used for the wet chemical etching of the silicon wafer.

The external bias (V_{ext}) of voltage ±1 V with a slew rate of 5 mV s^{-1} was applied between the working photoelectrode and a reference electrode. The current range was set to ±20 mA during the measurements. The study was carried out both in dark and under white light illumination of intensity (I_o) 50 mW cm^{-2}. The electrochemical impedance spectra (EIS) were carried out in a frequency range of 10 mHz to 25 kHz at a potential of 1 V under illumination. Electrical properties measurement of the working electrodes was carried out using the M-S investigation at a frequency of 1 kHz in the dark condition.

3. Results and discussion

In the metal assisted wet chemical etching method, Si substrate is immersed into the HF/$AgNO_3$. It results in faster etching of Si from the Si-metal interface in the presence of HF solution. This results in the dissolution of Si underneath the metal ions resulting in the fabrication of semiconductor nanostructures in the form Si nanowires, which were found to be decorated with the tree branches-like Ag nanowires (dendrites) having branches, sub-branches and leaves like structures (**Figure 2**) [1]. The potential advantage of such type of nanostructures is that the surface-to-volume ratio of the nanostructures increases significantly as compared to the vertical standing nanostructures over the substrate. This facilitates the nucleation site for the hydrogen production and hence photocurrent (as can be seen in this study). These nanostructures have been used in the photoelectrochemical method for the water splitting applications. Thus, water splitting using photoelectrolysis can be used for the production of hydrogen, which is an alternate source of renewable and sustainable solar energy.

The prepared dendritic heteronanostructures were utilized for the PEC splitting of the natural water. In the water splitting, PEC cell was considered with a three electrode gathering where the dendritic heteronanostructures were utilized as working electrode (photoanode). During PEC measurement, potential has been recorded versus the Ag/AgCl reference electrode. Here, it is to be noted that the E_g of the Si nanostructures is of the order so as to cause both the PEC oxidization $2H_2O + 4h^+ \rightarrow O_2 + 4H^+$ (working electrode and water interface) and reduction $2H^+ + 2e^{-1} \rightarrow H_2$ (counter electrode and water interface) reactions during the water

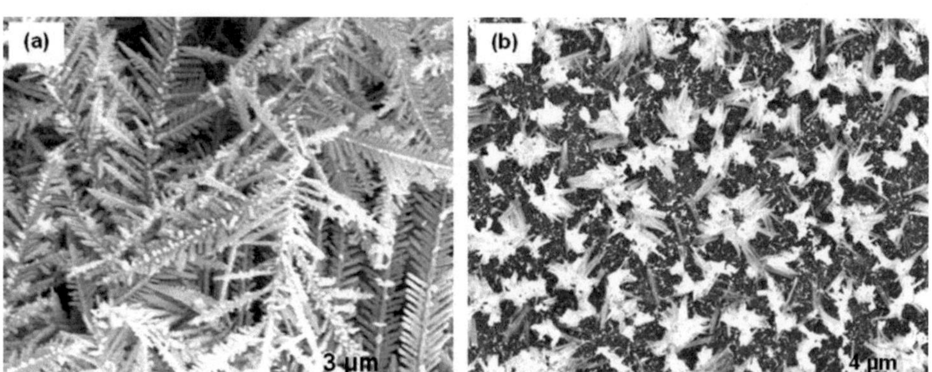

Figure 2.
Field-emission scanning electron microscopy (FE-SEM) images of the tree branches-like silicon dendritic nanowires. The nanostructures are synthesized using the method metal assisted wet chemical etching of silicon wafer in an aqueous solution consisting of 5 molar (M) HF (49%): 0.02 M $AgNO_3$ (1:1 by volume). The etching is carried out for 30 minutes at room temperature. (Dadwal et al., DOI: 10.1016/j.ijhydene.2018.10.163) [1]. Reprinted with permission from Dadwal et al., (2018), Elsevier.

splitting [1, 3, 8]. When the V_{ext} is applied between the photoanode and the reference electrode, photogenerated holes from the silicon nanowires oxidize the water, which result in photocurrent as can be seen from **Figure 3** [1].

It has been observed that the onset voltage at which the working electrode (dendritic nanostructures) reveals sharp rise in the measured photocurrent density (J_p) under illumination is about –1 V. From the onset voltage, J_p rises to 1.7 mA cm^{-2} with the increase in V_{ext}. It has been further observed that the rise in J_p beyond these values is negligibly small and gets saturated with Vext. The saturation photocurrent density (J_p^{sat}) approximately 1.7 mA cm^{-2} for the dendritic heteronanostructures is quite large as compared to the pristine silicon sample (**Figure 3**). This measured difference in J_p and saturation photocurrent density can be ascribed to the statement that under enlightenment by the energy of the incident light greater than the band gap, photogenerated electrons are excited more to the conduction band from the valence band of Si nanowires [1]. Consequently, the overall water splitting rate depends more upon the experimental arrangements of the PEC work station and on the behavior of dendritic heteronanostructures as photoanodes in determining the optimum photocurrent density.

The dendritic heteronanostructures and water interface impedance was explored using the electrochemical impedance spectroscopy, which shows the impedance (Z) of a PEC system through the frequency spectrum (Z_{real} versus Z_{imag}) (see **Figure 4**) [1, 29, 30]. The potential of the photoanode (working electrode) with respect to the reference electrode Ag/AgCl has been fluctuating at 5 mV. A DC bias of 1 V was applied between the working electrode and counter electrode. The impedance spectra have been noted in the frequency choice from 10 mHz to 25 kHz under dark and illumination conditions. **Figure 4** shows that the Nyquist plot in the low frequency region is assigned to the charge transfer resistance (R_{ct}) between the working electrode and water interface in the PEC water splitting [1, 31, 32].

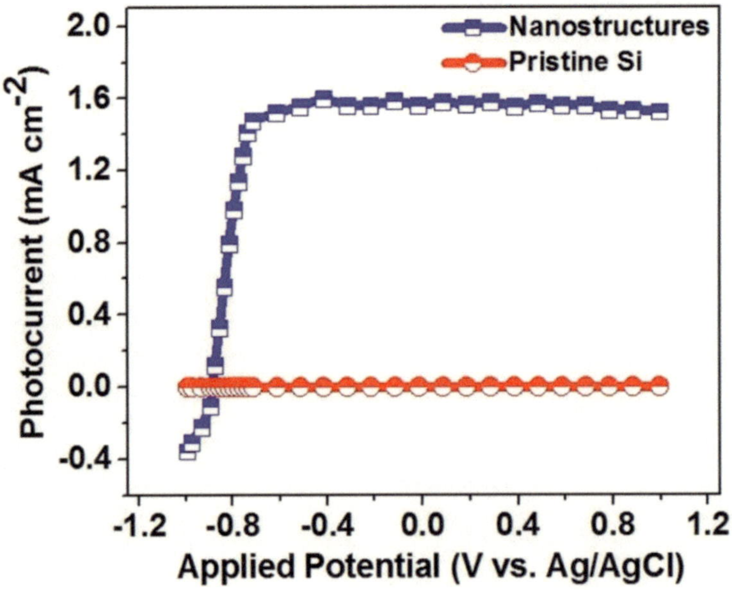

Figure 3.
Photocurrent density as a function of applied potential under a white light enlightenment 50 mW cm^{-2}. The photocurrent was produced by the PEC solar water splitting using natural water as an electrolyte and dendritic heteronanostructures as working electrodes (photoanode). The experiment was performed using a PEC cell in a three electrode arrangement (Dadwal et al., DOI: 10.1016/j.ijhydene.2018.10.163) [1]. Reprinted with permission from Dadwal et al., (2018), Elsevier.

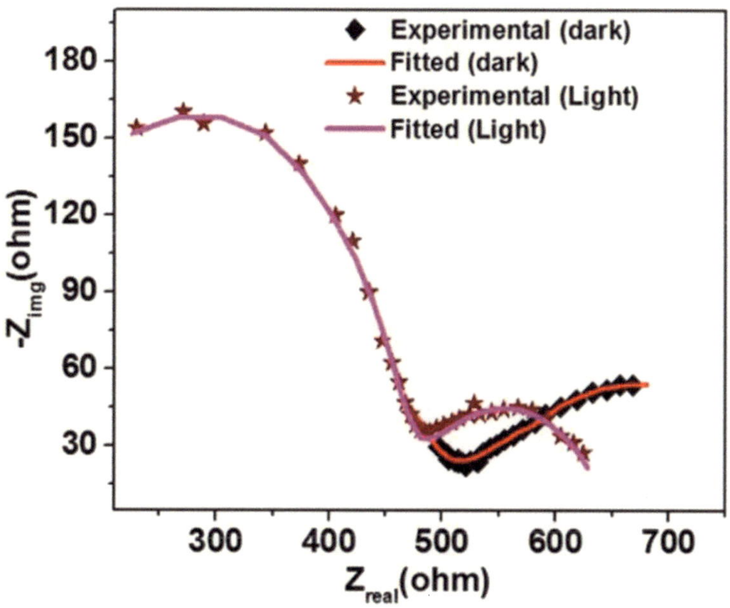

Figure 4.
Impedance measurement of the dendritic heteronanostructures as working electrodes obtained by the electrochemical impedance spectroscopy. The measurement was performed together in the light absence and under the white light enlightenment using a PEC cell in a three electrode assembly (Dadwal et al., DOI: 10.1016/j.ijhydene.2018.10.163) [1]. Reprinted with permission from Dadwal et al., (2018), Elsevier.

Here, R_{ct} is the difference between the high frequency and low frequency intercepts with the x-axis on a Nyquist plot. The R_{ct} value of 150 Ω was found to be decreased drastically under illumination [1]. As a consequence, the total impedance (|Z|) of the system gets reduced. It is to be noted that the importance of |Z| is considerable lower as compared to the previously reported results for the case of substances for instance Au-TiO_2 nanocomposite and α-Fe_2O_3 [33]. This means that R_{ct} of the dendritic heteronanostructures surface is sufficiently low to boost the charge carrier allocation at the dendritic heteronanostructures surface in the presence of V_{ext}. Low R_{ct} value can be accredited to the Ag dendrites role in the charge carrier allocation at the Si nanowires (photoanode) and water interface. Ag dendritic nanostructures can sweep away the photoexcited carriers from the Si nanowires/water interface which thereby decrease the photogenerated electron hole recombination. It thus made available carrier electrons from the external circuit to reach the photocathode (counter electrode). This effectively carries out PEC reduction of the water at the counter electrode [1]. That is why, a major rise in the eventual photocurrent has been observed during the present PEC solar water splitting experiment.

Further insight to the dendritic heteronanostructures assisted PEC water splitting reaction can be obtained through Mott-Schottky analysis at a frequency of 1 kHz in the dark condition. Mott-Schottky allows determining the internal electrical properties of the photoanode together with the flat-band potential (V_{fb}) and charge carrier concentration (N_D). The investigation has been carried out by computing the space charge section capacitance (C_{SC}) with external bias V_{ext}. C_{SC} procedures in the semiconductor near the photoanode and the water interface owing to equilibration of the chemical potential. Near the working electrode surface, for the n-type semiconductor, holes, locally come to be the majority carrier. In the presence of sufficient applied bias, deep depletion layer forms where the holes can counter with the hydroxyl ions in the water [36, 38]. Thus, at the onset voltage, efficient hole reaction (PEC oxidization) enables resulting a major rise in J_p, as observed [1]. For the semiconductor and water interface, C_{SC} is determined by on the V_{fb} according to the equation as $\frac{1}{C_{SC}^2} = \frac{2}{\varepsilon\varepsilon_O N_D e A^2}\left(V_{ext} - V_{fb} - \frac{k_B T}{e}\right)$ [31–33]. Here, ε is the dielectric constant of the silicon (11.7), ε_0 the permittivity of the free space, N_D represents semiconductor carrier or donor density, e denotes the fundamental charge, A denotes the effective surface area of the dendritic nanostructures for the PEC water splitting and $k_B T$ is the thermal energy. Using the Mott-Schottky expression, a plot between the $1/C_{SC}^2$ as a function of applied potential is shown in **Figure 5**. It showed that the slope of the graph is positive, indicating working electrode is n-type with electrons as the majority carriers. Further, from the slope of the straight line, N_D was calculated ~1×10^{20} cm^{-3} [1]. The existence of such a large carrier concentration endorses the role of metal (Ag nanostructures) which can reportedly act as a sink for the photogenerated carriers as well as donor dopants to subsequently increase the electron concentration. Ag dendrites donor dopants nature is due to the existence of Schottky barrier (ϕ_b) of ~0.2 eV (Mott-Schottky rule) between the Ag dendrites/Si junction. This consequence in the realization of satisfactory band alignment for the photogenerated carrier separation and their transferal from the Si nanowires to Ag dendrites heterostructures. Owing to this, a large value of the N_D has been found from the Mott-Schottky measurement. Since the charge carrier density is governed by linearly with the conductance, hence, improvement in the charge carrier concentration resulted in lowering of the charge transfer resistance among the working electrode and water boundary. Furthermore, for a large carrier concentration, it has been described that the metal atoms sweep away the photogenerated carriers from the working electrode and electrolyte interface [1]. Thus, there exists a definite possibility to enhance the lifetime of charge carriers resulting reduction in the electron and hole recombination process. Thus, the increase in charge carrier concentration and reduction in the electron and hole recombination

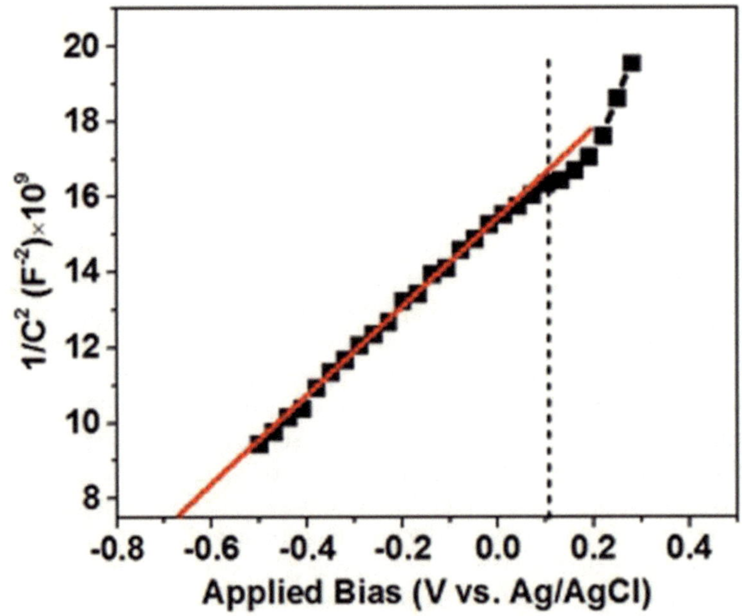

Figure 5.
Mott-Schottky plot of the dendritic heteronanostructures as working electrode at the frequency of 1 kHz. Solid red colored line corresponds to the linear fit to the data. The measurement was carried out in the dark using a PEC cell in a three electrode configuration (Dadwal et al., DOI: 10.1016/j.ijhydene.2018.10.163) [1]. Reprinted with permission from Dadwal et al., (2018), Elsevier.

in the company of Ag dendrites at the near surface region of Si nanowires, are believed to be the main subsidizing factor for enhancing the photocurrent density. On the other hand, the value of V_{fb} was found about −0.7 V and was acquired by extrapolating the interception of the straight line with the axis of the applied potential (**Figure 5**). It is to be noted that the V_{fb} is close to the onset potential which means that the photogenerated charge carrier separation efficiency is improved in the absence of band bending [1]. Such types of result have been also reported in the case of zinc oxide nanowire photoanodes with ultrathin Titania shells for the study of water splitting [34]. Thus, in this case, an improvement in the charge carrier separation can give a positive effect of increasing the carrier diffusion length and the eventual enhancement of the PEC water splitting.

In the PEC solar water splitting, the hydrogen gas evolution rate depends largely on the photocurrent density J_p and external bias V_{ext} between the photoanode with respect to the reference electrode. Thus, hydrogen gas evolution can be clearly observed from the Pt counter electrode. As the photogenerated current is given by the rate of change of the total charge (q) per second, the total number of carrier electrons (n) corresponding to the generated photocurrent, are calculated from the expression $n_e = \frac{J_p t}{q}$ per unit area per second. Where, t is the illumination time in the PEC water splitting. Since in the PEC water reduction reaction ($2H^+ + 2e^{-1} \rightarrow H_2$), two electrons correspond to the one hydrogen gas molecule. Thus, the total number of the evolved hydrogen gas molecules (n_{H_2}) can be found by dividing the above expression by two, that is, $n_{H_2} = \frac{J_p t}{2q}$. Results showed that at the onset voltage $J_p \sim 1.7$ mA cm^{-2}, n_{H_2} value was found to be 5.3×10^{15} per unit area per second. The value of n_{H_2} increased to 6.5×10^{15} per unit area per second for the saturation photocurrent density J_p^{sat} ~1.7 mA cm^{-2} under the illumination. In order to justify these results, hydrogen gas evolution as a function of illumination time has been analytically studied using the measured J_p (see **Figure 6**). The mole of the evolved hydrogen gas (N_{H_2}) during the PEC solar water splitting was calculated from the Faraday's laws of electrolysis $N_{H_2} = \frac{A}{2F}\int_0^t J_p \, dt$ [35]. Here, A = 3 cm^2 is the light receiving area and F = 96485 C/mole represents the Faraday constant (charge in coulomb carried by one mole of electrons). The evolved hydrogen gas during the PEC solar water splitting was calculated for 1 h. It is noted that the calculated amount of the hydrogen gas shows nearly a linear behavior over the illumination time indicating the good constancy of the dendritic heteronanostructures as working electrodes in the water. An average evolution rate of the hydrogen gas was found to about 1.7 μ mol min^{-1}. This generation rate is comparatively higher than the other semiconductor nanostructures that had been reportedly used as photoelectrodes for water splitting [14, 35–37].

The working electrode (photoanode) requires a V_{ext} versus the reference Ag/AgCl electrode in the presence of light illumination so as to derive the overall PEC solar water splitting reaction. The solar-to-hydrogen conversion efficiency (STH), defined as the quantity of the hydrogen gas evolved at the counter electrode under the light illumination with regard to the quantity of incident light photons and can be expressed as $STH(\%) = \Delta G^o \left[\frac{\{Hydrogen\ generation\ rate(moles^{-1})\}}{I_o \times Light\ receving\ area(m^2)} \right] \times 100$ [36]. Here, ΔG^o = 237 kJ mole^{-1} is the Gibbs free energy stored in the hydrogen molecules. The value of STH was derived for the dendritic heteronanostructures using water as electrolyte under the white light enlightenment of I_o = 50 mW cm^{-2} (**Figure 7**).

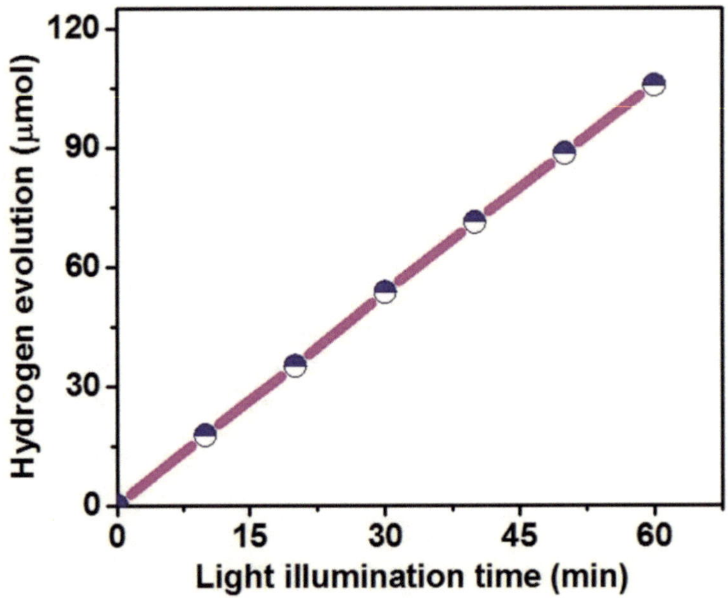

Figure 6.
Hydrogen gas evolution as a function of the white light illumination time. The evolution was calculated from the measured photocurrent current density under a white light illumination of intensity 50 mW cm^{-2}.

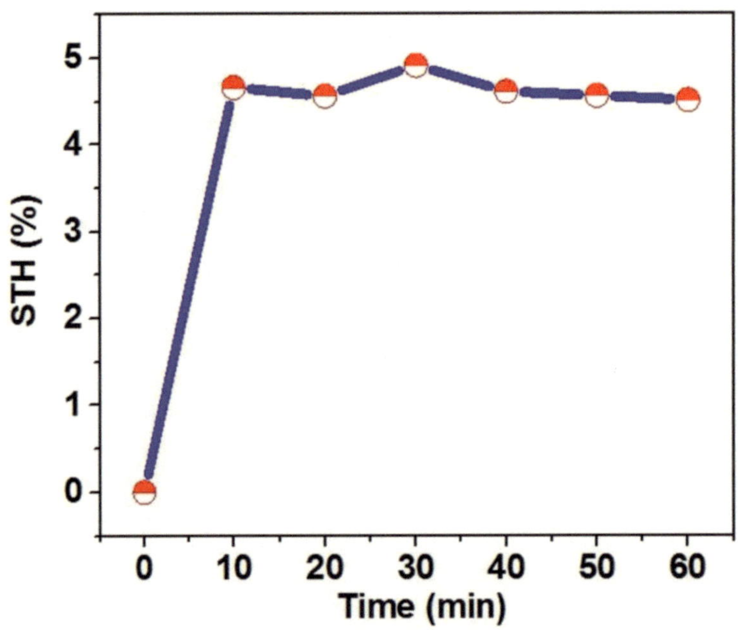

Figure 7.
Solar-to-hydrogen conversion efficiency (STH) of the dendritic heteronanostructures photoanodes as a function of applied bias. PEC solar water splitting was performed under a white light enlightenment of intensity 50 mW cm^{-2} using a PEC cell in a three electrode arrangement.

The maximum value of the STH ~4.5% near the J_p^{sat} ~1.7 mA cm^{-2} at V_{ext} = 0.4 V was found to be quite higher as compared to the earlier reports [36, 38]. This indicates the vibrant impact of the dendritic heteronanostructures on the PEC natural water splitting where the enhancement in STH is resulted from the rise in *Jp* owing to the lower value of R_{ct} between the photoanode and water boundary.

Thus, with the usage of silicon and lower cost metal assisted wet chemical etching method semiconductor dendritic nanostructures have been synthesized. The synthesized photoanode nanostructures are found to be suitable for the effective PEC splitting of natural water.

4. Conclusion

Metal assisted wet chemical etching of silicon resulted in the synthesis of semiconductor dendritic heteronanostructures. In the PEC water splitting, these dendritic heteronanostructures as working electrode showed reasonably good solar-to-hydrogen conversion efficiency of 4.5%. Photocurrent density was found to be enhanced due to the availability of (a) additionally large surface area of such type of nanostructures for the light absorption and (b) the presence of silver dendrites which increase the charge carrier transfer by minimizing the carrier recombination process. Low charge transfer resistance of the working electrode and water boundary and the higher carrier concentration resulted in efficient PEC water splitting. Hydrogen gas evolution showed a negligible decrease with an average rate of ~1.7 μ mol min^{-1} over the light illumination time. This indicates the good stability of the dendritic heteronanostructures as working electrodes in natural water. The solar-to-hydrogen conversion efficiency was found to be about 4.5% indicating that synthesized dendrites nanostructures was working as an efficient working electrode for water splitting.

Acknowledgements

The authors would like to thank the Ministry of Electronics and Information Technology (MeitY), Government of India, India, for the financial support to this work through a research project.

Author details

Uday Dadwal[1*] and Rajendra Singh[2]

1 Nanoscale Research Facility (NRF), Indian Institute of Technology Delhi, New Delhi, India

2 Department of Physics, Indian Institute of Technology Delhi, New Delhi, India

*Address all correspondence to: udadwal@gmail.com

References

[1] Dadwal U, Ali D, Singh R: Silicon-silver dendritic nanostructures for the enhanced photoelectrochemical splitting of natural water. Int. J. Hydrog. Energy. 2018;43:22815-22826. DOI: 10.1016/j.ijhydene.2018.10.163

[2] Shi Z, Wen X, Guan Z, Cao D, Luo W, Zou Z: Recent progress in photoelectrochemical water splitting for solar hydrogen production. Annals of Physics. 2015;358:236-247. DOI: 10.1016/j.aop.2015.04.005

[3] Li X, Zhao L, Yu J, Liu X, Zhang X, Liu H, Zhou W: Water splitting: From electrode to green energy system. Nano-Micro Lett. 2020;12:131. DOI: 10.1007/s40820-020-00469-3

[4] Liu D, Li L, Gao Y, Wang C, Jiang J, Xiong Y: The nature of photocatalytic "water splitting" on silicon nanowires. Angew. Chem. 2015;54: 2980-2985. DOI: 10.1002/anie.201411200

[5] Varadhan P, Fu HC, Kao YC, Horng RH, He JH: An efficient and stable photoelectrochemical system with 9% solar-to-hydrogen conversion efficiency via InGaP/GaAs double junction. Nat. Commun. 2019;10:5282. DOI: 10.1038/s41467-019-12977-x

[6] Cho Seungho, Jang JW, Lee KH, Lee JS: Research update: Strategies for efficient photoelectrochemical water splitting using metal oxide photoanodes. APL Materials, 2014;010703. DOI: 10.1063/1.4861798

[7] Nakamura A, Ota Y, Koike K, Hidaka Y, Nishioka K, Sugiyama M, Fujii K: A 24.4% solar to hydrogen energy conversion efficiency by combining concentrator photovoltaic modules and electrochemical cells. Appl. Phys. Express 2005;8: 107101. DOI: 10.7567/APEX.8.107101

[8] Riha SC, Vermeer MJ, Pellin MJ, Hupp JT, Martinson BF: Atomic layer deposition of a submonolayer catalyst for the enhanced photoelectrochemical performance of water oxidation with hematite. ACS Appl. Mater. Interfaces 2013;5:360-367. DOI: 10.1021/nn305639z

[9] Takeda H, Koike K, Inoue H, Ishitani O: Development of an efficient photocatalytic system for CO_2 reduction using rhenium(I) complexes based on mechanistic studies. J. Am. Chem. Soc. 2008;130:2023-2031. DOI: 10.1021/ja077752e

[10] Faraji M, Yousefi M, Yousefzadeh S, Zirak M, Naseri N, Jeon TH, Choi W, Moshfegh AZM: Two-dimensional materials in semiconductor photoelectrocatalytic systems for water splitting. Energy Environ. Sci. 2019;12,59-95. DOI: 10.1039/C8EE00886H

[11] Pavlenko M, Siuzdak K, Coy E, Załęski K, Jancelewicz M, Iatsunskyi I: Enhanced solar-driven water splitting of 1D coreshell Si/TiO2/ZnO nanopillars. Int. J. of Hydrogen Energy. 2020;45:26426-26433. DOI: 10.1016/j.ijhydene.2019.11.231

[12] Yidan Wang, Cheng Chen, Wei Tian, Weiwei Xu, Liang Li: Designing WO3/CdIn2S4 type-II heterojunction with both efficient light absorption and charge separation for enhanced photoelectrochemical water splitting. Nanotechnology. 2019;30:495402. DOI: 10.1088/1361-6528/ab4084

[13] Chen X, Shangguan W: Hydrogen production from water splitting on CdS-based photocatalysts using solar light. Front. Energy. 2013;7:111-118. DOI: 10.1007/s11708-012-0228-4

[14] Tang J, Durrant JR, Klug DR: Mechanism of photocatalytic water splitting in TiO_2. Reaction of water

with photoholes, importance of charge carrier dynamics, and evidence for four-hole chemistry. J. Am. Chem. Soc. 2008;130:13885. DOI: 10.1021/ja8034637

[15] Wang Y, AlOtaibi B, Chowdhury FA, Fan S, Kibria MG, Li L, Li CJ, Mi Z: Photoelectrochemical reduction of carbon dioxide using Ge doped GaN nanowire photoanodes. APL Materials. 2015;3:116106. DOI: 10.1063/1.4935307

[16] Huang Z, Geyer N, Werner P, Boor JDB, Gösele U: Metal-assisted chemical etching of silicon: A Review. Adv. Mater. 2011;23:285-308. DOI: 10.1002/adma.201001784

[17] Li M, Lia Y, Liua W, Yueb L, Lia R, Luoa Y, Trevora M, Jianga B, Baia F, Fua P, Zhaoc YZ, Shenc C, Mbengue JM: Metal-assisted chemical etching for designable monocrystalline silicon nanostructure. Materials Research Bulletin. 2016;76:436-449. DOI: 10.1016/j.materresbull.2016.01.006

[18] Leontis I, Botzakaki MA, Georga SN, Nassiopoulou AG: Study of Si nanowires produced by metal-assisted chemical etching as a light-trapping material in n-type c-Si solar cells. ACS Omega. 2018;3:10898-10906. DOI: 10.1021/acsomega.8b01049

[19] Srivastava SK, Kumar D, Schmitt SW, Sood KN, Christiansen SH, Singh PK: Large area fabrication of vertical silicon nanowire arrays by silver-assisted single-step chemical etching and their formation kinetics. Nanotechnology. 2014;25:175601. DOI: 10.1088/0957-4484/25/17/175601

[20] Ouertani R, Hamdi A, Amri C, Khalifa M, Ezzaouia H: Formation of silicon nanowire packed films from metallurgical-grade silicon powder using a two-step metal-assisted chemical etching method. Nanoscale Research Letters. 2014;9:574. DOI: 10.1186/1556-276X-9-574

[21] Valadez RFB, Aggarwal V, Pacholski C: Fabrication of porous silicon-based optical sensors using metal-assisted chemical etching. RSC Adv. 2016;6:21430-21434. DOI: 10.1039/c5ra26816h

[22] Cerfontaine S, Wehlin SAM, Elias B, Gautier LT: Photostable polynuclear ruthenium(II) photosensitizers competent for dehalogenation photoredox catalysis at 590 nm. J. Am. Chem. Soc. 2020;142:5549-5555. DOI: doi.org/10.1021/jacs.0c01503

[23] Venkataraman A, Amadi EV, Chen Y, Papadopoulos C: Carbon nanotube assembly and integration for applications. Nanoscale Research Letters. 2019;14:220. DOI: 10.1186/s11671-019-3046-3

[24] Kern W: The evolution of silicon wafer cleaning technology. J. Electrochem. Soc. 1990;137:1887-1892. DOI: 10.1149/1.2086825

[25] Tong QY, Schmidt E, Gösele U: Hydrophobic silicon wafer bonding. Appl. Phys. Lett. 1994;64:625-627. DOI: 10.1063/1.111070

[26] Mousavi BK, Behzadirad M, Silani Y, Karbasian F, Mousavi AK, Mohajerzadeh S: Metal-assisted chemical etching of silicon and achieving pore sizes as small as 30 nm by altering gold thickness. J. Vac. Sci. Technol. A. 2019;37:061402. DOI: 10.1116/1.5112776

[27] Um HD, Kim N, Lee K, Hwang I, Seo JH, Yu YJ, Duane P, Wober M, Seo K: Versatile control of metal-assisted chemical etching for vertical silicon microwire arrays and their photovoltaic applications. Scientific Reports. 2015;5:11277. DOI: 10.1038/srep11277

[28] Azeredo BP, Sadhu J, Ma J, Jacobs K, Kim J, Lee K, Eraker JH, Li X, Sinha S, Fang N, Ferreira P, Hsu K: Silicon nanowires with controlled

sidewall profile and roughness fabricated by thin-film dewetting and metal-assisted chemical etching. Nanotechnology. 2013;24:225305. DOI: 10.1088/0957-4484/24/22/225305

[29] García MID, Gómez R: Investigating water splitting with $CaFe_2O_4$ photocathodes by electrochemical impedance spectroscopy. ACS Appl. Mater. Interfaces. 2016;8:21387-21397. DOI: 10.1021/acsami.6b07465

[30] Niklasson GA, Qiu Z, Pehlivan IB, Edvinsson T: Impedance spectroscopy of water splitting reactions on nanostructured metal-based catalysts. IOP Conf. Series: Materials Science and Engineering. 2019;503:012005. DOI: 10.1088/1757-899X/503/1/012005

[31] Deng Z, Zhang Z, Lia Y, Liu J, Li J, Liu Y: Electrochemical impedance spectroscopy study of a lithium/ sulfur battery: Modeling and analysis of capacity fading. J. Electrochem. Soc. 2013;160:A553-A558. DOI: 10.1149/2.026304jes

[32] Lopes T, Andrade L, Formal FL, Gratzel M, Sivula K, Mendes A: Hematite photoelectrodes for water splitting: evaluation of the role of film thickness by impedance spectroscopy. Phys. Chem. Chem. Phys. 2014;16:16515-16523. DOI: 10.1039/ C3CP55473B

[33] Naseri N, Sangpour P, Moshfegh AZ: Visible light active Au:TiO_2 nanocomposite photoanodes for water splitting: Sol–gel vs. sputtering. Electrochim. Acta. 2011;56:1150-1158. DOI: 10.1016/j.electacta.2010.10.080

[34] Liu M, Nam C Y, Black CT, Kamcev J, Zhang L: Enhancing water splitting Activity and chemical stability of zinc oxide nanowire photoanodes with ultrathin titania shells. J. Phys. Chem. C. 2013;117:13396-13402. DOI: 10.1021/jp404032p

[35] AlOtaibi B, Harati M, Fan S, Zhao S, Nguyen HPT, Kibria MG, Mi Z: High efficiency photoelectrochemical water splitting and hydrogen generation using GaN nanowire photoelectrode. Nanotechnology. 2013;24:175401. DOI: 10.1088/0957-4484/24/17/175401

[36] Benton J, Bai J, Wang T: Enhancement in solar hydrogen generation efficiency using a GaN-based nanorod structure. Appl. Phys. Lett. 2013;102:173905. DOI: 10.1063/1.4803926

[37] Liao L, Zhang Q, Su Z, Zhao Z, Wang Y, Li Y, Lu X, Wei D, Feng G, Yu Q, Cai X, Zhao J, Ren Z, Fang H, Hernandez FR, Baldelli S, Bao J: Efficient solar water-splitting using a nanocrystalline CoO photocatalyst. Nat. Nanotechnology. 2014;9:69-73. DOI: 10.1038/nnano.2013.272

[38] Zhang RQ, Liu M, Wen Z, Jiang Q: Prediction of silicon nanowires as photocatalysts for water splitting: Band structures calculated using density functional theory J. Phys. Chem. C. 2011;115:3425-3428. DOI: 10.1021/ jp111182c

Chapter 12

Synthesis and Catalytic Activity Studies of Silver Nanoparticles Stabilized in Polymeric Hydro Gel

Jaya T. Varkey

Abstract

Silver nanoparticles (AgNPs) are one of the most vital and fascinating nanomaterials among several metallic nanoparticles that are involved in biomedical applications. But their stabilization towards agglomeration is a serious concern. Synthesized silver nanoparticles can be dispersed in polymeric hydrogel for stabilization and can be efficiently used in heterogeneous catalysis. Polystyrene crosslinked with 1, 6-hexanediol diacrylate can be suitably functionalized for catalytic activities. The nature of the support has a profound influence on the reactivity of the polymeric resin. A flexible support with optimum hydrophilic and hydrophobic balance enhanced the reactivity of the supporting system. Using this supported AgNPs catalytic reduction of Para-nitro phenol can be easily accomplished comparing to conventional method.

Keywords: Silver nano, polymer, hydrogel, catalyst, P-Nitrophenol

1. Introduction

Nanomaterials may offer novel answers for technological challenges in various areas, like medicine, solar energy conversion, catalysis, and water treatment. Nanoparticles exhibit improved activities compared to their parent metal atoms because of their increased surface area [1]. These usual activities may find applications in different fields such as catalysis, biosensing, imaging and drug delivery [2–8]. But significant problem with the synthesis of nanoparticles is their stabilization. Crosslinked polymers with suitable functionality can be used as an efficient support for incorporating nano particles. Polymer embedding represents the simplest way to protect the nanoparticles and providing a means of stabilization in handling and applications [9]. The incorporation of Ag NPs in the polymeric hydrogel matrix facilitates the recycling of catalyst without reducing the catalytic efficiency. In polymer supported catalysis, Styrene-divinyl benzene (PS-DVB) copolymer is the most used supporting system. Due to the hydrophobic and rigid nature of polymer backbone, PS-DVB systems show low metal ion uptake. Comparing to that 1, 6-hexanediol diacrylate (HDODA) cross-linked polystyrene (PS HDODA) possess optimum hydrophobic hydrophilic balance and is more flexible than PS–DVB system [10]. Thus PS-HDODA has been introduced as a more convenient catalyst support than PS-DVB system.

1.1 Silver nanoparticles

Silver nanoparticles are nanoparticles of silver of between 1 nm and 100 nm in size. AgNPs are one of the most crucial and interesting nanomaterials among various metallic nanoparticles that are involved in biomedical applications. AgNPs play a major role in nano chemistry, particularly in nanomedicine due to their antibacterial properties. Although several noble metals have been used for many applications, AgNPs have been concentrated on potential applications in cancer diagnosis and therapy [11]. Due to their distinct properties, they have been used for several applications, including as antibacterial agents, in industrial, household, and healthcare-related products, consumer products, medical device coatings, optical sensors, and cosmetics. Also found applications in pharmaceutical industry, diagnostics, drug delivery, as anticancer agents, and have enhanced the tumor-killing effects [12]. To accomplish the requirement of AgNPs for these applications, several methods have been adopted for its synthesis. But the conventional physical and chemical methods seem to be very expensive and hazardous. An efficient method for synthesizing and stabilizing silver nanoparticles is reported.

The challenges in the application of nano particles could be overcome by incorporation into hydro gels [13]. A hydrogel is a network of crosslinked polymer chains that are hydrophilic in nature. Such polymer supported catalyst systems can be separated from the products by means of simple filtration which led to operational flexibility [14]. Immobilized metal nano particles find application in several fields such as water treatment, food processing, catalysis etc. [15]. Silver nano particles stabilized in sulphonated 1,6-Hexanediol diacrylate (HDODA) cross-linked polystyrene were obtained by in situ reduction of silver nitrate in the hydrogel network. Hydro gel network is a carrier where the in situ reduction of $AgNO_3$ in the presence of $NaBH_4$ as reducing agent occurred.

1.2 Polymeric hydrogel

Hydrogels are cross-linked polymeric networks, which have the potential to carry water within the spaces available among the polymeric chains. The hydrogels have been used extensively in various biomedical applications like drug delivery, cell carriers and entrapment, wound management, and tissue engineering. The ability of hydrogels to hold water arises from the presence of hydrophilic functional groups attached to the polymeric backbone. Due to their crosslinking nature among polymer chains, they are insoluble in solvent systems. Many materials including both naturally occurring and synthetic, fit the definition of hydrogels. A polymeric material can act as a hydrogel when it exhibits the ability to swell and retain a significant fraction of water within its structure but will not dissolve in water. Hydrogels have received considerable attention in the past 50 years, due to their exceptional promise in wide range of applications [16]. They possess a degree of flexibility very similar to natural tissue due to their large water content. The ability of hydrogels to absorb water occurs from the hydrophilic functional groups attached to the polymeric backbone, while their resistance to dissolution attributes to cross-links between network chains. Polystyrene-DVB systems cannot act as efficient hydrogel systems due to their rigid and hydrophobic nature. But if DVB is changed to a flexible crosslinker like Hexanediol diacrylate (HDODA) the system could act as a hydrogel. This is due to the hydrophobic-hydrophilic balance and the flexible nature of the crosslinker (**Figures 1** and 2).

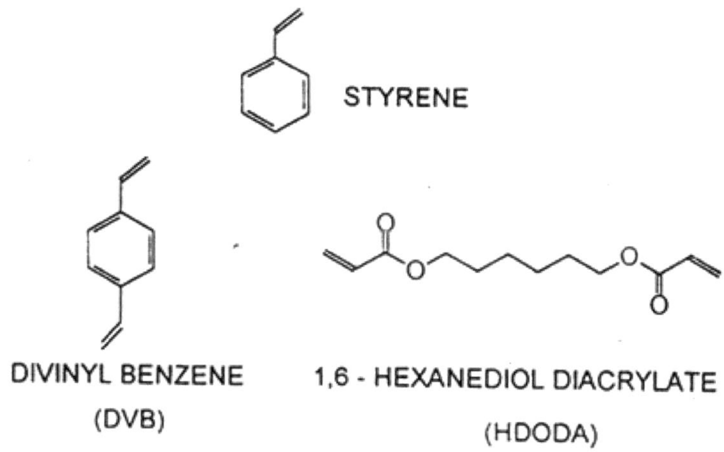

Figure 1.
Structural comparison of DVB and HDODA.

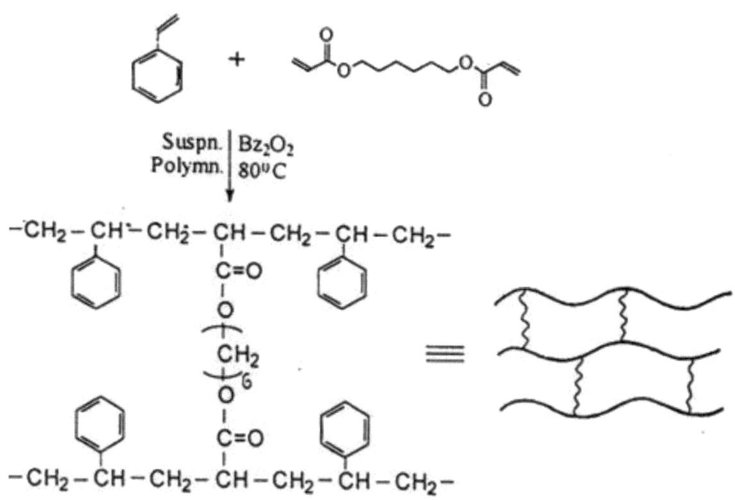

Figure 2.
Scheme of polymer preparation by suspension method.

1.3 Functionalization

Functionalized polymer networks [17] have gained great importance in many fields of scientific research as well as for industrial applications, which will still increase because of the variety of possible modifications of their chemical and physical properties. Macro porous polymer networks prepared from styrene (S) and divinylbenzene (DVB) can act as ion exchange resins after functionalization with sulfonic acid groups. This can play an important role as solid acid catalysts in heterogeneous catalysis. The large amount of work that has been done until now to investigate the catalytic properties of sulfonated PS-DVB polymer have clearly shown that both their catalytic activity and selectivity are dependent on different parameters which are interconnected in a complicated manner. Besides structure and crosslinking density of the polymeric network influences the swelling capacity of the polymeric systems. The catalytic activity depends on the distribution and the local concentration of the sulfonic acid groups in the polymeric network. Various methods for the synthesis of tailor-made functional networks are important for the development of new and improved catalysts. But the rate of sulphonation depends on the flexibility and hydrophilicity of the polymer support. The DVB-crosslinked polystyrene particles are so dense and hydrophobic that it is difficult for the hydrophilic sulfonation agent to enter the internal volume of the particles without the swelling agent. PS-HDODA support has more flexibility and optimum hydrophilic - hydrophobic balance than the conventional PS-DVB resin.

1.4 Sulfonation

Surface functionalization of porous polymer networks was investigated using the sulfonation. Functionalization of the polymer support can be done by using sulfuric acid. The rate of sulfonation depends on the composition of the sulfuric acid used. An increase in the concentration of sulfuric acid (**Figure 3**) leads to higher functionalization capacities within the same reaction time. PS-HDODA resin is functionalized by sulfonation and introduced both as an ion exchanger and as a catalyst support for heterogeneous catalysis. The degree of sulfonation can be controlled by varying the reaction time.

1.5 Ion-exchange capacity measurements

Ion-exchange capacity is the measure of the ability of an insoluble material like crosslinked polymeric supports to undergo shifting of ions attached to it which are loosely incorporated into its structure with oppositely charged ions present in the surrounding solution. Salt splitting titration is used for the determination of ion-exchange capacity of the sulfonated resin. Here the salts of carboxylic acids are decomposed to its acid and base compounds. In this process the salt is passed through ion exchange resins and the corresponding electrochemical reaction occurs and splitting takes place. The ion exchange membranes help to separate the metallic

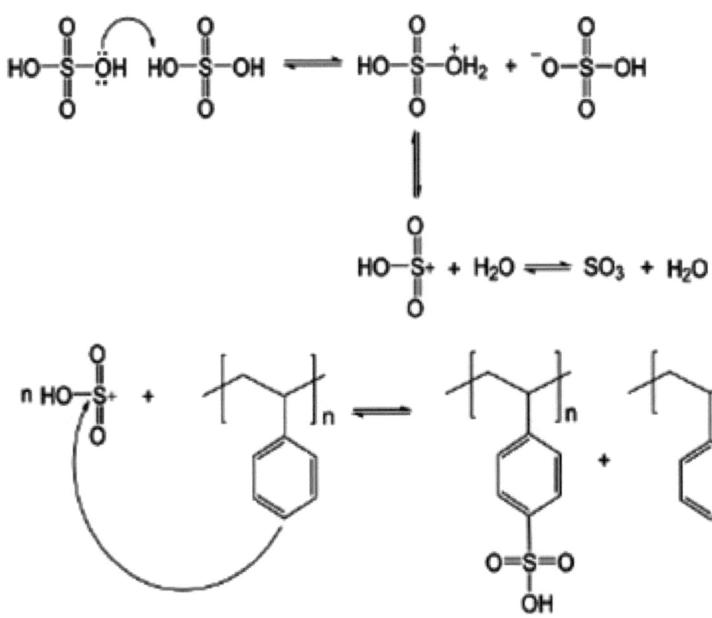

Figure 3.
Mechanism of Sulphonation reaction [18].

and acidic ions. The ion-exchange capacity of the sulphonated resin (IEC, meq/g) was calculated from the following equation.

$$IEC = c \times v / w \tag{1}$$

Where 'c' is the standardized concentration of sodium hydroxide, 'v' is the volume (ml) of the NaOH solution at an end point and 'w' is the weight (g) of determined sulphonated resin.

2. Metal loading to the polymeric support

Metal and metal oxide nanoparticles exhibit unique properties about sorption behaviors, magnetic activity, chemical reduction, ligand sequestration, among others. So, they can be effectively used in catalysis and biomedical applications. However, metal and metal oxide nanoparticles lack chemical stability and mechanical strength. Another serious problem is the agglomeration of nanoparticles which can reduce the surface area of the particles. This will lead to decrease in the surface to volume ratio of the particles and eventually bringing down the reaction rate. Proper distribution of nanoparticles into different polymer surfaces can reduce many of these shortcomings without affecting the parent properties of the nanoparticles. Moreover, the polymer support is not acting as an inert support. Proper selection of functional groups on the support may even lead to the enhancement of the properties of nanoparticles. These polymeric hydrogels can be used as microreactors to fabricate nanoparticles. Assembly of Silver nanoparticles into polymeric hydrogel system can be done by chemical reduction of silver ions inside the polymer network.

2.1 Attachment of silver nanoparticles to sulphonated PS-HDODA resin

Silver nanoparticles has found applications in many fields of nanoscience including medicine and catalysis. Also, silver is commonly used in consumer products due to its antibacterial properties. These properties of silver nanoparticles made them in immense number of studies. The various troubles developed while utilizing these nano particles could be controlled by incorporating these into hydro gels. Hydrogels are polymer supports with optimum hydrophobic-hydrophilic balance. Polymer supported catalyst systems can be easily recovered from the products after the reaction by filtration, which leads to operational flexibility. Polymer anchored metal nano particles find applications in various fields such as water treatment, food processing, medicine, and catalysis. A new polymer support, polystyrene crosslinked with HDODA was introduced, which after sulphonation can be effectively used as an efficient support. Silver nano particles incorporated into sulphonated HDODA cross-linked polystyrene were obtained by in situ reduction of silver nitrate in the hydrogel network. Sulphonated PS-HDODA is acting as a hydro gel network and as a carrier where the in situ reduction of $AgNO_3$ in the presence of $NaBH_4$ as reducing agent occurred.

2.2 Catalytic activity of silver nanoparticles stabilized in sulphonated PS-HDODA support

Nanomaterial-based catalysts are usually heterogeneous catalysts broken up into metal nanoparticles to enhance the catalytic process. Metal nanoparticles have high surface area, which can increase catalytic activity. Nanoparticle catalysts can be easily separated and recycled. The activity of catalytic system does not only depend upon size and shape of nanoparticles immobilized on different stabilizing systems, but it is also affected by stabilizing system. The nature of the support system has a profound influence on the reactivity of the nanoparticles anchored on it.

Para-nitro phenol (p-NP) is a harmful pollutant which is poisonous to living organisms [19]. These are commonly used in the manufacture of pesticides, insecticides, fungicides, and find applications in pharmaceuticals and synthetic dye industry [20]. Living things can get exposed to p-NP by many ways such as inhalation, ingestion, and dermal contact. Adverse effects of p-NP on health depends on duration, exposure dose, pathways, and individual characteristics such as gender, nutritional status, lifestyle etc. Exposure of p-NP by inhalation or ingestion can cause headache, drowsiness, nausea, cyanosis, and eye irritation [21]. One of the most toxic derivatives of the commonly used parathion insecticide is p-NP and it is carcinogenic and cytotoxic [22]. It is highly soluble in water bodies such as fresh water, marine environments and in industrial wastewater and that makes the water sources unhealthy. Because of this high solubility rate removal of p-NP from wastewater is not effective by traditional water purification methods. For water purification from contaminants catalytic approach plays an important role. Catalytic reduction of p-NP to p-Aminophenol which is an important chemical intermediate in the preparation of pharmaceuticals, dyes, pesticides, photographic developers is an efficient solution for the effective removal of p-NP [23–28]. Silver nanoparticles are widely used due to its size, shape, and antibacterial properties. It can be effectively used for the catalytic reduction of p-NP after stabilizing in polymeric PS-HDODA hydro gel. Reduction of p-NP using silver nanoparticles stabilized in sulphonated PS-HDODA resin is an efficient method for its successful removal. $NaBH_4$ can be as used as reducing agent. After catalysis, catalyst can be separated by simple filtration. The percentage catalytic efficiency can be easily calculated.

3. Morphology and Characterization

The surface morphology of the nanoparticles affects the uniform dispersion of NPs in the polymeric hydrogels. These surface properties are significant and that can affect the properties of the hybrid nanocomposite hydrogels. In this regard, the surface fracture of the hybrid hydrogels and the dispersion of Ag NPs in the nanocomposite were analyzed by SEM and TEM.

The surface morphology of the polymer beads was studied by Scanning electron microscopy. Bead forms of the copolymer retain morphology. It was found that the prepared PS-HDODA copolymer beads were having very smooth surface (**Figure 4**). But after sulphonation the external surface of the sulphonated copolymer beads showed scales and cracks indicating effective sulphonation. After the loading of silver nanoparticles, the external surface of the polymer beads became more and more rough, and the morphology has changed as evident from **Figure 4**.

The EDX image (**Figure 5**) of Silver nanoparticles loaded sulphonated PS-HDODA has clearly indicated the presence of metal loaded into the polymer.

The TEM image of sulphonated PS-HDODA resin showed very smooth surface (**Figure 6**). In metal loaded resin, the silver nano particles appeared as dark spot inside sulphonated PS-HDODA resin with the size in the range of approximately 30 nm.

Catalytic approach plays an important role for the purification of water from hazardous contaminants. Sodium borohydride reduction of p-NP using the silver nanoparticles loaded sulphonated PS-HDODA resin is a useful reaction to study

Figure 4.
SEM image of (a) PS-HDOD a resin, (b) Sulphonated PS-HDODA resin, (c) silver nanoparticles loaded sulphonated PS-HDODA resin.

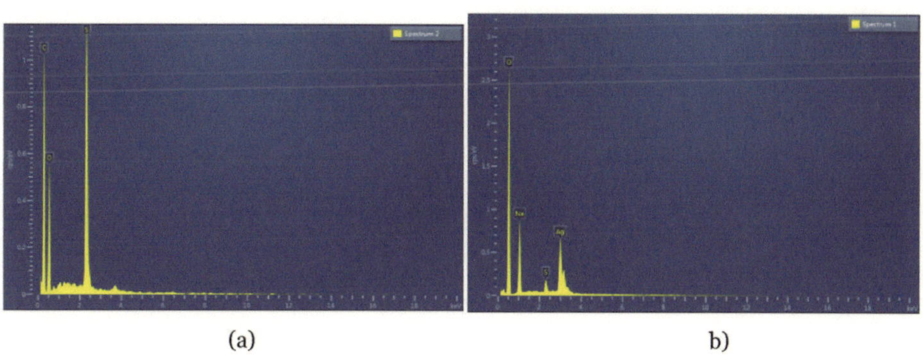

Figure 5.
EDX image of (a) Sulphonated PS-HDODA resin, (b) silver nanoparticles loaded sulphonated PS-HDODA resin.

the catalytic activity of supported silver nanoparticles (**Figure 7**). The percentage catalytic efficiency was calculated using the following equation.

$$\%\text{efficiency} = \text{Co} - \text{C} / \text{Co X } 100. \quad (2)$$

Where 'Co' is the initial absorbance of p-NP and 'C' is the absorbance of p-NP at the time t. From the percentage efficiency plot (**Figure 8**), as time increases the

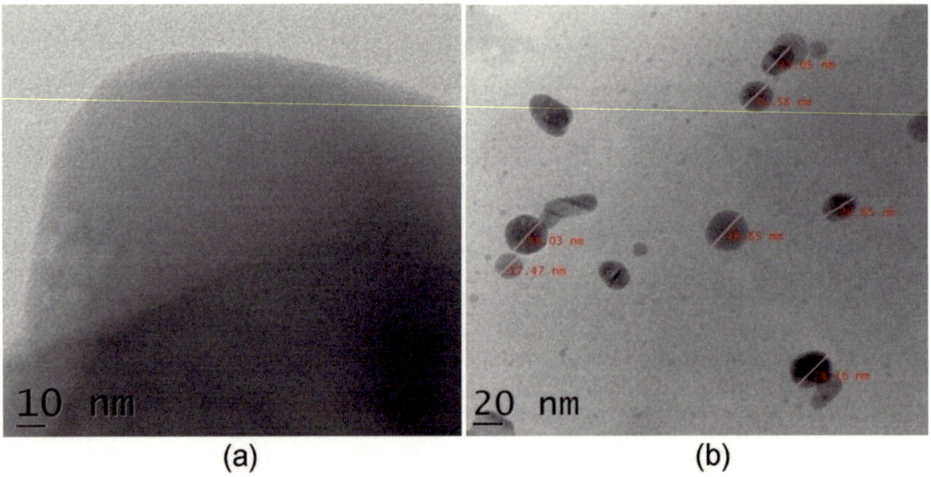

Figure 6.
TEM image of (a) Sulphonated PS-HDODA resin, (b) silver nanoparticles loaded sulphonated PS-HDODA resin.

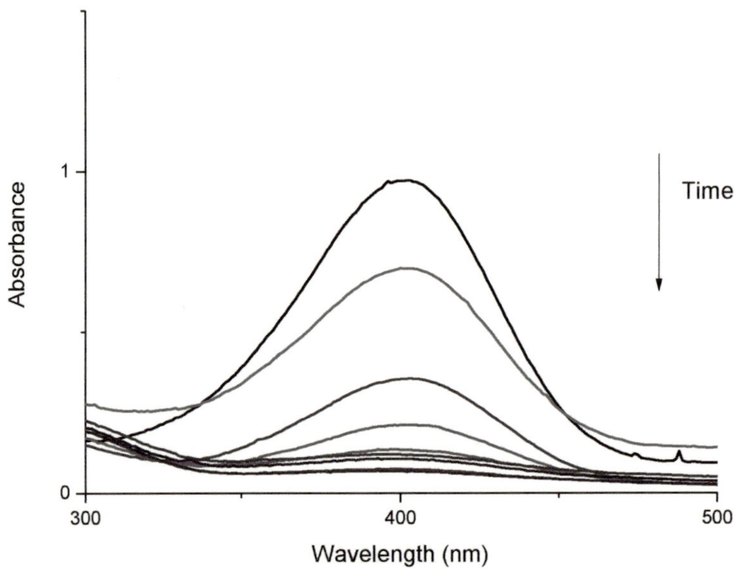

Figure 7.
UV- visible absorbance spectra for catalytic reduction of p-NP.

efficiency of the catalyst was also increased. From the experimental data the rate of the reaction corresponds to first order kinetics.

$$\ln C / Co = -kt. \quad (3)$$

Where Co is the initial absorbance of the reaction system, C is the absorbance at time t. From kinetic curve rate constant k was calculated for the reaction system (**Figure 9**).

In the FTIR spectrum the resins exhibited a band around 1720–1724 cm^{-1} indicating carbonyl stretching frequency, which is an indication HDODA cross-linking to the polymeric support system. Bands at 1598–1657 cm^{-1}, 1473–1493 cm^{-1} and 1443–1451 cm^{-1} corresponds to phenyl ring stretching vibrations of it C=C-C

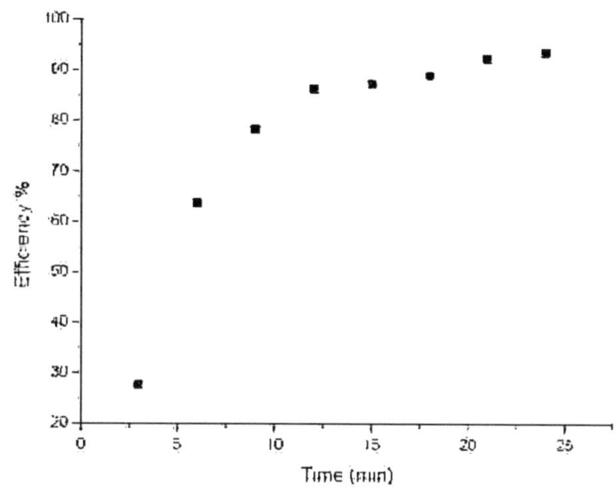

Figure 8.
Efficiency % with irradiation time for catalytic reduction of p-NP.

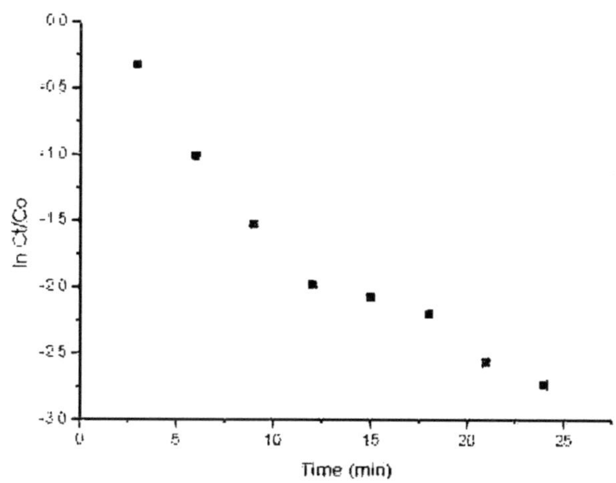

Figure 9.
Kinetic curves for p-NP reduction.

conjugative system. The frequency in the range 1166 cm^{-1} was attributed to S=O symmetric stretching vibration. When silver metal nanoparticles loaded the S=O stretching frequency lowered to 1130 cm^{1}. 1033–1078 cm^{-1} and 907 cm^{-1} represented phenyl in-plane and out of plane C-H bending vibrations respectively. 700–759 cm^{-1} phenyl ring wagging and twisting, respectively.

4. Conclusion

Polymers found application in many fields because of their highly useful properties such as high strength, low weight, and low cost of preparation. Although polymers are essential to modern society, several major challenges block their uses in many fields. One of the major problems is that almost majority of world's supply of polymeric materials is derived from fossil resources, which is a resource that is consuming rapidly and affects the environment significantly. A potential solution to both challenges is to replace fossil fuel-derived polymers with those produced from renewable resources.

The polymer support obtained by the copolymerization of Styrene and HDODA can be effectively used in polymer supported reactions. Suspension polymerization can be used for the efficient synthesis of copolymer beads. PS-HDODA copolymer beads were converted into hydro gel by sulphonation of the cross-linked polymer. Metal loaded to this polymeric support was converted into nano scale by sodium borohydride reduction. Preparation of silver nano particles stabilized in this polymeric hydro gel can be used as an efficient catalyst system for heterogeneous catalysis.

PS–HDODA copolymer beads can be synthesized by suspension polymerization and functionalized by sulphonation using con. Sulfuric acid, and the ion exchange capacity of the obtained resins were determined by salt splitting titration. Metal ions can be loaded into the polymeric support by equilibrating the polymeric ligand with the metal salt solution. The metal loaded to the polymeric support was reduced into the nano scale by sodium borohydride. Sulphonated resin showed swelling in polar solvents indicating the presence of hydrophilic group in the polymeric system. Before sulphonation PS-HDODA copolymer beads showed no ion-exchange capacity due to the absence of sulphonic acid group.

The catalytic activity of prepared silver nanoparticles stabilized in polymeric hydrogel can be tested against p-NP reduction reaction. The catalyst showed 93% efficiency in the reduction reaction within 24 min from the start of the reaction. Sulphonated PS-HDODA system can be successfully utilized as supporting material for preparation and storage of stable silver nanoparticles and the supported catalyst system can be efficiently used for the treatment of wastewater containing p-NP.

Conflict of interest

The author declare that there is no conflict of interest regarding the publication of this chapter.

Author details

Jaya T. Varkey
Department of Chemistry, St. Teresa's College (Autonomous), Ernakulam, Kochi, Kerala, India

*Address all correspondence to: jayavarkey@yahoo.com

References

[1] Schmid, G. In nanoscale Materials in Chemistry, Klabunde, K. J., Ed.; Wiley-Interscience: New York 2001; pp 15-59.

[2] Sudeep, P.K.; Ipe, B. I.; Thomas, K.G.; George, M.V.; Barazzouk, S.; Hotchandani, S.; Kamat, P. V. Nano Lett. 2002, 2, 29-35.

[3] Merschdorf, M.; Pfeiffer, w.; Thon, A.; voll, S, Gerber, G. Appl. Phys. A 2000, 71, 547-552.

[4] Zhong, C.-J.: Maye, m. M. Adv. Mater. 2001, 13, 1507-1511.

[5] Li, X. G.; Takahashi, S. Watanabe, K.; Kikuchi, Y.; Koishi, M. Nano Lett. 2001, I, 475-480.

[6] Sun, X.; Gutierrez, A.; Ycaman, M. J.; Dong, X.; Jis, S. Mater. Sci.Eng. A 2000, 286, 157-160.

[7] Kim, Y.; Johnson, R. C.; Hupp, J. T. Nano Lett. 2001, I. 165-167.

[8] Ruiz, A.; Arbiol, J.; Cirera, A.; Cornet, A.; Morante, J. R. Mater. Sci. Eng.C 2002. 19, 105-109.

[9] L. Nicolaisb and G. Carotenuto, Metal- Polymer Nanocomposites, Wiley, Hoboken, 2005.

[10] Shiney Baby and KS Devaky*, Effect of polymer characteristics on Beckmann Rearrangement of polymer supported ketoximes, Polym Int, 2002, 51:1411-1416.

[11] Gurunathan S., Park J.H., Han J.W., Kim J.H. Comparative assessment of the apoptotic potential of silver nanoparticles synthesized by *Bacillus tequilensis* and *Calocybe indica* in MDA-MB-231 human breast cancer cells: Targeting p53 for anticancer therapy. Int. J. Nanomed. 2015; 10:4203-4222.

[12] Chernousova S., Epple M. Silver as antibacterial agent: Ion, nanoparticle, and metal. Angew. Chem. Int. Ed. 2013; 52:1636-1653

[13] Praveen Thoniyot, Mein Jin Tan, Anis Abdul Karim, David James Young, and Xian Jun Loh* Nanoparticle – hydrogel composite: Concept, Design, and applications of These Promising, Multi-Functional Materials, Wiley-VCH Verlagb GmbH& Co. KGaA, Weinheim, Adv. Sci. 2015, 2, 1400-1010

[14] Maura, M.R.; Kumar, M.; sweta Sikarwar, B. React Funct Poly 2006, 66, 808.

[15] Ruparelia, J.P.; Chatterjee, A; Duttagupta, S. P.; Mukherji, S. Strain specificity in antimicrobial activity of silver and copper nanoparticles. Acta Biomater. J. A. Moore (Ed.), 4, 2008, 707-716.

[16] Ahmed, E. M., Hydrogel: Preparation, characterization, and applications: A review, Journal of Advanced Research, 6, 2, 2915, 105-121.

[17] Ferruti, P., Functionalization of polymers, Reactions on polymers, 1973, 73-101.

[18] Mulijani, S.K. Dahlan and A. Wulanawati, Sulphonated Polystyrene Copolyer: Synthesis, Characterisation, and Its Application of Membrane for Direct Methanol Fuel Cell (DMFC), International Journal of Materials, Mechanics and Manufacturing, 2014, Vol. 2, No.1.

[19] Sadanand Pandey, Shivani B. Mishra, Catalytic reduction of p-nitrophenol by using platinum nanoparticlesstabilised by guar gum, Carbohydrate polymers 2014, 113, 525-531.

[20] Agency for Toxic substances and Disease Registry (ATSDR), 1990.

[21] U.S Department of Health and Human Services (USDHHS), 1993.

[22] Banik, R. M., Prakash, M.R., & Upadhyay, S.N Microbil biosensor based on whole cell of pseudomonas sp. for online measurement of p-nitrophenol. Sensors and Actuators B: Chemical, 2008. 131, 295-300

[23] Marais, E., & Nyokong, T. Adsorption of 4-nitrophenol onto Amberlite IRA-900 modified with metallophthalocyanines. Jouranal of hazardous materials, 2008, 152, 293-301

[24] O'Connor, A. O., & Young L.Y. Toxicity and anaerobicbiodegradability of substituted phenols under methanogenic conditions. Environmental Toxicology and Chemistry, 1989, 8, 853-862.

[25] Dieckmannand, M.S., & Gray, K.A., A comparison of the degradation of 4-nitrophenol viadirect and sensitised potocatalysis in TiO_2 slurries Water research, 1996, 30. 1169-1183

[26] Bo, L. L., Zhang, Y.B., Quan, X., & Zhao, B. Microwave assisted catalytic oxidation of p- nitrophenol in aqueous solution using carbon supported copper catalyst. Journal of hazardous materials, 2008, 153, 1201-1206.

[27] Modirshahla, N., Behnajady, M. A., & Mohammadi- Aghdam, S. Investigation of the effect of different electrodes and their connectionson the removal of efficiency of 4-nitrophenol from aqueous solution by electrocoagulation. Journal of Hazardous materials, 2008, 154, 778-786.

[28] Canizares, P., Sez, C., Lobato, J., Rodrigo, M. A. Electrochemical treatment of 4-nitrophenol containing aqueous wastes using boron dopped diamond anodes. Industrial & Engineering Chemistry Research, 2004, 43, 1944-1951.

Chapter 13

Silver Nanoparticles in Various New Applications

Ainil Hawa Jasni, Azirah Akbar Ali, Suresh Sagadevan and Zaharah Wahid

Abstract

The use of silver in antimicrobial management is very ancient. Silver-based materials have proven interesting, practical, and promising for various applications. Silver nanoparticles (AgNPs) have been one of the nanostructures most studied and investigated over the past several years. AgNPs have greater specific properties depending on their size and form. These noble synthesised metrics have numerous optical, electrical, catalytic, and optical characteristics. These properties are ideal for many fields, depending on their size and shape. The outbreak of multiple infectious diseases has been a major strain on global economies and the public health sector. Extensive treatments have been suggested for disease control in environments containing infectious diseases through advanced disinfectant nanomaterials. This chapter investigates the application and mechanism of silver nanoparticles in certain nanobiotechnology sectors as a useful nanomaterial. In the sense of the market statistical survey research, AgNPs are emerging as one of the fastest developing product groups in the nanotechnology industry, providing a wide variety of nanosilver products in various applications. Lastly, due to the massive use of AgNPs in products recently, there are many concerns about AgNPs toxicity and safety had also been discussed.

Keywords: nanotechnology, metal-nanoparticles, silver nanoparticles, nanomaterials and applications

1. Introduction

The phrase 'nanotechnology' was proposed in 1974 by Norio Taniguchi, a researcher at the University of Tokyo, who pointed to the possibilities of treating nanometric materials with dimensions between 1 and 100 nm. In this variety of sizes, particles show new physical and chemical properties that can be used in various fields of science, such as medicine, food, textiles, chemicals, and energy, to name a few. Nanotechnology is a dynamic interdisciplinary science consisting of nanochemistry, nanophysics, nanomaterials science, nanoelectronics, optoelectronics and nanoengineering, nanobionics and nanometrology, nanodevice-building and nano-craft [1].

Metal nanoparticles are currently starting to be used as silver (Ag), gold (Au), and copper (Cu) with biochemical, optical, and magnetic properties, of which AgNPs are the most studied because of their antimicrobial ability for microbes, viruses, and fungi [2]. AgNPs have been used in nanomedicine and also being

IntechOpen

known as antimicrobial agents and disinfectants without adverse effects. AgNPs have been used against both aerobic and anaerobic microbes. As correlated with other compounds, great interest in AgNPs increased because it shows more efficacy [3]. Its antimicrobial properties can be linked to many internal systems such as communicating with the S–S disulfide bond of the metabolic enzymes, cleaving cells, and disrupting the respiratory chain [4].

Antimicrobial resistance is a public health problem increasing day by day as it is caused by microorganisms that gradually became resistant to drug therapy. Population overcrowding, urbanisation, inadequate water supply, and lack of environmental hygiene are also the major factors of rapid disease outbreaks. Many scientists had developed new effective antimicrobial reagents against the increasing number of microbial resistances with various antibiotics, due to clinical limits on antibiotic prescription. The use of smaller-size AgNPs with a larger surface area to volume ratio as they can enter cell cytoplasm rapidly and kill microorganisms effectively was proven compared to larger AgNPs. At very low concentrations of 5 mg/ml only, AgNPs act as effective antimicrobial activity against both gram-positive and negative bacteria. Furthermore, it has been revealed that AgNPs solution showed an anti-fouling effect on certain bacteria strains [5].

Fungal infections are also very common in immunosuppressed patients. Overcoming fungal diseases is a boring process due to the small number of antifungal drugs available in the current scenario. There is a solution to the urgent and inevitable development of antifungal agents that would be non-toxic, biocompatible, and eco-friendly. At this part, AgNPs play an important role as antifungal agents against various fungal diseases. AgNPs were tested towards different phytopathogens like *Fusarium solani*, Curvularia lunata, and showed efficient antifungal activity [6].

Nowadays, the AgNPs synthesis process is cost-effective, economical, energy-efficient, and offers healthy workplaces to societies. Hence, health and environmental safety surely leading to less waste and safer goods. The utilisation of the plant method can be useful in comparison to other biological processes. The plant-based method can shorten the time-consuming processes compared to microbes and retain their capacity during AgNPs synthesis.

1.1 New emerging applications of AgNPs

The most promising nanomaterials for current commercial applications are AgNPs. They function as antibacterial agents and are used in textiles, electronics products, the medical industry, environmental applications, coatings, food preservation, and other applications. AgNPs are widely used for different uses from home appliances and are used as a disinfectant for sewage treatment and surgical equipment for sterilisation [7]. The most promising nanomaterials for current commercial applications are AgNPs are used as a coating for cardiovascular implants and central venous and neurosurgical catheters, in latex membranes where AGNPs are applied as a biomaterial for a skin regeneration treatment and monitor the delivery rate of the nanoparticle membranes [8].

AgNPs are used in biosensors, where the silver nanoparticles content is used for quantitative detection as biological tags. AgNPs are also incorporated in clothes, shoes, paints, wound dressings, appliances, cosmetics, and plastics. AgNPs are used to improve conductivity in conductive tinting and are incorporated into composite systems. AgNPs are used in opical applications for the effective collection and enhanced optical spectroscopy, including improved metal fluorescence (MEF) and superficial Raman dispersion (SERS). With regard to nanoelectronics, optoelectronics and nanoengineering, innovative technological processes, nanomotors, nanoactuators, nanodevices, micro-opto-electro-mechanical systems (MEMS, MOEMS), ultra-large integrated circuits (ULCI), nano-robots, etc. were the new applications. **Figure 1** is the various applications of AgNPs meanwhile **Figure 2** are the listed advantages and disadvantages of AgNPs utilisation.

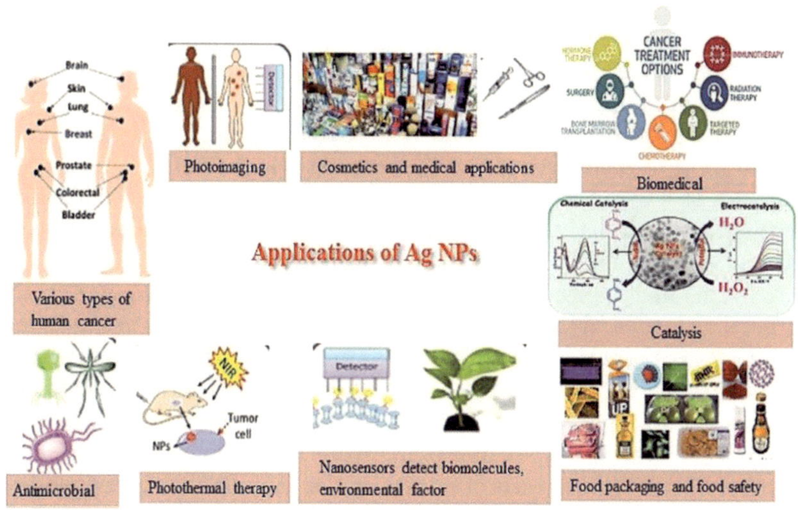

Figure 1.
Various applications of silver nanoparticles.

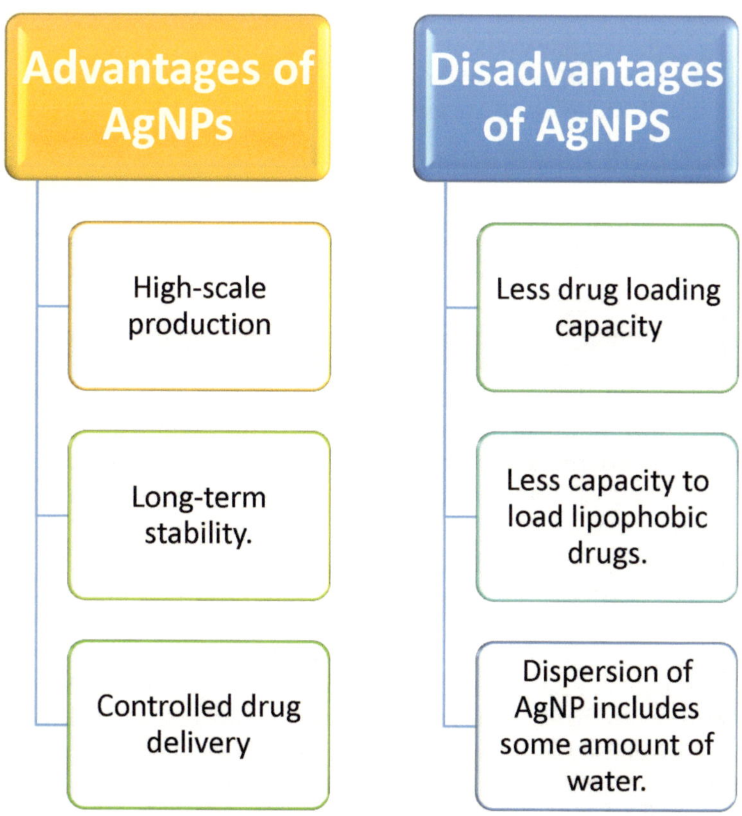

Figure 2.
The advantages and disadvantages of AgNPs utilisation.

2. Nanobiotechnology related applications

Many research techniques are introduced in various physical and chemical methods by scientists to develop metal nanoparticles. However, these approaches for various synthetic compounds are costly. They may result in the presence on the surface of nanoparticles of toxic chemical organisms, which have potentially harmful impacts in diverse biological and biomedical uses [9]. The need to develop environmentally sustainable methods for synthesising nanoparticles by "green synthesis" is increasing.

AgNPs are among the most thoroughly studied nanomaterials and the most common target of the above listed 'green' methods, which fascinate scientists. The research fields of research and the development of nanoparticles with plant extract are emerging. Intracellularly and extracellularly, the plant structures can be used to manufacture various metal nanoparticles [10]. Intracellular processes in nanoparticles' processing include seed plants in high-metal media and hydroponic solutions, such as metal-rich soils. While extract leaves, prepared by boiling or moulding leaves in water or ethanol, are an extracellular processing nanoparticles method [11]. *Medicago sativa* is a plant that can synthesise silver and gold nanoparticles by exploiting its biomolecule which is the first plant recorded used for the extracellular synthesis of a nanoparticle [12]. Since then, plants have gained considerable attention as a medium for the synthesis of nanoparticles. AgNPs have been identified for use in the treatment of wounds, burns, in the development of nano-containing materials for bone implants, dental materials, and antibacterial, antifungal, antiviral, anti-protozoans, anti-arthropods, anti-larvicidal anti-cancerous agents. $AgNO_3$ was used clinically as an important antibacterial agent until the invention of AgNPs [13]. The use of nanostructured materials in nanotechnology has been growing quite rapidly in the last few years. It is used for promising biomedical applications to detect and prevent different forms of diseases. AgNPs can be easily coated with titanium or titanium alloys and are used for dental titanium implants [14].

2.1 Wound healing

Wound healing is a complex biochemical pathway, including several cells that work to regenerate functional skin, including skin cells and immune cells. The use of silver in wound management is very ancient. Pharmaceutical companies and scientists are searching for new antibacterial due to infectious disease attacks and the development of antibiotic resistance. Silver-based products with more complex shapes and increased effectiveness than conventional dressings have been patented and commercialised in wound dressing applications. In wound care, biopolymers combined with a bioactive antimicrobial, antibacterial and anti-inflammatory nanoparticles have great potential to promote wound healing, particularly in the management of diabetic foot ulcers (DFUs), which still pose a huge problem and are associated with high amputation rates and clinical costs [15].

Silver plays an important function in wound healing and, along with its distinctive role in avoiding infection, AgNPs may also facilitate the transformation of fibroblasts into myofibroblasts, which in turn facilitates wound contracting, accelerates the pace of healing, and induces keratinocyte proliferation and relocation [16]. As reported by [17], the effect of nanosilver was tested in vitro and in vivo, and the result showed that, at a concentration of 10 ppm, AgNPs facilitated fibroblast migration, which also demonstrated higher levels of the alpha-smooth muscle actin (alpha-SMA) marker, signalling silver's capacity to turn fibroblasts into myofibroblasts and speed up the healing phase. Several laboratory animal models, such as rat (*Rattus norvegicus*), rabbit (*Oryctolagus cuniculus*), and pig (*Sus scrofa*), are used to test AgNP antibacterial effectiveness in wound healing [18, 19]. Studies in the Sprague–Dawley rat and pathogen-free domestic pigs found that AgNP-loaded dressing speeds up excisional wound healing, while after treatment, the wounds suffer from bacterial infection [18]. Pig explant culture model and mouse experiments have reported that AgNP dressing prevents epidermal cells' proliferation and re-epithelialisation of wounds [20]. The caudal fin of Zebrafish, including the epidermis, blood vessels, nerves, pigment cells, and fibroblast-like cells, has a relatively basic but symmetric structure [21]. The major wound healing mechanisms in mammals include the regenerative adult zebrafish fins [22]. Furthermore, the fin regeneration of adult zebrafish is an ideal model for studying the effects of chemicals on tissue regeneration and wound healing.

2.2 Antimicrobial applications

The interest and understanding of the antibacterial potency of AgNPs have been highlighted with the advancement of nanotechnology. A comparative study of AgNPs, $AgNO_3$, and AgCl find that the antibacterial efficacy of AgNPs is greater than that of free silver ions [23]. The AgNPs have known to be an active bactericide against several bacteria, both Gram-negatives and Gram-positives, including several highly pathogenic bacterial strains [13]. As reported by [24], an experiment was carried out using a model of both Gram-positive (*Staphylococcus aureus*) and Gram-negative (*Escherichia coli*) bacteria to explore the antibacterial properties of AgNPs. *E. coli* development is inhibited at a very low concentration of AgNPs (3.3 nM), which is ten times lower than the minimum inhibitory concentration (MIC) of *S. aureus* (33 nM). The robust antibacterial efficacy and improved stability of AgNPs (10–15 nm) against various drug-resistant bacterial strains were recorded in another study. There are several records that the dose-dependent antibacterial activity of AgNPs is more prevalent against Gram-negative than Gram-positive bacteria [25]. Furthermore, the AgNPs also found potential antibacterial activity against multidrug-resistant gram-negative such as *Klebsiella pneumonia* and *E. coli* and for gram-positive bacteria *like S. aureus*, which isolated from human pathogens. Different parameters impact antibacterial behaviour AgNPs including size and shape, time of exposure [26]. Silver concentration, compound forms, and microorganisms of the targets [27].

Silver reagent ($AgNO_3$, AgCl) has to be high enough to inhibit bacterial cell growth. In the case of aquaculture, one of the greatest problems was preventing diseases caused by viruses, microbes, fungi, and parasites. Traditionally, antimicrobials have been used to eliminate bacterial infections in aquaculture. However, excessive application of these chemicals has caused resistant strains, rendering the treatments not so effective. A previous study by [28] analysed resistant strains in fish farmers in 25 countries showed that tetracycline was the antibiotic most widely used. The isolated tilapia bacteria show a wide variety of antibiotic resistance, such as tetracycline, erythromycin, and streptomycin. Resistant strains were *Aeromonas salmonicida, Photobacterium damselae, Yersinia ruckeri, Listeria sp, Vibrio sp, Pseudomonas* sp, and *Edwardsiella sp.* AgNPs were synthesised to control Vibrio harveyi by *Camellia sinensis* in infected *Feneropenaeus indicus* organisms. In vivo tests have shown that the concentration of 10 μg mL-1 inhibited 70 percent bacterial growth [29]. *Bacillus subtilis*, a non-pathogenic organism used to synthesise nano compounds, was evaluated for its antimicrobial effect on *V. parahaemolyticus* and *V. harveyi* in infected *Litopenaeus vannamei*. The results showed a survival rate of 1% in the control group, but a survival rate of 90% with nano compounds [30]. AgNPs encapsulated with starch and applied in immersion baths (20 minutes) at 10 ng of nanoparticles' concentrations, infected by *Ichthyophthirius multifiliis* and *Aphanomyces* parasites, anti-parasitic and antifungal impact. The findings revealed that the fish recovered after three days without any toxic impact on the use of AgNPs [31]. Much research has been done to prove the effectiveness of silver nanoparticles as disease control in aquaculture (**Table 1**).

2.3 Antifungal applications

The efficacy of AgNPs compared to free silver against a wide range is mostly found in fungi like *Aspergillus, Candid*a, and *Saccharomyces,* and it is as fungicide properties also reported by [32]. Antifungal action of AgNPs in conjunction with heterocyclic compounds like Pyrazolo, thiazolidine, tetrazole, phthalazine, and pyridazine derivatives have been tested against *Aspergillus falvu*s and *C. albicans.* Results have reported enhanced antifungal efficacy of AgNPs combined with heterocyclic compounds in contrast to heterocyclic compounds alone. AgNP antifungal properties against commonly found fungal strains and recorded MIC of AgNPs vs. *Candida albicans, Candida glabrata* varied from 0.4 to 3.3 μg/ml. Additionally, emerging viral diseases are the main threat to human and veterinary sector. **However, AgNPs have established tremendous attention for their antimicrobial properties, but the antiviral properties for AgNPs remain an unexplored**

AgNPs characteristics	Microorganisms	Results
AgNPs with chitosan	*Aliviibrio salmonicida*	MIC, 50 μg mL^{-1} and 2-(3-cyanophenyl)-1H-Benzimidazole (MCB), 100 μg mL^{-1}
Commercial nanoparticles of Al_2O_3, Fe_3O_4, CeO_2, ZrO_2, MgO,	*Aeromonas hydrophila, Bacillus subtilis,Vibrio harveyi,Vibrio parahaemolyticus and Serratia sp.*	The CeO_2 Naps show higher antibacterialthe effect when 10 μg mL-1 concentration wasused.
Nanocid®	*Streptococcus iniae, Lactococcus garvieae, Yersinia ruckeri, Aeromonas hydrophila*	*S. iniae* MBC of 5 to 0.15 μg mL^{-1}, *L.garvieace* MBC of 10 μg mL^{-1} to 0.62 μg mL^{-1}, *A. hydrophila* MBC of 0.31 μg mL^{-1} to <0.15 μg mL^{-1}, *Y. ruckeri* MBCs of 2.5 to 0.62 μg mL^{-1}
Nanoparticles of CuO, ZnO,Ag, TiO_2	*Aeromonas hydrophila, Edwardsiella tarda, Pseudomonas aeruginosa, Flavobacterium branchiophilum, Vibrio spp Staphylococcus aureus, Bacillus cereus and Citrobacter spp.*	Show antibacterial effect in tested strains.
Synthesised nanoparticles with leaves of *Mangifera indica*, *Eucalytus terticonis, Carica papaya* and *Musa paradisiaca* plants	*Aeromonas hydrophila*	Synthesised nanoparticles with *Carica papaya* show antimicrobial activity with 153.6 μg mL^{-1} concentration.
Synthesised AgNPs with *Boerhaavia diffusa*	*A. hydrophila, P. fluorescens and F.branchiophilum.*	AgNPs concentration (50 μg mL^{-1}) was demonstrated higher zones of inhibition (15 mm) for *F. ranchiophylum*, 14 mm for *A. hydrophilla* and (12 mm) for *P. fluorescen.*

Table 1.
The application of silver nanoparticles used as pathogen control in aquaculture.

area. Therefore, a new, unique and develop technique needed to overcome the problem of antiviral resistance. As a result of their possible antiviral efficacy, AgNPs are emerging as a better approach to treating viral infections in HIV-1 and influenza at a certain stage.

3. Food and personal care industry

AgNPs in the food industry, [33] reported for antibacterial and conservation-free action, are commonly used. Most viruses and bacteria are harmless for humans but lethal, so they are used for sanitising food and drink every day, and they are immune to drug infections. Nanosilver fresh food bag is one of the bags that are ready for sale on the market. Because of its anti-fungicidal and anti-bactericidal activity, AgNPs are commonly used in everyday goods such as soap, fruit, plastics and pastes. Nano-based smart and active foods provide many advantages over standard packaging approaches to increase mechanical efficiency, barrier properties and antimicrobial films for nano-sensitive pathogens identification and alert customers to food protection [34]. "Active content" nanocomposites for packaging and material food packaging may also be strengthened with coating. Many researchers are aware of and important to organic compounds' antimicrobial characteristics, such as essential oils, organic acids and bacteriocins as an antimicrobial packaging, polymeric matrices. However, these compounds cannot be used for different processing phases in food susceptible to physical environments that require high temperature and pressure. Inorganic nanoparticles' use makes it possible to obtain good antibacterial activity in low concentrations and become more soluble under intense conditions. It is also noteworthy that these nanoparticles are recently used in antimicrobial food packaging [35]. Control of food quality has become a global, very common problem. AgNPs have unique physicochemical properties that can be used to produce, package and store food. AgNPs could additionally develop into detectors that are capable of recognising or responding to changes in environmental conditions, such as molecular contaminants, microbial quality organisms, gases. Another popular method used is electrochemical detection based on nanomaterial sensors in the food sector [36].

AgNPs had been added into skincare (**Figure 3**), body wash, cosmetics, deodorants, and others with the advancement of personal care technology. This had urbanised the use as antibacterial agents of AgNPs. AgNPs are added as an antiseptic into personal care products for skin problems. Interestingly, AgNPs also have been incorporated in make-up remover cloth, textiles, gloves, bath towels and cleaning fabrics. Such an application's relevance is silver's natural ability to remove

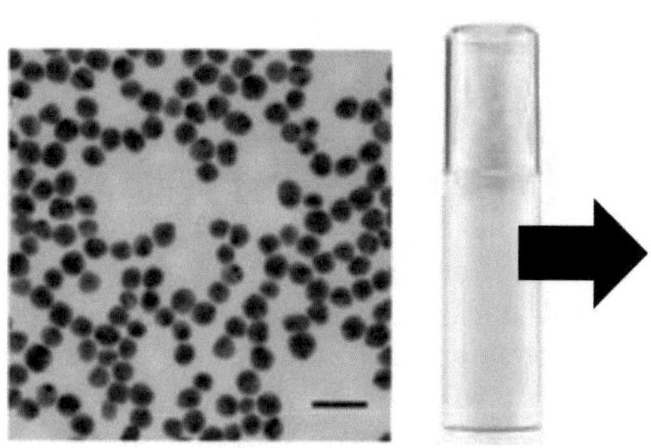

Figure 3.
AgNPs image with 20 nm diameter under a transmission electron microscope (TEM) in a skincare product.

bacteria as a major cause amount of infections in the skin. Silver ions also prevent their persistence and eliminate them very easily on the microfiber fabric. AgNPs are very helpful in maintaining the optimal balance of the skin condition as they preserve a stable pH by preventing usage of other chemicals in a wide range of items. Regardless of gender or age, AgNPs have shown excellent results, including people with the most sensitive skin. It was tested that consumer goods that contain small quantities of AgNPs do not have major effects on health.

4. Nanophysics applications

Nanophysics is an area which studies more the artificial assembly and fabrication of nanostructures as well as research of their external size effects. To identify and measure them, Nanophysics develops different devices and instruments. A range of manufacturing techniques, including e-beam lithography, focused-ion-beam milling, nano-manipulation, and self-assembly, are used to create novel materials, structures and devices [37] Novel nanostructured architectures have been developed in nanomaterial science. In addition, functional nanomaterials and intelligent nanocomponents with special properties have also been manufactured with the integration of AgNPs [38].

4.1 Fabrication of antennas

There are two ways of silver-based antenna fabrication. The first fabrication is with nano-metallic silver, and the other one contains micrometre-sized grains. These two samples are prepared through a thick-film fabrication procedure. The metal powder is used as a material for preparation; the metal oxide is used as an inorganic metal binder and an organic vehicle that evaporates through the initial stages. These samples are usually characterised for electrical performances. It was found that in the lower frequency range, of both samples had similar behaviour with a loss in an electrical parameter and linear increase in the frequency range (from 0.1 dB/mm/GHz up to 80 GHz), but beyond 80 GHz frequency, the prepared AgNPs had a lower electrical loss, for the entire frequency range. The lower loss from the AgNPs and the broad range in loss per wavelength does not rely on the frequency. Hence, it has been summarised that the AgNPs fabricated conductors showed a very less electrical loss with higher frequency range which in turn assigned to the roughness present in the lower side of the nanoparticles because of better packing may allow opportunities for low-temperature fabrication of antennas and sub-THz metamaterials with the improved performance [39].

4.2 In electronically conductive adhesives

AgNPs can be utilised as a silver paste in the electrodes due to their high conducting nature. It is also used as conductive fillers in electronically conductive adhesives (ECAs). A few of the research groups have prepared the AgNPs by reducing the silver nitrate with ethanol in polyvinylpyrrolidone (PVP) [40]. Different reaction conditions have been tuned like PVP concentration, reaction time, and reaction temperature. In this method, PVP prevents aggregation; and also increases the nucleation rate spontaneously and simultaneously decreases the size of the silver nanoparticles. Ethanol is used as the reducing agent to adjust the viscosity of the ECAs. The produced AgNPs with the chemical reduction method showed very fine dispersion and narrow size distribution. Ethanol is also used for the re-dispersing of AgNPs. The absorption peak was recorded at 410 nm, which is an indicator of the quantum size effect of the AgNPs, which is occurring in the absorption property of AgNPs. It is also summarised that the size of the AgNP has been decreased with an increase in the concentration of silver nitrate, the increase in reaction temperature, and the reaction time.

Researchers have synthesised AgNPs, silver nanorods, and epoxy resins containing high-performance electrically conductive adhesives (ECAs) using a novel preparation method. The synthesised nanoparticles and nanorods were dispersed well, with the absence of agglomeration in the matrix. The volume electrical resistivity tests determine the volume electrical resistivity of the ECA was close correlates with the various sintering temperatures and time and time, and the ECA could achieve the volume electrical resistivity of (3–4) × 9 10–5 Ω after sintering at 160 °C for 20 min. Hence, they found that the synthesised and prepared ECA to tend to achieve low-temperature sintering and showed excellent electrical, thermal, and mechanical properties [40]. **Nano metal particles such as Ag, Cu, Zn, and Au are particularly useful for electrical circuitry development because nano-sized metal particles can be shielded from inks and can also be used to boost electrical conductivity. Uniformly formed AgNPs are capable of exhibiting improved electrical conductivity, which makes electronic stuff beneficial. Films with sufficiently smooth and continuous surface morphology may be manufactured by adding AgNPs at a level of 0.05 g/100 g (0.05 per cent) in the CS-GL matrix. It has also demonstrated desirable mechanical strength for industrial packaging. CS-GL-AgNPs at 0.1 g/100 g (0.1 per cent) amount of film proved to be a promising protective packaging material that could increase the shelf life of red grapes by 14 days.**

A study [41] reported a simple method to manufacture surface improved fluorescence spectroscopy (SEFS) substrates based on highly sensitive superhydrophobic Ag nanorods (AgNR) arrays using the glancing angle deposition method at 133 K substrate temperature and resultant coating of molecules of heptadecafluoro-1-decanethiol (HDFT). SEFS substrates display more than 3 times greater fluorescence signal amplification than traditional AgNR films in the HDFT coated superhydrophobic AgNR arrays (In **Figure 4**).

4.3 Ink-jet printing

Ink-jet technology for electronic circuit manufacturing at very low costs has been used, and additional applications in this desirable technology have been noted. It is very interesting to create powerful inks for the versatile display of electronic devices using ink-jet technology. Researchers have prepared the AgNPs through chemical reduction from the silver nitrate using triethylamine to reduce and protect agents. The nanoparticles have been sintered through washing it with

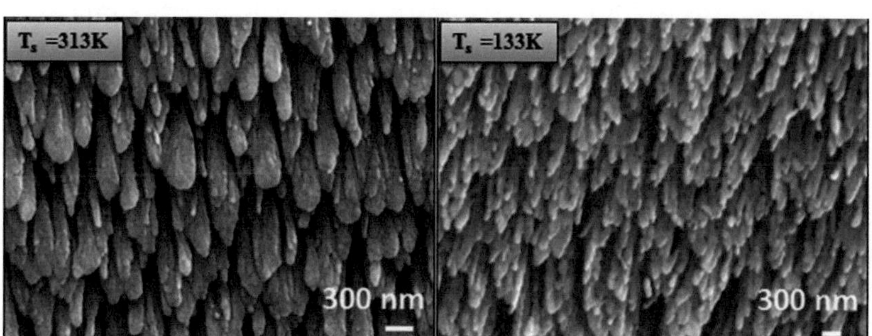

Figure 4.
SEM image of silver nanorods grown on the Si substrate at (a) 313 K and (b) 133 K [41].

acetone and deionised water to remove the particles and organic contaminants present on the surface; after cleaning the film, it was treated with ozone by UVO-100 UV ozone for 30 min. These AgNPs suspensions were spin-coated with 500 rpm, 15 seconds on the polyimide substrate and dried at room temperature to remove the solvent. The resulting AgNPs on the polyimide substrate was heated from 100 to 200 °C and held at 200 °C for 1 h to convert to silver films. The synthesised AgNPs were sintered at different temperatures, and it was shown that the resistivity of the silver film sintered at 150 °C for 1 h was close to the resistivity of bulk silver. Upon the results, the prepared nanoparticles showed a low sintering temperature. Therefore, silver nanoparticles would be used to construct the electronics through ink-jet printing [42].

4.4 Fillers

The micro-sized silver particle fillers are seen as full density silver flakes; they had demonstrated to be highly porous agglomerates. Conjugated polymers with different nanoscale filler inclusions **have been investigated for** sensor applications, including gas sensors, biosensors, and chemical sensors. **Nanofill**ers used include nanowires of metal oxide, nanotubes of **carbon and nanoscale** silver [43]. In the coming decades, nanofiller-based silver nanoparticles and their related products are very promising. In the future, we will face many risks and challenges, in particular energy problems, and research into sustainable energy conversion is expected to explode, both theoretically and experimentally, and polymer nanocomposites will not stand out from this trend, such as self-cleaning or "easy-to-clean" coatings, coated on the surface of a building, protective substrates and glass can help to save energy and water in facility cleaning. In contrast, insulating nanocomposite coatings can help save billions of dollars in energy savings to keep homes in winter.

4.4.1 AgNPs as inorganic filler

Mineral or metallic fillers are reviewed as inorganic fillers. Clay, nano clay [montmorillonite (MMT)], silver nanoparticles (AgNPs), and calcium carbonate ($CaCO_3$) are the most common inorganic reinforcement material-based composites. AgNPs is also a popular metallic filler due to its physicochemical properties, including optical, electrical, catalytic, antimicrobial, and therapeutic properties. It can be synthesised either chemically or biologically. Conventional physical and chemical methods are more costly and toxic than biological methods that can produce nanoparticles in high yields, with good solubility and stability [44]. The biological method is most preferred and very simple, cost-effective, fast, environmentally friendly, and non-toxic. The green approach is preferred for synthesising AgNPs because it uses bacteria, plant extracts, fungi, vitamins, and amino acids far from chemical methods. Few researchers had analysed these nanoparticles via TEM. The resistivity was also evaluated through various levels of filler loading. The AgNPs prepared and synthesised were spheres and nano-sized approximately 50–150 nm in diameter and micro-sized particles with diameters of 5–8 μm, and flakes of silver of 10 μm in length. The data concluded that it was not easy to find the cross-linkage of particles, and there are very lesser chances of different contact and contact area. Through the resistivity measurements, it has been demonstrated that the conductivity of micro-sized silver particle-filled adhesive is influenced by constriction resistance, the AgNPs-containing adhesive is controlled by tunnelling and even thermionic emission. Therefore the particular AgNPs were succeeded to increase the electrical conductivity [45].

4.5 Solar cell optimisation

In recent years, the emerging technology in plasmonic effects in thin-film silicon solar cell is widely studied. It has been a promising application in solar cell fabrication industries due to the nanoscale properties of the silver nanoparticles established in the interface between the metal and dielectric contacts which in turn enhance the light-trapping properties of thin-film silicon solar cells through an increase in absorbance capacity and production of hot electrons that enhance the photocurrents in the solar cell [46]. Silver indium sulphide quantum dots (QD) supported by glutathione (GSH) and polyethylene (PET) as dual-ligands have been synthesised by [47] using an environmentally friendly and reproducible aqueous method. The resulting silver indium sulphide quantum dots present surprisingly long lifespans of 3.69 ls, excellent fluorescent stability and low cytotoxicity, making them the ideal candidate for real-time bioimaging. Thus, they can effectively passivate the surface trap centres and thus reduce non-radiative emissions in **Figure 5**.

4.6 Biosensor fabrication

Few groups have tried to fabricate nano-enzymatic glucose biosensors by depositing AgNPs using the in-situ chemical reduction method on TiO_2 nanotubes, which is synthesised using an anodic oxidation process. Scanning electron microscopy was used to determine the structure, morphology, and mechanical activity of the electrode. It was present both inside and outside of TiO_2 nanotubes whose length and diameter were about 1.2 μm and 120 nm. The composition was constructed and used as an electrode of a non-enzymatic biosensor for glucose oxidation. The electrocatalytic properties of the prepared electrodes for glucose oxidation were investigated by cyclic voltammetry (CVs) and differential pulse voltammetry (DPV). Nano enzymatic glucose biosensors have exciting applications in catalysis and sensor areas [48].

In [49] work, their silver nanostructures were significantly stabilised by the Mercaptopropionic Acid (MPA), and Self Assembled Monolayer (SAM) biosensor detection limit for endotoxin *E.coli* was estimated at 340 pg./ml. We investigated biosensor selectivity in different experiments, taking endotoxin *B.abortus* as the second form of endotoxin contamination in our target samples (HBs-ag developed

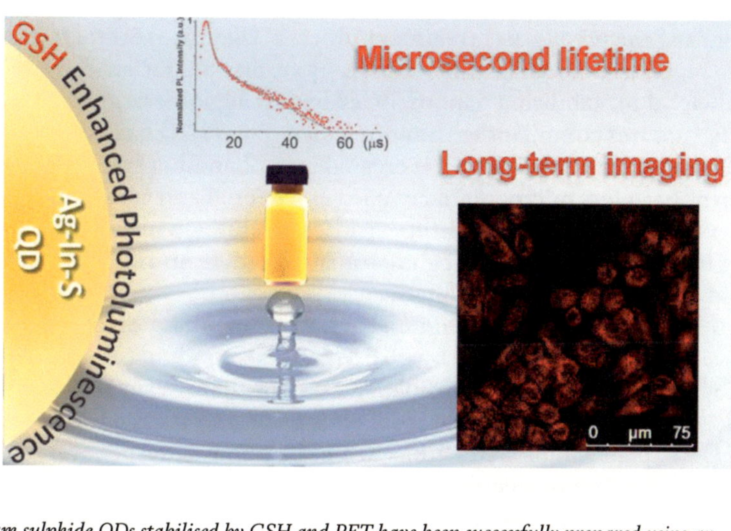

Figure 5.
Silver indium sulphide QDs stabilised by GSH and PET have been successfully prepared using an environmentally friendly and reproducible aqueous route. The resulting QDs have an absolute quantum yield (QY) photoluminescence (PL) of up to 37.2 percent [47].

at the Institute Pasteur, Iran). Overall, this study suggests that Localised Surface Plasmon Resonance (LSPR) biosensing can be considered in quality control laboratories as a sensitive, simple, and label-free method for detecting endotoxins. They showed that this biosensor could selectively detect both forms of endotoxins compared to other biological organisms. The construction of a silver-based LSPR biosensor for endotoxin detection was described. They used the Glancing Angle Deposition (GLAD) method to obtain reproducible nanocolumns of silver.

5. Environmental impacts, toxicity and disposal

The cytotoxicity of AgNPs must also be taken into account, particularly if used to treat chronic wounds. The wound healing and tissue repair mechanisms appear to be complex, and it is unknown if there is a temporary window on the exposure of AgNPs [15]. The toxicity of AgNPs depends on a variety of factors, such as concentration, duration, dosage and particle size. In comparison to (Michigan Cancer Foundation) MCF-7 cell culture, toxicity is dose-dependent and to cause cell damage in Human Epidermoid Larynx (Hep-2) cell line by Reactive Oxygen Species (ROS) formation. AgNPs biogenically synthesised from *Podophyllum hexandrum* leaf extract showed cytotoxicity and apoptotic influence, possibly through caspace-cascade activation and loss of mitochondrial integrity [50].

Metallic nanoparticles in general and AgNPs, in particular, are growing warning triggers for human and environmental managers due to the growing introduction of AgNPs into consumer goods. The enormous use of AgNPs in the products points out so many questions about the toxicity and safety [13]. Release of AgNPs from consumer goods is required to shift in land-based environments, but their fate and transition are very complicated and little are understood about their effect on the climate. AgNPs textile and cosmetic compounds in Europe have the highest environmental exposure when washing and rinsing water in wastewater treatment plants is treated. The resulting silver release from these plants is predicted to be low in the soil and surface waters.

However, owing to the widespread use of silver-based goods, the public is concerned about various aquatic lives when AgNPs reach the sea. Another concern with AgNPs is that they do not differentiate between helpful and harmful microorganisms. Free silver or AgNPs found in urban wastewater have been greatly converted into Ag_2S during the wastewater treatment process. There are several techniques available to regulate and reduce their toxicity by surface alteration. Surface modification is helpful in stabilising nanoparticles against agglomeration and keeping these nanostructures compliant with another step. The WHO has estimated 50% of the biological pollution indoor air is created by air handling systems and the production of harmless microorganisms such as bacterial and fungal pathogens has been observed in air filters. Most of these pathogens develop mycotoxin, which is harmful to human health, decreasing microbial growth in air filter by incorporating antimicrobial Ag-NPs through air filters.

The antimicrobial effect of Ag-NPs on bacterial contamination of activated carbon filters (ACFs) was investigated by [51]. The study found that Ag-deposited "ACF filters" effectively eliminated bioaerosols. Analysis of the antibacterial activity of the Ag-coated "ACF filters" revealed that there were two bacteria called "*Bacillus subtilis*" and *E. coli* is completely inhibited within 10 and 60 minutes, respectively. Polymer air filters made of polypropylene and silver nitrate ($AgNO_3$) were examined for bacterial survival [52]. The results indicated that the addition of antibacterial agent $AgNO_3$ to filter effectively prevented bacteria from colonised filter. The presence of the antimicrobial compound $AgNO_3$ in the filtration systems reduces the number of bacteria observed in Gram-negative and Gram-positive bacterial strains *Micrococcus luteus, Micrococcus roseus, Bacillus subtilis* and *Pseudomonas luteola*.

The significant reduction in bacterial pathogens' growth in silver-treated filters has made the antimicrobial filters therapy technology quite important for the future. Water is one of the most valuable substances on Earth and is necessary for all living beings. About 70 percent of the world is saturated with water, but just 0.6 per cent is suitable for human use. Access to clean water is a major health and social challenge in many developed countries. There has been considerable interest in water disinfection using AgNPs. Chemicals formed by nanosilver (chem-Ag-NPs) can be uniformly decorated on porous ceramic materials to form Ag-NPs-porous ceramic, composite materials using 3-aminopropyltriethoxysilane (APTES) as a molecule interaction [53]. Since silver and AgNP are commonly used in clinical fields, more study is required into the cytotoxic effects of normal cells and cancer cells. AgNPs have been observed as toxic to some of their mammalian cells. The IC50 was of 5 lg/mL (GT-AgNP) and 272.14 ± 0.09 lg/mL (C-AgNP), for example, in AgNPs (C-AgNP and GT-AgNP), synthesised with coffee and green tea extract. Besides, the susceptibility of NIH/3 T3 to the AgNPs was improved from cervical cancer, with IC50, 14.26 ± 0.05 lg/mL (GT-AgNP). AgNPs demonstrated cytotoxic effects by cell line and by the exposure concentration. Significantly cytotoxic activity against the cells of B16F10 (mouse melanoma; IC501/26.43 ± 3.4108 g/mL), SKOV3 (human ovarian, carcinoma; IC501/416.24 ± 2.4808 g/mL), A549 (human lung adenocarcinoma; IC501/43 ± 33978 g/mL) and PC3 (human prostate cancer; IC501/41 ± 28 g/mL) were observed in AgNPs synthesised with fungal *Pestalotiopsis microspora* (mL) [54]. The silver accumulation's major organs are the liver and kidney, irrespective of whether the AgNPs are intravenous, oral or nasal [36]. AgNPs from diverse resources has different impacts on the wellbeing of mammals. In vivo experiments in rats treated with 10 mg/kg AgNPs synthesised with *Ficus religiosa* leaf extract showed an accumulation of silver in the liver, brain and lungs at a concentration of 4.77, 3.94 and 3.043 μg/g of tissue respectively (In **Figure 6**) [55].

AgNPs are not only detrimental to humans; they are also harmful to fish. The toxicity in Zebrafish is measured. Suggested findings The LD50 for silver nitrate was 100 lg/L, but the LD50 was 80 and 400 lg/L for chemical and plant synthesis AgNPs, respectively [36]. Fungal-derived AgNPs did not significantly impact *Poecilia reticulate* after 48 h of exposure to 3.41 mg/L. This may mean that fungal-derived AgNPs are the least toxic among the numerous AgNPs, whereas chemical-derived AgNPs are usually more toxic than silver nitrates. AgNPs also possess size-dependent toxicity to Zebrafish. For example, smaller size AgNPs appear to cause higher mortality rates and lower hatchability rates, resulting in higher embryotoxicity than larger particle sizes [56]. The cytotoxicity of AgNPs is influenced by many factors, including shape, size, composition, surface charge, and capping molecule or coating. **Uncoated Ag NPs are more harmful than Ag NPs, which are coated. AgNPs should be kept in the dark at a temperature of 4 °C as they are photosensitive. AgNPs that are not properly stabilised can rapidly oxidise and easily mix in solutions** [57].

5.1 Future concerns

Nevertheless, more research is still needed to explore the potential risks of inorganic metallic filler (e.g., AgNPs) towards ecological, human, and animal activities. Inorganic fillers are not biodegradable, and it varies with different synthesis approaches and methods for the quality. Hence, clay and calcium carbonate is the most common inorganic fillers used in research since they are usually environmental-friendly and inexpensive.

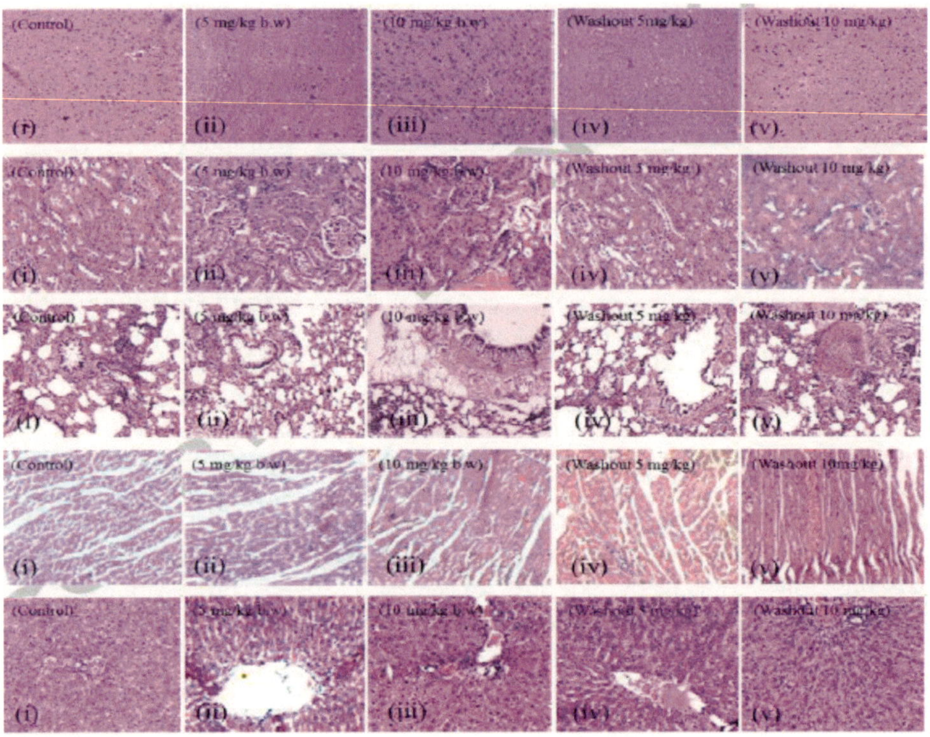

Figure 6.
No histopathological changes have been observed in the kidneys, brain, heart, lungs, and spleen from rats treated with FRAgNPs of 5 and 10 mg/kg b.w on day 29 as well as on day 89 [55].

6. Conclusion

The use of AgNPs as antibacterial agents must, therefore, be extended. Moreover, extensive research and development towards applying for multifunctional AgNPs and study on human health and the environment impacts are much needed. Then, the development of technology to prevent systemic side effects is also required. Numerous studies showed that AgNPs could slow down the viability of the antiviral activity's exact mechanism remains elusive and demand further testing. In future, the environmentally sustainable and advance approach of AgNPs synthesis could provide a viable alternative method to traditional physical and chemical methods.

Author details

Ainil Hawa Jasni[1]*, Azirah Akbar Ali[2], Suresh Sagadevan[3] and Zaharah Wahid[4]

1 Department of Biotechnology Engineering, Faculty of Engineering, International Islamic University Malaysia, Kuala Lumpur, Malaysia

2 School of Biological Sciences, Universiti Sains Malaysia, Penang, Malaysia

3 Nanotechnology and Catalysis Research Centre, University of Malaya, Kuala Lumpur, Malaysia

4 Department of Science in Engineering, Faculty of Engineering, International Islamic University of Malaysia, Kuala Lumpur, Malaysia

*Address all correspondence to: a.hawa.jasni@gmail.com

References

[1] Pokropivny V, Lohmus R, Hussainova I, Pokropivny A, and Vlassov S. Introduction to Nanomaterials and Nanotechnology, Ukraine,: Tartu University Press, 2007.

[2] Marquez JCM, Partida AH, Dosta, MDCM, Mejia, JCM and Martinez, ABSilver nanoparticles applications (AgNPs) in aquaculture.," International Journal of Fisheries and Aquatic Studies., 2018, 6(2) 05-11.

[3] Shah A, Lutfullah G, Ahmad K, Khalil AT, Maaza M. Daphnemucronata-mediated phytosynthesis of silver nanoparticles and their novel biological applications, compatibility and toxicity studies. Green Chem Lett Rev. 2018; 11(3):318-33.

[4] Ilangovana R, Subhaa V, Earnest Ravindranb RS, Kirubanandanb Sand Renganatha S. In Sabu T. and Balakrishnan P. Chapter 2: Nanomaterials: Synthesis, physicochemical characterisation, and biopharmaceutical applications., Nanoscale Processing, Elsevier Book. Amsterdam, Netherlands. 2021, 36-42.

[5] Manish D., Biosynthesis of Silver Nanoparticles (AgNPs) and Its Applications. Center of biotechnology, Slideshare. 2018.

[6] Abdel-Aziz SM, Prasad R, Hamed AA, Abdelraof M Fungal nanoparticles: A novel tool for a green biotechnology? Fungal Nanobionics: Principles and Applications, Springer. 2018; 61-87.

[7] Yaqoob AA, Umar K., and Ibrahim MNM. Silver nanoparticles: various methods of synthesis, size affecting factors and their potentials- a review. Applied Nanoscience. 2020.

[8] Kumar, Jaanbee Shaik and Anitha C. In Sabu T. and Balakrishnan P. Chapter 11: Nanostructures for biomedical devices., Nanoscale Processing, Elsevier Book. Amsterdam, Netherlands. 2021, 299-316.

[9] Mittal AK, Bhaumik J, Kumar S, and Banerjee, UC. Biosynthesis of silver nanoparticles: elucidation of prospective mechanism and therapeutic potential. Journal of Colloid Interface Science 2014(415): 39-47.

[10] Renugadevi K, and Aswini RV. Microwave irradiation assisted synthesis of silver nanoparticles using *Azadirachta indica* leaf extract as a reducing agent and In vitro evaluation of its antibacterial and anticancer activity. International Journal of Nanomaterials and Biostructures 2012; 2: 5-10.

[11] Parashar V, Parashar R, Sharma B, and Pandey AC. Parthenium leaf extract mediated synthesis of silver nanoparticles: a novel approach towards weed utilisation. Digest Journal of Nanomaterials and Biostructures. 2009(4): 45-50.

[12] Gardea-Torresdey JL, Gomez E, Peralta-Videa JR, Parsons JG, Troiani H, and Jose-Yacaman M. Alfalfa sprouts: a natural source for the synthesis of silver nanoparticles. Langmuir 2003(19): 1357-1361.

[13] Anjum S, Abbasi BH and Shinwari ZK. Plant-Mediated Green Synthesis of Silver Nanoparticles For Biomedical Applications: Challenges And Opportunities. Pakistan Journal of Botany 2016; 48(4): 1731-1760.

[14] Sameer J Nadafa, Sandip Bandgar, Indrayani D Raut, In Sabu T and Balakrishnan P. Nanostructures for antimcrobial therapy., Nanoscale Processing, Elsevier Book. Amsterdam, Netherlands. 2021, 362-375.

[15] Pang S, Gao Y, Wang F, Wang Y, Cao M, Zhang W, Liang Y, Song M, and

Jiang G. Toxicity of silver nanoparticles on wound healing: A case study of zebrafish fin regeneration model. Science of the Total Environment 2020(717): 137-178.

[16] Paladini F, and Pollini M. Antimicrobial Silver Nanoparticles for Wound Healing Application: Progress and Future Trends. Materials 2019; 12, 2540.

[17] You C, Li Q, Wang X, Wu P, Ho JK Jin R, Zhang L, Shao H, and Han C. Silver nanoparticle loaded collagen/chitosan scaffolds promote wound healing via regulating fibroblast migration and macrophage activation. Scientific Reports 2017(7); 10489.

[18] Lin YH, Hsu WS, Chung WY, Ko TH, and Lin JH. Silver-based wound dressings reduce bacterial burden and promote wound healing. International Wound Journal 2016(13); 505-511.

[19] Masood N, Ahmed R, Tariq M, Ahmed Z, Masoud MS, Ali I, and Hasan A. Silver nanoparticle impregnated chitosan-PEG hydrogel enhances wound healing in diabetes-induced rabbits. International Journal of Pharmaceutics, 2019; 559, 23-36

[20] Burd A, Kwok CH, Hung SC, Chan HS, Gu H, Lam WK and Huang L. A comparative study of the cytotoxicity of silver-based dressings in monolayer cell, tissue explant, and animal models. Wound Repair Regeneration. 2007; (15) 94-104.

[21] Poss KD, Keating MT and Nechiporuk A. tales of regeneration in Zebrafish. Developmental Dynamics 2003; 226 (2): 202-210.

[22] Sonnemann KJ and Bement WM. Wound repair: Towards understanding and integration of single-cell and multicellular wound responses. Annual Review of Cell and developmental biology 2011(27): 237-263.

[23] Choi O, Deng KK, Kim NJ, Ross L, Surampalli RY and Hu Z. Theinhibitory effects of silver nanoparticles, silver ions, and silver chloride colloids on microbial growth. Water Research, 2008 (42): 3066-3074.

[24] Kim JS, Kuk E, Yu KN, Kim JH, Park SJ and Lee HJ. Antimicrobial effects of silver nanoparticles. Nanomedicine: Nanotechnology, Biology and Medicine, 2007 (3):95-101.

[25] Kumar VP, Pammi S, Kollu P, Satyanarayana K, and Shameem U. Green synthesis and characterisation of silver nanoparticles using. *Boerhaavia diffusa* plant extract and their antibacterial activity. Industrial Crops and Products, 2014, (52): 562-566.

[26] Sadeghi B, Garmaroudi FS, Hashemi M, Nezhad H, Nasrollahi A and Ardalan S. Comparison of the antibacterial activity on the nanosilver shapes: nanoparticles, nanorods and nanoplates. Advanced Powder Technology. 2012 (23): 22-26.

[27] Guzman M, Dille J and Godet S. Synthesis and antibacterial activity of silver nanoparticles against gram-positive and gram-negative bacteria. Nanomedicine: Nanotechnology, Biology and medicine. 2011(8): 37-45.

[28] Tuševljak N, Dutil L, Rajić A, Uhland FC, McClure C, St-Hilaire S et al. Antimicrobial use and resistance in aquaculture: findings of a globally administered survey of aquaculture-allied professionals. Zoonoses and public health, 2013; 60(6):426-436.

[29] Vaseeharan B, Ramasamy P, Chen JC. Antibacterial activity of silver nanoparticles (AgNPs) synthesised by tea leaf extracts against pathogenic *Vibrio harve*yi and its protective efficacy on juvenile *Feneropenaeus indicus*. Letters in applied microbiology. 2010; 50(4):352-356.

[30] Sivaramasamy E, Zhiwei W, Li F, Xiang J. Enhancement of Vibriosis Resistance in *Litopenaeus vannamei* by Supplementation of Biomastered Silver Nanoparticles by *Bacillus subtilis*. Journal of Nanomedicine Nanotechnology. 2016; 7(352):2.

[31] Barakat KM, El-Sayed HS, Gohar YM. Protective effect of squilla chitosan–silver nanoparticles for *Dicentrarchus labrax*. International Aquatic Research. 2016; 8(2):179-189.

[32] Ramya M and Subapriya MS. Green synthesis of Silver Nanoparticles. International Journal Pharmacy, Medical and biology Sciences 2012.

[33] Cushen M, Kerry J, Morris M, and Cruz-Romero ME. Cummins, Nanotechnologies in the food industry recent developments, risks and regulation. Trends Food Science Technology 2012 (24): 30-46.

[34] Mihindukulasuriya SDF and Lim LT. Nanotechnology development in food packaging: A review. Trends Food Science Technology. 2014(40): 149-167.

[35] Verma P and Maheshwari. Application of Silver nanoparticles in diverse sectors. International Journal Nano Dimension, 2019, 10 (1); 18-36.

[36] Zhao X, Zhou L, Rajoka MSR, Yan L, Jiang C, Shao D, Zhu J, Shi J, Huang Q, Yang H and Jin M. fungal silver nanoparticles: Synthesis, Application and Challenges. Critical Reviews in Biotechnology, 2017.

[37] Mohd H R and Nur A I. In Sabu T. and Balakrishnan P. Chapter 1: Basic concepts and processing of nanostructure materials. Chapter 11: Nanostructures for biomedical devices., Nanoscale Processing, Elsevier Book. Amsterdam, Netherlands. 2021, 1-20.

[38] Bhusan B. Introduction to nanotechnology, in Springer Handbook of Nanotechnology, Berlin, Heidelberg, Springer, 2017, 1-19.

[39] Alshehri AH, Jakubowska M, Mlozniak A, Horaczek M, Rudka D, Free C, Enhanced electrical conductivity of silver nanoparticles for high-frequency electronic applications. Applied Materials and Interfaces. 2012;4: 7007-7010.

[40] Chen D, Qiao X, Qiu X, Chen J. Synthesis of electrical properties of uniform silver nanoparticles for electronic applications. Journal of Material Science. 2009; 44:1076-1081.

[41] Kumar, Samir; Goel, Pratibha; Singh, Dhruv P; Singh, J P. Highly sensitive superhydrophobic Ag nanorods array substrates for surface enhanced fluorescence studies. Applied Physics Letters, 2014, 104(2), 023107.

[42] Wu JT, Hsu SL. Preparation of triethylamine stabilised silver nanoparticles for low temperature sintering. Journal of Nanoparticle Research. 2011; 13(1):3877-3883.

[43] Neethumol Va, Tania F, Meril Sb, and Ajalesh B N In Sabu T. and Balakrishnan P. Chapter 14: Nanocomposite of polymer matrices, Nanoscale Processing, Elsevier Book. Amsterdam, Netherlands 2021, 384-401.

[44] Zhang XF; Li, ZG; Shen, Wei; Gurunathan, Sangiliyandi. Silver Nanoparticles: Synthesis, Characterisation, Properties, Applications, and Therapeutic Approaches. International Journal of Molecular Sciences,2016, 17(9), 1534.

[45] Ye L, Lai Z, Liu J, Tholen A. Effect of silver particle size on the electrical conductivity of isotropically conductive adhesives. Transactions on Electronics Packaging Manufacturing.1999; 22(4):299-302.

[46] Sanago R, Maity S, Mehta RK. Plasmonic effect due to silver nanoparticles on silicon solar cell. Procedia Computer Science. 2016; 92:549-553.

[47] Jiao M; Li, Yun J, Yuxiu L, Chenxi; Bian, Hao; Gao, Liting; Cai, Peng; Luo, Xiliang. Strongly emitting and long-lived silver indium sulfide quantum dots for bioimaging: Insight into co-ligand effect on enhanced photoluminescence. Journal of Colloid and Interface Science, 2020, 565, 35-42.

[48] Li Z, Zhang Y, Ye J, Guo M, Chen J, Chem W. Nanozymatic glucose biosensors based on silver nanoparticles deposited on TiO2 Nanotubes. Journal of Nanotechnology. 2016.

[49] M. Zandieh, S. N. Hosseini, M. Vossoughi, M. Khatami, S. Abbasian and A. Moshaii. Label-free and simple detection of endotoxins using a sensitive LSPR biosensor based on silver nanocolumns. Analytical Biochemistry,. 1830-1933, 2018.

[50] Jeyaraj M, Rajesh M, Arun R, Mubarak AD, Sathishkumar G and Sivanandhan G. An investigation on the cytotoxicity and caspase-mediated apoptotic effect of biologically synthesised silver nanoparticles using *Podophyllum hexandrum* on human cervical carcinoma cells. Colloids Surf B Biointerfaces, 2013(102): 708-717.

[51] Invalid reference

[52] Yoon KY, Byeon JH, Park CW and Hwang J.Antimicrobial Effect of Silver Particles on Bacterial Contamination of Activated Carbon Fibres. Environmental Science Technology 2018 (42): 1251-1255.

[53] Miaskiewicz-peska E, and Lebkowska M. Effect of antimicrobial air filter treatment on bacterial survival. Fibres Textiles Eastern Europe 2011 (19): 73-77.

[54] Lv Y, Liu H, Wang Z, Liu S, Hao L, Sang Y, Liu D, Wang J, and Boughton RI. Silver nanoparticle decorated porous ceramic composite for water treatment. Journal Membrane Science 2009(331): 50-57.

[55] Netala VR, Bethu MS, Pushpalatha B, Baki VB, Aishwarya S, Rao JV and Tartte V. Biogenesis of silver nanoparticles using endophytic fungus P*estalotiopsis microspora* and evaluation of their antioxidant and anticancer activities. International Journal of Nanomedicine 2016(11)5683-5696.

[56] Nakkala JR, Mata R, Sadras SR. Green synthesised nanosilver: synthesis, physicochemical profiling, antibacterial, anticancer activities and biological in vivo toxicity. Journal of Colloid Interface Science 2017(499)33-45.

[57] Mosselhy DA, He W, Li D, Meng Y and Feng Q. Silver nanoparticles:in vivo toxicity in zebrafish embryos and a comparison to silver nitrate. Journal Nanoparticle Research 2016;(18) 222.

Section 4

Health Concerns

Chapter 14

AgNano, the Construction of Occupational Health Standards: A Status Update

Guillermo Foladori and Noela Invernizzi

Abstract

The regulation of chemical substances involves a negotiation between social actors to translate controversial scientific evidence about risks into legal norms. This chapter addresses the discussion elicited by a public consultation on a voluntary regulation guide on silver nanoparticles (AgNP) in workplaces. It examines the comments made from 2016 to 2018 by diverse social actors – business representatives, non-governmental organizations (NGO), and independent researchers – to two successive draft versions of a Recommended Exposure Limit (REL) in working environments with AgNP. The REL is a voluntary guideline on permissible exposure limits elaborated by the NIOSH in the U.S. The methodology used was a qualitative content analysis, structured upon a historical and sociotechnical contextualization of nanotechnologies carried out through literature review. The findings show how different social actors position themselves in the controversy, revealing a pattern of behavior consistent with their position in the research, production, and commercialization of this new nanomaterial. While a group of actors, aligned with the interests of AgNP producers, proposed the restriction of mandatory and AgNP-specific regulation, another group of more heterogeneous actors, identified with the interests of workers and consumers, demanded more scientific and technical information and stricter health protection measures.

Keywords: nanosilver, risks, recommended exposure limits, regulation, occupational health

1. Introduction

The regulation of chemical substances involves a difficult negotiation between social actors to translate often controversial scientific evidence about risks and safety into legal norms. When the regulation faces chemical substances with uncertain risk, as in many of the nanomaterials, the difficulties increase.

This article addresses the public discussion of a proposal for a voluntary guide to regulate the exposure limits to silver nanoparticles (AgNP) in workplaces in the United States. The draft guide, known as *Recommended Exposure Limits* (REL), was prepared by the *National Institute for Occupational Safety and Health* (NIOSH) on-demand from the *Centers for Disease Control and Prevention* (CDC), and went through two stages of discussion and rework during 2016–2018. We examine the public online discussion of both drafts by different social actors, basically academics, business organizations, and non-governmental organizations (NGOs).

Analyzing this discussion required placing nanotechnologies in their historical and socio-technical context. Nanotechnology is the intentional manipulation of matter to form new structures with a dimension smaller than 100 nanometers. Nanoparticles have particular physical–chemical properties (electrical, optical, magnetic, thermal, mechanical) and are different from the same material on a larger scale [1]. The interaction of nanoparticles with biological systems is highly unpredictable and their use may involve unknown risks [2].

From mid-2000s there was an explosion of nanotechnology products in the market. Although there are no detailed records, StatNano [3] registered 8 452 products until November 2018, present in almost all economic sectors.

The development of this emerging technology coincides with the wake-up call by the World Health Organization and the United Nations Environment Program on the global pandemic caused by toxic chemicals [4]. These organizations indicate that about five million people die annually from the exposure and handling of chemical substances and contact with consumer items that contain them [5, 6].

Silver is a metal known both for its toxicity and for its healing effects since ancient times [7]. Currently, its use in the form of nanoparticles is blossoming. The inventory of nanotechnology products of the *Woodrow Wilson International Center for Scholars* identified 442 using AgNP, and reports that silver is the most commonly used nanomaterial in the whole set of products [8, 9]. The antibacterial properties of AgNP justify its use in textiles, food packaging, paints, toys, environmental technologies, cosmetics, implants, and other medical devices; they are also used in the electronics industry (semiconductor printing, radio frequency identifiers, flexible circuits, solar panels) [10–12]. The United States produced 20 tons of AgNP in 2010; in 2014 between 450 and 542 tons were produced at a word level [13].

Toxic effects of AgNP on the human organism have been detected when certain exposure levels are exceeded [14]. In the workplace, the AgNP enter the body mainly through inhalation. The final destination within the organism is uncertain. Whereas there was a consensus to consider that were the lungs, more recent research has showed that they can move from the lungs to the liver, and eventually to the spleen and kidneys, accumulating [12], thus exposing workers to a variety of potential health threats. These characteristics of AgNP have raised the concern of CDC of the United States, which has recommended NIOSH to develop a voluntary guide (REL) of permissible exposure for AgNP [15].

The toxicity of silver in larger sizes, when certain exposure limits are exceeded, is already widely known, causing, for example, argyria, and there are safety regulations in this regard [16]. With the increasing use of AgNP, a debate arises about whether, in smaller sizes, such as nanoparticles, the toxicity of silver remains the same, as some of the actors who participated in the public consultation analyzed here argue, or if toxicity manifests itself differently, as other actors claim.

Regarding occupational safety, there are mandatory regulations and voluntary guides. In the United States, a chain of regulations can be identified. The first is the *Occupational Exposure Limits* (OEL), which are scientific studies about the maximum acceptable limit of particles in workplaces of hazardous substances of certain material or class of materials. The OELs are established considering functional categories (exposure period time according to the degree of concentration, maximum exposure, and an emergency category when danger is imminent).

Based on OEL, mandatory workplace standards called *Permissible Exposure Limit* (PEL) are developed. These are prepared by the *Occupational Safety and Health Administration* (OSHA). Voluntary standards, such as the Recommended Exposure Limit (REL) examined in this chapter can also be developed, often based on OELs. These are elaborated by NIOSH.

On December 18, 2015, the NIOSH put out a first REL draft, entitled *Current Intelligence Bulletin: Health Effects of Occupational Exposure to Silver Nanomaterials* [13] for public consultation. This received comments from different institutions, organizations and scientists, from which a second draft was prepared [13] and made public on August 24, 2018. The latter also underwent public consultation, which ended in November 2018. This article examines the two drafts with their corresponding comments available on the *website* of CDC (https://www.regulations.gov/docket?D=CDC-2016-0001). The analysis considers only the comments from the public, since the comments from peer reviewers asked by the agency are anonymous and not publicly available.

The antecedent of this draft REL is the existence of a PEL based, in turn, on a 1988 OEL, which controls the exposure to silver in the workplace. The OSHA imposes a PEL of 10 $\mu m/m^3$ for soluble and powdered silver. Both, the OEL study and the mandatory exposure limit established in the PEL, refer to silver in larger sizes; no standard exists, be it an OEL or a PEL, for nanosized silver. What is under construction, and is discussed here, is a voluntary guide, a REL. It is important to note that, at the beginning of this process, when NIOSH based the first draft of the REL for silver in nanosize on the existing OEL that referred to silver in larger sizes, implicitly proposed an equivalent regulatory treatment for silver and nanosilver. However, as the consultation process advanced, and critical comments were made on this point, NIOSH responded with a second draft that distinguishes nanoparticles in the air, establishing a much lower maximum exposure of 0.9 $\mu m/m^3$ and leaving the original exposure limit of 10 $\mu m/m^3$ for particles in dust, smoke, and soluble compounds. As will be seen, the public consultation evidenced that there are opposing positions regarding whether the OEL for silver is sufficient to elaborate a REL for nanosilver [11, 17].

The commentaries correspond to the following social actors: PISC, PBNS, NIA, CTA, SNWG, Faustman, Oberdörster, and Fox, briefly described in what follows.

- *PETA International Science Consortium Ltd* (PISC) is an international consortium aimed at promoting strategies to replace the use of animals in experiments [18].

- *Pennsylvania Bio Nano Systems* (PBNS) is a one-person company that advises nanotechnology companies regarding technical regulatory matters [16, 19].

- *Nanotechnology Industries Association* (NIA) is an association of companies and other entities related to the production and commercialization of nanotechnology products. It advises national and international institutions and organisms, such as the OECD and the ISO, and has the goal of promoting the use of nanotechnologies [20].

- *International Center for Technology Assessment* (CTA) is an NGO oriented to assess the social impacts of technologies [21].

- *Silver Nanotechnology Working Group* (SNWG) is an enterprise organization that promotes scientific knowledge production and public information regarding the beneficial use of silver nanoparticles in industrial products and final consumption [22].

- *Elaine Faustman* is an investigator for the Institute *Risk Analysis and Risk Communication*, and the *Department of Environmental and Occupational Health Sciences*, at the *University of Washington*, WA [23].

- *Guenter Oberdörster* is a recognized scientist specialized in the toxicology of nanomaterials at the *Department of Environmental Medicine*, of the *University of Rochester*, NY [17].
- *Mary A Fox* is an assistant professor at *Johns Hopkins Bloomberg School of Public Health*, and *Co-director of the Risk Sciences and Public Policy Institute* [24].

2. Methodology

This qualitative investigation was elaborated in four stages, that go from the general to the particular, placing the problem to analyze, which is the regulation of AgNP, within a broader historical and socio-technical perspective. The first three stages were built upon a revision of literature based on consultations to the Scopus, Web of Science, and PubMEd databases, and the analysis of documents and databases of products that use nanotechnologies. The last stage was developed through content analysis of the interventions of the different actors in the public consultation of the REL document.

The first stage consisted of tracing the historical and socio-technical context in which nanotechnologies arose, which allowed us to underscore two aspects. First, the fast growth of nanotechnology products since the beginning of the century, most of them without regulation. Second, the entry of new chemical products such as silver nanoparticles into the market, without prior assessment of their risks, occurred in the context of the pandemic caused by toxic chemicals used in everyday consumer goods, as stated by the Organization World Health (WHO) and UNDP (United Nations Development Program).

The second stage was aimed at identifying the main characteristics of AgNP, both in technical terms and concerning their potential risks.

The third stage was to describe the object of study, that is, the voluntary guideline for AgNP regulation in preparation. This required explaining the main aspects and restrictions of the preparation of a voluntary guide such as a REL, which led to the identification of the contradictory nature of the process of transforming technical-scientific risk criteria into legal standards. Next, the actors (organizations or individuals) who commented on the drafts were identified.

The fourth stage consisted of the analysis of the NIOSH draft document and the comments made by the various actors in the public consultation. The content analysis was based on the information obtained in the previous stages. A voluntary guide such as the REL adapts scientific-technical information regarding the hazards/risks of a work environment with AgNP, to legal norms that involve different actors: The State, private companies, and workers. The first actor creates (and enforces when it is the case) the regulation; the second is the target group of the regulation, and the last group is the main subject of risk and beneficiary of the regulation. From the two fields in interaction, scientific-technical and legal, three dimensions of content analysis were derived, which were formulated as three specific questions to examine the positions defended by the different actors who participated in the discussion of the voluntary guide drafts.

First, based on the available methods, techniques and information, the standards, both mandatory (PEL) and voluntary (REL), conform to a given state of knowledge, feasible for further extension or revision, which, in turn, would lead to the updating of the rules. This raises the problem of how and when to regulate and limit (or expand) the production and marketing of new chemicals. In the case at hand, the question is stated as follows:

- At what point, regarding the progress in research and development (R&D), the production and commercialization of AgNP should a REL and/or a PEL be elaborated?
  Second, since the knowledge on hazard/risk is always incomplete and subject to controversies,

- How is the conflict between insufficient knowledge and administration of risks solved?
  Third, being the main involved actors, namely the State, the companies, and the workers,

- What are the actors' opinions regarding the level of responsibility on risk (the producer, the regulatory organism, or the worker), and what should be the degree of access to the relevant information (confidential or public)?

3. Results and discussion

When examining the social actors who commented on the document, clear differences in the relationship they maintain with the subject under discussion emerge. NIA and SNWG are industrial representatives and PBNS a business advisor. This group of three actors has a conflict of interest regarding the subject because their final goal is the production and incorporation of AgNP in consumer products and their commercialization. PISC is an animal rights defense organization; therefore, it has a conflict of interest regarding the *in vivo* methods of risk assessment. Oberdörster, Faustman, and Fox are researchers from research centers who have declared no conflict of interest in articles published on the subject. CTA is an NGO, based in the United States, aimed at assessing and advising society on the economic, ethical, social, environmental and political impacts that result from the application of technologies, without a declared conflict of interest, although manifestly biased towards workers and consumers. This different location of the actors regarding the subject necessarily determines their perspectives.

In the following subsections, we examine the arguments deployed by the actors in their commentaries to the document under consultation, organizing them around the three questions formulated in the methodology.

I. At what point should REL and/or PEL be elaborated regarding the stage of R&D, production and commercialization of the AgNP?

Despite the enormous variety of nanoparticles, and that each one can imply different health risks for workers, there is an element in common to all of them: the matter in nanoscale shows different biological and physicochemical properties compared to the same matter on a larger scale. Moreover, the same material behaves differently within the range of 1 to 100 nm, depending on its shape, crystallography, number of dimensions in the nanoscale and other characteristics. Nanoparticles' behavior also varies according to the route through which they enter the organism and the exposure time. Regarding silver, and without considering the nano size, several studies indicate different toxicity depending on the way it is presented (dust, soluble, etc.) [16], which already warns that size is associated with distinct toxic effects. In this context, it is relevant to ask: why do new chemicals enter the market without toxicity analysis, or assessed on the basis of methodologies developed for the matter in a larger size, as in the case of nanoparticles? Given the current pandemic caused by toxic chemicals, the uncertainty about the risks derived from the properties of the nano-sized matter, and the existence of sophisticated risk assessment techniques, it is necessary to understand why there is such a temporary lag between entry of products with AgNP to the market and their regulation.

In the REL discussion, NIOSH presented a first draft of the document in 2016, then corrected it in 2018. In none of the versions, mention was made on the contradiction between the elaboration of a REL while AgNP continued to enter the market in various products without any specific regulation. In this way, the NIOSH is trying to "manage the existing situation", adopting an *effective risk management* approach, without any mention to the possibility of modifying the path of production and consumption by controlling the market. There is also no mention of the fact that the regulation under discussion was already delayed, considering the increasing commercialization of products with AgNP since mid-2000s.

Within the commentators, only the CTA refers to this issue and advocates for a moratorium on the commercialization until there is enough data confirming the safety of AgNP:

> *No data should mean no new production [...] companies should stop manufacturing unapproved nanosilver products [21].*

Except for this actor, the NIOSH proposals, and the commentators, take as a natural fact the marketing of products without sufficient data regarding their safety. In doing so, they promote an *ex-post* safety policy, instead of a preventive policy.

II. How is the conflict between insufficient knowledge and administration of risk solved? The question of uncertainty.

The discussion evidences three conflictive areas regarding knowledge and uncertainties on the dangers and risks of AgNP. The first one refers to the relationship between size and toxicological effects. The second relates to the way in which scientific data is interpreted and transformed into legal rules. The last conflictive point has to do with the validity of scientific methods and their limitations.

i. The first area of controversy is the distinction between the effects of silver *vis a vis* nanosilver. The industrial actors and their advisors affirm that nanosilver has the same toxicological behavior as silver in a larger size and that there is already a PEL issued by OSHA on silver, which would make unnecessary the elaboration of a specific REL on AgNP. As already said, the first draft of the NIOSH proposal also considered the risks of AgNPs and silver in larger size equivalent. Later on, NIOSH changed this perspective and acknowledged the different risks of silver in nanoscale and larger sizes [15].[1] The industry working group commented:

> *SNWG is extending support of the Agency's recommendation that effective risk management control practices be implemented so that worker exposures to all forms of silver, including silver nanomaterials, do not exceed the NIOSH REL of 10 μg/m3 (8-hour time-weighted average) for silver metal dust, fume, and soluble compounds measured as a total airborne mass concentration. [...] workers will be more than adequately protected from any potential harmful exposures to all forms of silver, including nanosilver. [...] In light of some of the uncertainties concerning nanosilver, the SNWG believes that the toxicity of nanosilver is not significantly different from bulk or dissolved Ag (colloidal silver) [22].*

[1] Here is not the place to develop on this issue, but the reader must consider both the economic and political power of the chemical industry and the neoliberal phase of capitalism that has been replacing the control of the State over private enterprise by business self-responsibility, a transition from regulation to governance, from *hard to soft law* [25].

The industry working group begins supporting NIOSH's recommendation, but the support is limited to the first draft of the document, where 10 μg/m3 was suggested as a limit, the same limit that OSHA uses for silver in larger size. In the following sentence, the commentator explicitly identifies silver and nanosilver as equal for risk analysis; and, in the last one, emphasizes that the uncertainties are the same for silver and nanosilver. In summary, SNWG argues that there is no need for a specific standard for nanosilver.

From the same opinion is the NIA,

> *[...] the Association insists that silver nanomaterials do not present a different toxicological profile to other forms of silver, including colloidal silver. The antimicrobial action of silver, and therefore its toxicological profile, originates in silver ions (Ag+) and may not be attributed to the nanoparticles themselves [20].*

PBNS, an industry consultant entity, considers argyria as the final point in the organism of silver potential health risks, and argues that the maximum permissible contemplates all types of particles, so there would be no difference between nano and large-scale silver, and, since there is an OSHA PEL for the larger size, the NIOSH should not insist on the specificity of the nanosize. However, contradictorily, PBNS recognizes that nanomaterials can present "unexpected properties"; but, if the NIOSH understands that the endpoint of silver is argyria, there would not be, according to PBNS, a novel effect, and using the nano concept would be incorrect:

> *In selecting argyria as the valid endpoint, there is then no novel use, nor first time exposure nor unexpected property. Yet, using the term nanomaterial implies that there should be a particle size dependence [19].*

The entire PBNS comment goes in the direction of invalidating the specificity of nano and suggesting to follow the OSHA's already approved criteria based on silver in a larger size [19].

It is worth noticing that the three industrial actors' arguments do not rebut the scientific articles published over the past two decades, which provide evidence about the different behavior of AgNP and silver in a larger size -- see, for example, the systematic review of Akter et al. [14]. In doing so, they are simply ignoring the available scientific information that does not fit their interests. Neither they refer to the current uncertainty involving the risks of the matter at the nanoscale, an aspect which is considered a crucial issue in the NIOSH first draft, which demands attention to the likely different risks associated with AgNP in the air, in the solid and liquid forms, due to the different routes of introduction into the organism.

The animal defense NGO (PISC), for its part, only emphasizes the need to replace analysis *in vivo* with *in vitro* and *in silicon;* and by not questioning the equalization of risks in nano and larger scales stated in the first draft of the document, reinforces the business position.

The remaining actors, the NGO and the independent researchers, recognize that nanosilver implies a different risk than silver in a larger size. Independent researchers, for example, explicitly call attention to the specific risks associated with nano size. Faustman argues that:

> *While an OEL for micro-sized silver dust and silver fumes of 10 μg/m3 is in place, we believe that the physicochemical properties of AgNPs allow for additional health risks not observed from exposure to micro-sized particles [23].*

Oberdörster also stresses the specificity of AgNP by emphasizing the different risks of inhaled nanoparticles, as well as indicating the liver as the final point of destination in the organism of the nanosized silver [17]. CTA, meanwhile, shows that there is a much wider variety of AgNP on the market than the NIOSH draft recognizes; and that each of these varieties may have different risks, so a specific REL is necessary for each case [21]. Fox points out the need to specify when referring to pure AgNP, and, perhaps, it would be necessary to establish different RELs for soluble and insoluble nanoparticles [24].

ii. The second area of controversy has to do with the degree of correspondence between the scientific references provided by the NIOSH document (bibliography) and its normative conclusions; that is, between the scientific-technical information and its legal adaptation.

NIA calls to reduce the scope of the regulation to a specific form of nanoparticles, spherical not covered; just because the bibliography of the NIOSH draft only includes this modality.

> *[...] document scope should be revised to reflect the data presented in the Draft Bulletin. While NIOSH mentions the ISO definition of a nanomaterial, which includes particles, plates and wires, studies mentioned in the Draft Bulletin mostly address spherical silver nanoparticles. In addition, the studies in the document mostly focus on uncoated silver nanoparticles. As a result, the Draft Bulletin should explicitly focus on health effects of uncoated spherical silver nanoparticles [20].*

There is a huge variety of nanoparticles, and the regulation cannot deal with them one by one, but the industry takes refuge in this limitation of the cited literature to avoid or reduce the scope of the regulation. The industrial consultant PNBS takes a similar stand, asking for restricting the scope of the REL to strictly adjust it to the literature displayed in the document:

> *Narrow the current REL (10 μg/m3) to substantively spherical primary particles, their aggregates and agglomerates, and caution that the REL does not extend to shapes with high aspect ratios [...] Narrow the current REL to uncoated silver-metal- particles [19].*

In the opposite direction, proposing to expand the scope of the REL, the CTA claims that the intended maximum of 100 nanometers established by the REL should be extended to 1,000 nm [21] and, to that end, introduces the argument that other governmental agencies, such as the FDA, have extended the analysis of AgNPs to 1000 nm, when it merits [26]:

> *WHAT SIZE IS NANO? This review simply uses the narrow US government definition for "nano," i.e. 1-100 nm. The NIOSH definition would be enhanced if it used the expanded standard used by the FDA, i.e. companies are asked to report as "nano" any change in size below 1000 nm that changes the properties of the chemical [21].*

The industry supports the 100 nm limit for nanosize and considers that existing analyses of silver in a larger size are appropriate to assess risks. SNWG, for example, supports the use by NIOSH of the categories of exposition of the existing PEL issued by OSHA. In doing so, it also agrees with equivalent toxicity assessments between nano-size and larger-size as stated in the NIOSH first draft [22].

Therefore, while independent scientists and the environmental NGO claim the necessity to expand the bibliographic references to better assess the risks of AgNP, and raise questions on the ones used [24], in order to construct a broader regulation, companies prefer to keep the existing bibliographic references and seek to restrict the regulation scope.

iii. Finally, the third area of controversy over risk uncertainties regards the validity of scientific methods and their limitations. Currently, most risk analyzes include various techniques, *in vitro, in vivo, in silicon*. The analysis *in silicon* has expanded due to its speed, economy, and the possibility of standardizing the procedures, and also because of the ethical concerns regarding tests on animals. PISC, for example, suggests that NIOSH should replace analysis *in vivo* with *in vitro* and *in silicon* [27] and justifies this demand not only because of ethical but also methodological reasons, particularly regarding the uncertainties of extrapolating information from animals to humans:

> *The dissolution of silver nanoparticles (AgNPs) in different physiological environments can be addressed using alternative methods (including in vitro and ex vivo), which are considered a vital tool in understanding AgNP behavior in vivo.*
>
> *[...] Of note here is that there are many uncertainties in extrapolating toxicity outcomes from animals to humans, including variations in responses to chemicals in different species and strains of animals, gender differences within species of animals, as well as different toxic thresholds between species including humans [27].*

Some independent researchers have criticized NIOSH's preference for the application of the PBPK method to AgNP, rather than relying on research with *in vivo* methods. While the PBPK method is *in silicon*, the one used as the basis in independent studies extrapolates results from an analysis *in vivo* [11], which, among other things, suggests the liver instead of, or in addition to, the lungs, as the toxicological endpoint, as the first draft of the NIOSH suggests [23]. Moreover, the methods *in silicon* have been criticized by many epidemiologists because even when using several variables, they are always restricted compared to the number of variables present in a living organism [28]. Computerization also implies that the selection of the variables to be considered may be subject to manipulation [29, 30]. Oberdörster suggests that PBPK should not be used due to a lack of reliability:

> *REL are not well justified, because of either questionable PBPK modeling using disputed data or of rather simplistic unscientific extrapolation [17].*

In the opposite position, SNWG applauds the use of the PBKP on which the NIOSH relies:

> *In light of these standards based on argyria, the endpoint of concern, the SNWG applauds the use of the Bachler et al., 2013 PBPK model for silver nanoparticles to evaluate the potential adverse effects of working lifetime exposure to silver nanoparticles at the current NIOSH REL for silver (10 ug/m3, 8-hr TWA concentration of soluble or insoluble silver, total airborne particle mass sampling). This PBPK model was developed based on data in rats, extrapolated to humans, and validated with limited bioassay data in humans [22, 31].*

Many of the arguments deployed in the comments to the public consultation are not based on scientific information, but reveal how participants use the inconsistencies of the draft to limit, extend, or reject conclusions. This is evident, for instance, when the toxicological effects of silver in nano and larger scales are considered equivalents in spite of mounting evidence indicating they are not; or when it is suggested to restrain the regulation to the form of nanoparticles mentioned by the bibliography of the first draft, instead to enlarging the references to include the as much scientific information on diverse nanoparticle forms as possible; or when the limitations of scientific techniques are not discussed in terms of their implications for the effective protection of the workers' health. Such examples reveal the intricacy of the process of regulation, in which scientific information is subjected to diverse interpretations from the perspectives and interests of different social actors.

III. What are the perspectives of the actors regarding the hierarchy of responsibility on risk (producer, regulatory body, worker), and the degree of access to information (confidentiality or disclosure)?

Risk analysis considers the potential hazard and the degree of exposure of the worker [32, 33]. Exposure can be reduced by an uncontaminated environment or using protective equipment. The legislation aims to avoid hazards, maintaining a pollution-free environment in the first instance, and when this is not possible, using personal protective equipment [34]. The REL draft reproduces this hierarchy of controls in its recommendations. Although this hierarchy of protection procedures is a widely established legal fact, the emphasis on one or another alternative is significant in the position of the different actors. Thus, for example, CTA is explicit in emphasizing hazard control: "*workplace controls, not respirators are needed*" [21], and at large:

> *NIOSH, however, needs to stress even more strongly that in the absence of sufficient data on the inhaled toxicity of nanosilver products, that it is EXTREMELY important that workplaces implement a hierarchy of controls that keep workers from breathing any nanosilver. NIOSH needs to strengthen its risk management control practices to note that respirators will not be adequate to protect workers and that avoiding exposures is the best way to protect workers [21].*

The claim is valid because the REL is a voluntary guide, and, as long as there is no PEL from which the State can impose a firmer measure, the different approaches on how to avoid hazards lead to different responsibilities. Maintaining the work environment without risk is the responsibility of the employer, while the use of personal protective equipment places the responsibility on the worker. This criticism was assumed by NIOSH in the second draft:

> *The revised document recommends using the hierarchy of controls, encouraging the elimination or substitution of silver nanomaterials before employment of engineering controls, with PPE, including respirators, being the final and least preferable control [35].*

Responsibility for risks is intricately linked to the availability of information. If workers do not have information about the materials they handle, their hazards, and the risk to which they are subjected, they can hardly adopt a preventive attitude towards illnesses and accidents. The publicity or confidentiality of the information that the companies handle is a point of contention. CTA asks NIOSH to use information about the effects of AgNP available at other government agencies, such as the EPA and the FDA, information that these agencies have because they have authorized the entry of products with AgNP into the market [21]. The given answer reveals that there are confidentiality clauses that frequently prevent this circulation of information:

> *NIOSH collaborates with other Federal agencies when possible on chemical assessments to avoid a duplication of effort [35].*

SNWG insists on the confidentiality of potential requests for information by the NIOSH:

> *In regard to the research needs discussed in Section 8 of the NIOSH document, one of the functions of the Silver Nanotechnology Working Group is to identify, gather and consolidate industry data in an anonymous manner to protect CBI (Confidential Business Information). If such a mechanism is needed by NIOSH to bring forth needed data as listed on p. 120–121 of the External Review Draft [3] in a manner consistent with CBI, the SNWG would be glad to serve in such a capacity [22].*

The analytical answer to the third question leads to a similar conclusion to the previous two. The three industrial actors agree to reduce the available scientific information, or raise doubts about its relevance, to ensure the confidentiality of data on the materials used in production. In the opposite position are the independent researchers, who insist on expanding the range of literature and methods related to the subject, and on sustaining the differences between silver and nanosilver. In addition to this, the environmental NGO demands to consolidate the employer's responsibility instead of the workers and asks for the dissemination of technical information.

4. Conclusions

The analysis of the voluntary guide regarding workers' exposure to the risks of AgNP in occupational environments, as well as the comments made by several actors, allow us to draw some conclusions. The first and most general is that, except for one commentator, the issue under discussion is considered within a broader context in which the entrance of novel materials into the market, without assessment of their safety, is taken as a given fact. Therefore, the proposals are limited to administering the state of affairs, that is, the production and marketing of AgNPs and the commodities that incorporate them, notwithstanding the existence of scientific evidence of risks for the workers operating in its production or handling. Regulation faces an economic dynamic that overcomes it and aims just to lessen its side effects.

The second conclusion is that the commentators, despite responding individually to the draft, can be grouped analytically into two large groups, according to the coincidence of opinions. The first group responds to the interests of the producers and marketers of AgNP. Their views agree in restricting as much as possible the advent of mandatory regulatory measures. This is explicit in their arguments about the equivalent risks of AgNPs and silver in larger sizes. It is also evident when their spokesmen raise doubts about potential risks of AgNP; when they prioritize confidentiality over information on production processes, and when they derive the responsibility on risk control on workers. The second group, with less cohesion, demands to broaden the spectrum of scientific-technical information and to limit the production and marketing of commodities with AgNP until they are proved safe, in order to protect workers and consumers. The first group, more compact and convergent in their opinions, is clearly identified with business interests. The second, more dispersed, represent the interests of workers and consumers, as well as independent intellectuals who demand further investigation.

The third conclusion regards the role, largely transparent and responsible, of the government agency that conducted the process, the NIOSH. The transparency approach lies in opening the draft for public comments, as well as the flexibility demonstrated in the changes made in the second version of the document responding to the comments received. For example, a modification from the originally proposed exposure limit of 10.0 μg/m ^{3} for all particulate forms of AgNP to a 0.9 μg/m3 in the specific case of AgNPs in the air. It is also relevant to highlight that the second draft included a specific mention of the hierarchy of risk control, placing the producer in the first position and secondarily the worker, specifying that the priority is to avoid hazards in the work environment, and only as a last resource, individual protection equipment must play its role.

Author details

Guillermo Foladori[1*] and Noela Invernizzi[2]

1 Autonomous University of Zacatecas, Zacatecas, Mexico

2 Federal University of Parana, Brazil

*Address all correspondence to: gfoladori@gmail.com

References

[1] Khan I, Saeed K, Khan I. Nanoparticles: Properties, applications and toxicities. Arabian Journal of Chemistry [Internet] 2017 [cited 2018 Sep 25];Available from: http://www.sciencedirect.com/science/article/pii/S1878535217300990

[2] Department of Health and Human Services, CDC. Silver Nanoparticles (AgNP); information and comment request. Federal Register2012;75169-71.

[3] StatNano. Nanotechnology Products Database (NPD) [Internet]. 2018 [cited 2018 Sep 24];Available from: http://product.statnano.com/

[4] UNEP. Costs of Inaction on the Sound Management of Chemicals [Internet]. United Nations Environment Programme; 2013 [cited 2013 Feb 17]. Available from: http://www.unep.org/hazardoussubstances/Portals/9/Mainstreaming/CostOfInaction/Report_Cost_of_Inaction_Feb2013.pdf

[5] Di Renzo GC, Conry JA, Blake J, DeFrancesco MS, DeNicola N, Martin JN, et al. International Federation of Gynecology and Obstetrics opinion on reproductive health impacts of exposure to toxic environmental chemicals. International Journal of Gynecology & Obstetrics 2015;131(3):219-25.

[6] Prüss-Ustün A, Vickers C, Haefliger P, Bertollini R. Knowns and unknowns on burden of disease due to chemicals: a systematic review. Environmental Health [Internet] 2011 [cited 2013 Mar 20];10(9). Available from: http://www.ehjournal.net/content/10/1/9

[7] Wadhera A, Fung M. Systemic argyria associated with ingestion of colloidal silver. Dermatology Online Journal 2005;11(1):1.

[8] Vance ME, Kuiken T, Vejerano EP, McGinnis SP, Hochella MF, Rejeski D, et al. Nanotechnology in the real world: Redeveloping the nanomaterial consumer products inventory. Beilstein J Nanotechnol 2015;6:1769-80.

[9] WWICS. A nanotechnology consumer products inventory project on emerging nanotechnologies [Internet]. Washington DC: WWICS (Woodrow Wilson International Centre for Scholars); 2015 [cited 2012 Jun 20]. Available from: http://www.nanotechproject.org/inventories/consumer/

[10] Thamilselvi V, Radha KV. A Review On The Diverse Application Of Silver Nanoparticle. IOSR Journal Of Pharmacy 2017;7(1):21-7.

[11] Weldon BA, M Faustman E, Oberdörster G, Workman T, Griffith WC, Kneuer C, et al. Occupational exposure limit for silver nanoparticles: considerations on the derivation of a general health-based value. Nanotoxicology 2016;10(7):945-56.

[12] Wiemann M, Vennemann A, Blaske F, Sperling M, Karst U. Silver Nanoparticles in the Lung: Toxic Effects and Focal Accumulation of Silver in Remote Organs. Nanomaterials 2017;7(12):441.

[13] NIOSH. Revised External Review Draft - Current Intelligence Bulletin: Health Effects of Occupational Exposure to Silver Nanomaterials. By Kuempel ED, Roberts JR, Roth G, Zumwalde RD, Drew N, Hubbs A, Dunn KL, Trout D, Holdsworth G. Cincinnati, OH: U.S. Department of Health and Human Services, Centers for Disease Control and Prevention, National Institute for Occupational Safety and Health. 2018;

[14] Akter M, Sikder MdT, Rahman MdM, Ullah AKMA, Hossain KFB,

Banik S, et al. A systematic review on silver nanoparticles-induced cytotoxicity: Physicochemical properties and perspectives. Journal of Advanced Research 2018;9:1-16.

[15] CDC, NIOSH. Health Effects of Occupational Exposure to Silver Nanomaterials; Notice of Public Meeting; Availability of Document for Comment. Federal Register2016;3425.

[16] Drake PL, Hazelwood KJ. Exposure-Related Health Effects of Silver and Silver Compounds: A Review. Ann Occup Hyg 2005;49(7):575-85.

[17] Oberdörster G. Regulations.gov / Comments [Internet]. Comment on the Centers for Disease Control and Prevention (CDC) Notice: Draft Current Intelligence Bulletin: Health Effects of Occupational Exposure to Silver Nanomaterials; Notice of Public Meeting; Availability of Document for Comment; Extension of Comment Period2016;Available from: https://www.regulations.gov/document?D=CDC-2016-0001-0015

[18] PISC. Regulations.gov - Comment [Internet]. Comment on the Centers for Disease Control and Prevention (CDC) Notice: Draft Current Intelligence Bulletin: Health Effects of Occupational Exposure to Silver Nanomaterials; Notice of Public Meeting; Availability of Document for Comment; Extension of Comment Period2016 [cited 2018 Sep 22];Available from: https://www.regulations.gov/document?D=CDC-2016-0001-0020

[19] PBNS (Pennsylvania Bio Nano Systems). Regulations.gov - Comment [Internet]. Comment on the Centers for Disease Control and Prevention (CDC) Notice: Draft Current Intelligence Bulletin: Health Effects of Occupational Exposure to Silver Nanomaterials; Notice of Public Meeting; Availability of Document for Comment; Extension of Comment Period2016 [cited 2018 Sep 22];Available from: https://www.regulations.gov/document?D=CDC-2016-0001-0004

[20] NIA. regulations.gov / Comments [Internet]. Comment on the Centers for Disease Control and Prevention (CDC) Notice: Draft Current Intelligence Bulletin: Health Effects of Occupational Exposure to Silver Nanomaterials; Notice of Public Meeting; Availability of Document for Comment; Extension of Comment Period2016;Available from: https://www.regulations.gov/document?D=CDC-2016-0001-0018

[21] CTA. Regulations.gov / Comments [Internet]. Comment on the Centers for Disease Control and Prevention (CDC) Notice: Draft Current Intelligence Bulletin: Health Effects of Occupational Exposure to Silver Nanomaterials; Notice of Public Meeting; Availability of Document for Comment; Extension of Comment Period2016;Available from: https://www.regulations.gov/document?D=CDC-2016-0001-0019

[22] SNWG. regulations.gov / Comments [Internet]. Comment on the Centers for Disease Control and Prevention (CDC) Notice: Draft Current Intelligence Bulletin: Health Effects of Occupational Exposure to Silver Nanomaterials; Notice of Public Meeting; Availability of Document for Comment; Extension of Comment Period2016;Available from: https://www.regulations.gov/document?D=CDC-2016-0001-0017

[23] Faustman E. regulations.gov / comments [Internet]. Comment on the Centers for Disease Control and Prevention (CDC) Notice: Draft Current Intelligence Bulletin: Health Effects of Occupational Exposure to Silver Nanomaterials; Notice of Public Meeting; Availability of Document for Comment; Extension of Comment Period2016;Available from: https://www.regulations.gov/document?D=CDC-2016-0001-0005

[24] Fox MA. Regulations.gov - Comment [Internet]. Comment on the Centers for Disease Control and Prevention (CDC) Notice: Draft Current Intelligence Bulletin: Health Effects of Occupational Exposure to Silver Nanomaterials; Notice of Public Meeting; Availability of Document for Comment; Extension of Comment Period2018 [cited 2018 Nov 30];Available from: https://www.regulations.gov/document?D=CDC-2016-0001-0020

[25] Dunoff JL, Pollack MA. Hard and Soft Law [Internet]. In: Interdisciplinary Perspectives on International Law and International Relations: The State of the Art. New York: Cambridge University Press; 2013. page 197-221.Available from: https://books.google.com.br/books?id=0BbLHGzaekIC&pg=PR5&hl=pt-BR&source=gbs_selected_pages&cad=2#v=onepage&q&f=false

[26] FDA C for FS and A. Guidance Documents - Guidance for Industry: Safety of Nanomaterials in Cosmetic Products [Internet]. U.S. Food and Drug Administration2014 [cited 2018 Sep 26];Available from: https://www.fda.gov/cosmetics/guidanceregulation/guidancedocuments/ucm300886.htm

[27] PISC. Regulations.gov - Comment [Internet]. Comment on the Centers for Disease Control and Prevention (CDC) Notice: Draft Current Intelligence Bulletin: Health Effects of Occupational Exposure to Silver Nanomaterials; Notice of Public Meeting; Availability of Document for Comment; Extension of Comment Period2016 [cited 2018 Sep 22];Available from: https://www.regulations.gov/document?D=CDC-2016-0001-0020

[28] Grant RL, Combs AB, Acosta Jr. Experimental models for the investigation of toxicological mechanisms. In: McQueen CA, editor. Comprehensive toxicology. Amsterdam: Elsevier; 2010.

[29] Brown V, Grossman E. Why the United States Leaves Deadly Chemicals on the Market. In These Times [Internet] 2015 [cited 2015 Nov 3];Available from: http://inthesetimes.com/article/18504/epa_government_scientists_and_chemical_industry_links_influence_regulations

[30] Brown V, Grossman E. Triumph of "Digital Toxicology": Why the US won't Regulate Deadly Chemicals [Internet]. The Ecologist2015;Available from: https://theecologist.org/2015/nov/27/triumph-digital-toxicology-why-us-wont-regulate-deadly-chemicals

[31] Bachler G, von Goetz N, Hungerbuhler K. A physiologically based pharmacokinetic model for ionic silver and silver nanoparticles. International Journal of Nanomedicine 2013;8:3365-82.

[32] Howard J. Setting Occupational Exposure Limits: Are We Livin in a Post-OEL World? University of Pennsylvania Journal of Business Law 2005;7(3):513.

[33] Schulte P, Geraci C, Zumwalde R, Hoover M, Kuempel E. Occupational risk management of engineered nanoparticles. J Occup Environ Hyg 2008;5(4):239-49.

[34] Hierarchy of Controls | NIOSH | CDC [Internet]. 2020 [cited 2020 Sep 30];Available from: https://www.cdc.gov/niosh/topics/hierarchy/default.html

[35] NIOSH. NIOSH Responses to Public Comments on: Current Intelligence Bulletin: Health Effects of Occupational Exposure to Silver Nanomaterials [Internet]. 2016;Available from: www.cdc.gov/niosh/docket/archive/docket260A.html